BUGATTI

BUGATTI

••• THE MAN AND THE MARQUE •••

Jonathan Wood

Foreword by
Barrie Price

The Crowood Press

First published in 1992 by
The Crowood Press Ltd
Ramsbury, Marlborough
Wiltshire SN8 2HR

www.crowood.com

This impression 2018

© Jonathan Wood 1992

All rights reserved. No part of this publication may be reproduced or transmitted in any form or by any means, electronic or mechanical, including photocopy, recording, or any information storage and retrieval system without permission in writing from the publishers.

British Library Cataloguing in Publication Data

A catalogue record for this book is available from the British Library.

ISBN 978 1 85223 364 8

Typeset by Inforum Typesetting, Portsmouth
Printed and bound in India by Replika Press Pvt. Ltd.

Contents

	Acknowledgements	6
	Foreword	7
1	Milan's Young Maestro	9
2	Opportunity in Germany	31
3	The Move to Molsheim	62
4	A Flight of Folly	92
5	Breakthrough at Brescia	111
6	Enter the Eight	138
7	Masterpiece	164
8	Four, Eight and Sixteen Cylinders	199
9	The Royale Rebuffed	243
10	Twin Cam Twilight	278
11	Much More than a Car Factory	333
12	Reversals, then Renaissance	352
	Appendix 1: Owners' Clubs	373
	Appendix 2: Type Numbers	376
	Selected Bibliography	379
	Index	381

Acknowledgements

I must begin by setting down a debt of gratitude to the works of the late Hugh Conway, whose *Bugatti 'Le pur-sang des automobiles'* and *Grand Prix Bugatti* are required reading for anyone undertaking a project such as this one. Conway was unquestionably the greatest authority on the marque and he was also instrumental in setting up the Bugatti Trust which has given me every help with this book. Another work on which I have drawn is the stimulating *Bugatti by Borgeson*, essayed by that dedicated seeker after truth, Griffith Borgeson. Likewise WF Bradley's *Ettore Bugatti, Portrait of a Man of Genius* is another useful source, provided that it is referred to with care.

I would also set down my thanks to a number of individuals who have helped me with this work though I should stress that I alone am responsible for any conclusions drawn. First and foremost my thanks must go to the Bugatti Trust which provided me with every assistance. I am most grateful to its chairman, Barrie Price, for contributing the Foreword and also to its curator, David Sewell, who offered endless help, advice and encouragement. My thanks to him and Jennifer for their kind hospitality during my visits to Prescott. I should also like to thank Dr Norbert Steinhauser, who is making an exhaustive study of Bugatti's all-important pre-1919 activities, for his advice. Also Malcolm Jeal, editor of *The Automobile*, Dudley Gahagan, Doug Nye, Cyril Posthumus and 'Wilkie' Wilkinson, who all provided help in their respective spheres. No request was too much trouble for Lynda Springate and her helpful staff at the library of the National Motor Museum. I am also grateful to Betty and Paul Sheldon of the St Leonard Press, publishers of *A Record of Grand Prix and Voiturette Racing*. Where would we be without these marvellous volumes?

Photographs, of course, play an important part in any book such as this. The majority have been supplied by the Bugatti Trust and my thanks to its photographer, Tim Curr, for producing the many prints with such speed and efficiency. The balance have come from the National Motor Museum's photographic library and I am also particularly grateful to David Hodges and Cyril Posthumus for digging into their collections. Thanks are similarly due to David Burgess-Wise, Brian Hatton and Robin Townsend of Jarrot Engines of Chalford, Gloucestershire.

Neill Bruce, and his Peter Roberts Collection, is responsible for the magnificent colour photographs of the pre-1939 Bugattis and Neill also provided additional help and assistance. Some of the cars were provided by the Midland Motor Museum and my thanks to EA Stafford-East for making his 1935 Bertelli-bodied Type 57 and Type 59 available for photography.

There is one further debt I wish to acknowledge, though its creditor, alas, died some twenty years ago. My fascination for Bugattis was fired during my school days when a friend lent me a copy of *The Kings of the Road*, the work of that marvellous American journalist and Bugatti owner, Ken Purdy. Its first chapter, *The Fabulous Bugatti*, still delights today. Unfortunately when I first read the book, it was out of print and, being unable to obtain a copy in Britain, I wrote to Purdy, care of his publishers in America. Almost by return of post came a signed copy which is still in my library. I hope that Purdy would have approved of what follows.

Jonathan Wood

Foreword

One might be excused for believing that yet another book on any of the great makes of motor car, Rolls-Royce, Bentley, Bugatti, Mercedes-Benz and so forth would be superfluous. Those of us who are closely associated with the 'Bugatti World', however, are constantly amazed by the discovery of fresh material which surfaces almost daily and often after the passage of 100 years and more. Enthusiasts for other makes will no doubt experience the same state of affairs. Fresh facts tend to shift one's previous notions and impressions of the cars and personalities concerned and thus history needs continual revision.

Those who are familiar with the writings of Jonathan Wood will know that he undertakes most painstaking and detailed research, and this book contains the result of his labours into the life and works of the entire Bugatti family. We learn many hitherto unknown facts, concerning Ettore Bugatti's forebears and his early life. This information is put into the perspective of Italian and Central European politics and life at the time.

The motor cars are placed in perspective with other contemporary makes in more detail than has been attempted before and the Author enquires into the mind of Bugatti and explores the probability that he may have been influenced by the work of this or that designer. Interesting to contemplate, but one should bear in mind that engineers would all be pondering the same problems at the same time and were all bound by the same laws of physics. It is thus a fact that different firms often came to similar conclusions quite independently of each other.

There is no doubt whatever that Ettore Bugatti was a genius, unique in character, larger than life, of magnetic personality and one of the most important individuals of modern times. His works represent the combination of art and engineering in its highest form, and are unlikely to be surpassed.

My researches cannot find any individual or any firm who remotely approaches the Bugatti commitment; we cannot find for instance, another senior designer/proprietor who chose to live inside his factory. Imagine the scene as, morning and evening, one thousand workers walked or cycled through the gates and immediately passed by Madame Bugatti's kitchen window while 'Le Patron' and daughter Lidia were probably entertaining the King of the Belgians in the garden room next door.

We are grateful to Jonathan Wood for this book which not only extends our knowledge further but is likely to become the standard reference work on this fascinating and entertaining subject.

Barrie Price
Chairman of the Bugatti Trust
and former chairman of the
Bugatti Owners' Club.

'I shall never starve, whatever happens to me, for I can do everything.'
Ettore Bugatti (1881–1947)

1
Milan's Young Maestro

'In the motor car field Bugatti is very well known; son of an artist, he is himself an artist at heart.'

Gazzetta dello Sport, Milan, 10 May 1901

It was the summer of 1990 at a test track at Ladoux in central France. The circuit, wedged between the number nine national highway on one side and the Paris to Clermont-Ferrand railway on the other, belongs to Michelin, which is the world's largest tyre manufacturer. Amongst its multifarious activities, the company had been developing an ultra high performance tyre for four-wheel drive supercars and was playing host to what would be its first recipient. The latter was an exclusive and costly model, powered by a purpose-designed 3.5-litre V12 engine, which employed no less than four turbochargers. Although this light-coloured, mid-engined experimental coupe was put through its paces at the Michelin track, there was little clue to the true appearance of what was, in reality, a heavily disguised prototype.

After completing a demanding series of tests, the enigmatic four-wheel drive coupe was returned to its recently completed factory at Campogalliano, near Modena in northern Italy. But the low-built two-seater was not just another hardworking experimental mule. It represented the rebirth of one of the most legendary and respected marques in the history of the motor car, with an intended official launch date of 15 September 1991, which was the 110th anniversary of the birth of the make's creator. For after a close on forty years' absence, the Bugatti was back.

The choice of Italy for the new factory is singularly appropriate because it is only through a quirk of historical fate that the Milan-born Ettore Bugatti did not build the cars that bore his name in his native city and, instead, spent much of his life the 400 or so kilometres (250 miles) to the north on the other side of the Alps in Alsace. There at the small town of Molsheim, from 1910 until the outbreak of World War II in 1939, he only built about 7,800 cars. Every one of them, from the diminutive Type 13 of the pre-World War I era to the fabulous and mighty Royale, carry his indelible, individual stamp.

Perhaps the greatest expression of his abilities was the Type 35 of 1924, when Bugatti produced what can be numbered amongst the most successful racing cars in the world, and also one of the most beautiful.

For although Ettore Bugatti was a cautious, intuitive engineer he was, above all, a supreme artist who endowed both his cars and his engines with a degree of visual precision and proportion which has rarely been equalled; and when it was, the rival design was invariably the work of a fellow countryman.

To begin to comprehend why Bugatti designed his cars as he did, we must follow his own advice. At a time when his cars were sweeping all before them on the motor racing circuits of Europe, he wrote: 'In order to explain the strange development of my career I must first describe my environment during my childhood and what my life was like as a youth.'[1]

Throughout much of his life and particularly during his first twenty formative years, Ettore was exposed to the influence of Carlo, his multi-talented artist father, who was to have a profound impact on the way in which his son created and built his cars. This manifested itself not only in the cars' exclusive appeal, but in that they embodied some features and exquisite proportions of the type that had first appeared in Carlo's furniture, while the inspiration of the very spirit of their construction is to be found rooted in the Milan workshops of the elder Bugatti.

When Ettore Arco Isidoro Bugatti was born at the Castello Sforzesco in Milan on Thursday 15 September 1881, the Italian state was a mere twenty years old. Prior to the country's unification in 1861, in the words of Austrian Chancellor Klemens Metternich, 'Italy was no more than a geographic expression.' Under the direction of Count Comillo di Cavour of Piedmont, the strong, industrially dominant northern states were united with those of the poorer south and the Kingdom of Italy declared with Turin as its capital. It ceded this position to Rome in 1871 which completed the country's integration.

Then, as now, Milan, capital of Lombardy, is the north's largest city. In the nineteenth century, this most European of Italian communities was to expand dramatically through the growth of its commerce and industry. Milan had a population of 320,000 in 1881 which had nearly doubled to 560,000 in 1906. This followed Austria's cession of Lombardy in 1859, after the Battle of Magenta, that had paved the way to the Italian alliance of two years later. So, when Ettore's father Carlo was born in the city, on 2 February 1856, for the first three years of his life he was a citizen of the Austrian empire.

Alas, little is known of Carlo's sculptor father, Giovanni Luigi Bugatti, and his mother, who was born Amalia Salvioni. We do know, however, that he had a sister, named Luigia after her father. What few fragments of biographical information we have come from Ettore's daughter, L'Ébé, who was so named after the initials of her father's name. She tells us that he had '. . . been interested in science and architecture. In the mid-nineteenth century [he] had sculptured monumental chimney-pieces which were much admired for their elegant form, but he had also poured all his money into solving the secret of perpetual motion . . .'[2]

It is not clear to what stage in Luigi's life this refers, but it would appear that he had sufficient funds to pursue this elusive chimera. In the late 1870s, his son Carlo was able to study art at the city's Brera Academy and he later attended the Beaux Arts in Paris – for Carlo had decided to follow his father into an artistic career. It seems likely that he displayed a versatile and precocious ability, because his granddaughter reveals that he was known as 'The Young Leonardo', a reminder that the great and multifariously talented da Vinci spent most of his working life in Milan. It was a family characteristic to be inherited by Carlo's sons, Rembrandt and Ettore, and the latter's son, Jean, all of whom displayed strong, creative abilities before they were twenty years old.

L'Ébé Bugatti remembers that her grandfather was '. . . of medium build and very handsome; he had bright blue eyes which sparkled with intelligence and mischief, and his keen glance was quite intimidating yet winning at the same time. When still quite young he grew a beard which was silky and very fair, and he developed a habit of stroking and twisting it. In later life it turned grey, then white with a yellow stain from ruminative pipe smoking.'

Carlo also adopted a distinctive form of dress, a trait which he also bequeathed to Ettore. L'Ébé tells us that Carlo suffered from stomach trouble as a young man, which he treated with salts of San Pellegrino. This prompted him to wear flannel belts but even they proved to be too restrictive and

---

[1] 'Confidences de Monsieur Bugatti' (1926) in *The Bugatti Story* by L'Ébé Bugatti (Souvenir Press, 1967). First published in French as *L'Épopée Bugatti* (1966).

L'Ébé was the oldest of Barbara and Ettore's children, and, according to Hugh Conway, she was insistent that the accent on the first E of her name be scrupulously observed, even though it was not shown on her birth certificate.

[2] L'Ébé Bugatti, *The Bugatti Story*.

Carlo Bugatti (1856–1940), father of Ettore and a multi-talented artist, as this self-portrait testifies.

Following his father's example, Carlo initially pursued an architectural career, though, despite undertaking a number of projects ('a pyramid-shaped war memorial and arab-style house, bear witness to the originality of his approach')[3], there is no evidence to suggest that any were actually executed. Then, probably in 1880, came a change of emphasis, when Carlo started to design and manufacture furniture. This was destined to change his life and was to culminate in him becoming an internationally renowned figure. It was to lead the critic Maxime LeRoy to perceptively describe him in 1903 as 'an isolated genius whose flair for the bizarre defies classification.'[4]

In 1880 when he was twenty-four, Carlo's sister, Luigia, married the painter Giovanni Segantini, who had had the misfortune to have been abandoned by his father at the age of five, following the death of his mother. Romantically, he is said to have been brought up by shepherds in the mountains outside his native Milan and, such were the artistic talents he displayed in his youth, they raised money for him to attend the Brera Academy though he soon rebelled against its orthodoxy.

resulted in him creating his own idiosyncratic wardrobe.

His trousers were replaced by a far more suitable type of dungarees which were, accordingly, supported by his shoulders instead of the waist. 'Over this garment he wore a frock-coat which fell straight to his knees; it had split pockets at the sides, and the 2-inch, upright collar was fastened under the chin by two glass buttons the size of hazel-nuts; they were like cat's eyes set in gold, and were linked by a thin gold chain.'

On top, Carlo wore a dark, well-tailored coat though during the summer it was made of a silk-like material. The outfit was completed by a fine straw hat for everyday use but, on special occasions, Carlo wore a low-crowned silk one with the flat brim edged with a corded ribbon.

Carlo's Furniture

As a wedding present, Carlo designed and probably himself built, a bed and matching side cabinet for the couple. This is his first known surviving work. It is painted and decorated with a Japanese floral design, and displays the strong geometrical themes which recur in much of his later work. For the next twenty-five years, Carlo would devote himself to the design and construction of his extraordinary, eccentric furniture. But before examining its evolution, it is appropriate to pause and consider the state of applied arts in the new Italy and the part that Carlo played in their execution.

The country's international influence in this field

[3] Philippe Garner, 'Carlo Bugatti', in *The Amazing Bugattis* (The Design Council, 1979).

[4] Alastair Duncan, *Art Nouveau Furniture* (Thames and Hudson, 1982).

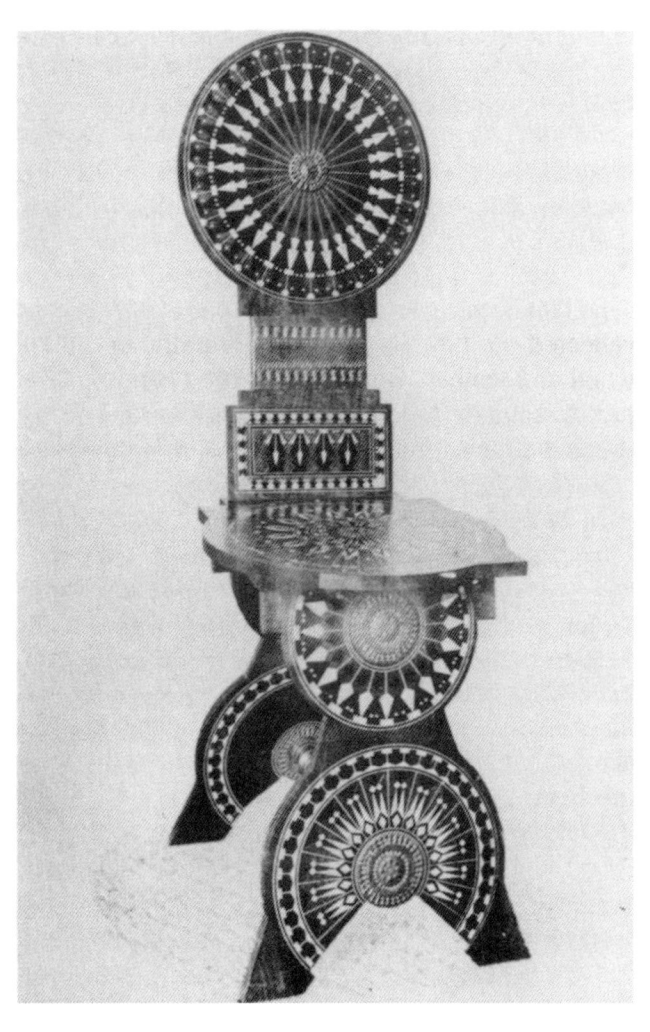

A beautifully inlaid chair by Carlo Bugatti, with geometrical themes readily apparent.

An Arabesque cabinet by Carlo Bugatti; a fine interplay of differing woods, metal and velum.

did not manifest itself until after World War II so, during the nineteenth century, it can be seen to have been virtually non-existent. What styles did exist were strongly influenced, from the 1850s, by French fashion. There, 'Images of an idealized version of the land of *A Thousand and One Nights* had known great popularity. Orientalist painters were in vogue . . .'[5] It is not surprising, therefore that Carlo's furniture soon reflected Oriental and Arabesque themes.

Only eight years after completing his first known

[5] Garner, *op. cit.*

work, Carlo was displaying his furniture at the Italian exhibition held in London in 1888. *The Queen* of July of that year devoted space to his work which it described as 'Quaint Furniture Exhibited by Carlo Bugatti, Milan.' By this time, his style had evolved and become more elaborate. Featured were a chair of ebonized wood with inlaid white metal and another brown painted one decorated with a Japanese floral design. There were also wall brackets and a table fringed with silk and leather with chamois panels. Carlo was awarded a first prize, a Diploma of Honour, for the originality of his designs which, it should be said, were like nothing else in Europe.

It was then that a reviewer in the *Journal of Decorative Art*, claimed that Bugatti was a true disciple of John Ruskin.

Arts and Crafts

Carlo's furniture was both expensive and exclusive. It can be seen as part of the Arts and Crafts movement, which began in Britain and soon spread to Europe and America. This movement represented a revulsion to the mass-produced, machine-made artefacts of the type displayed with such aplomb at the Great Exhibition held in London at the Crystal Palace in 1851. One of the movement's formative influences was the nineteenth-century critic, John Ruskin. The latter, in his voluminous writings, advocated that art should reflect a truth to nature and he also venerated the virtues of medieval craftsmanship. Writing in 1853 of the west fronts of many of Europe's gothic cathedrals, he thundered in praise of the old sculpture; the statues, goblins and associated monsters, he declared: 'Do not mock at them, for they are the signs of the life and liberty of every workman who struck the stone . . . it must be the first aim of all Europe at this day to regain for her children.'[6]

Artist craftsman William Morris, a disciple of Ruskin, positively responded to these exhortations in 1861 by establishing, in London, the firm of Morris, Marshall, Faulkner and Company. For Morris the machine-made article was an anathema. Instead, craftsmen were employed who were 'Fine Art Workmen in Painting, Carving, Furniture and the Metals'. Its impact was such that historian Nikolaus Pevsner would write that the 'event marks the beginning of a new era in Western Art.'[7]

Although Morris spoke of real art being made 'by the people and for the people' the reality was that the firm's hand-crafted furniture, wallpaper and fabrics were only accessible to a few discriminating and wealthy connoisseurs. It was what Morris described (with his socialist hat firmly in place) as art 'for the swinish luxury of the rich'.

In 1888 Carlo Bugatti moved his Milan workshop to 6 Via Castelfidardo. Although we do not have any detailed description of its form, it is likely that skilled craftsmen were employed there to produce Carlo's furniture, while he might have also relied on home-based artisans whose expertise was reflected in the time-honoured crafts of metal, wood and leather working. Yet, the twentieth century was not excluded: from 1883, Milan was one of the first cities in Europe to adopt electric power for street lighting and trams. It also permitted workers to use lathes and associated tools in their homes.

Such practices could not have failed to influence the young Ettore and when, in 1910, he began building cars under his own name there was never any attempt to produce them in quantity. Instead, the cars were built in small, hand-assembled batches. Significantly, when Ettore did create a design suitable for mass production, what became the Bébé Peugeot in 1912, he sold it.

Although machine tools were employed, Bugatti relied, just as his father had done, on the skills and craftsmanship of local labour. Little wonder that when *The Autocar*'s W F Bradley visited Molsheim in 1929, writing of Ettore, he maintained that: 'John Ruskin's soul [he had died in 1900] would have been delighted at this example of the artist-artisan; of the engineer who seeks to vent his artistic temperament . . . of the man who has happily combined an artistic home with production methods, and who has succeeded in transmitting this joyous creative spirit to all those who labour with him.'[8]

A Growing Reputation

Carlo Bugatti's next international presentation was at the 1900 Paris Exhibition, when Italy's stand proved an uneasy combination of the styles of Siena Cathedral's black-and-white stripes and the strong, Byzantine themes of St Marks, Venice.

[6] John Ruskin, 'The Nature of Gothic' in *The Stones of Venice* (George Allen & Unwin, 1853).

[7] Nikolaus Pevsner, *Pioneers of Modern Design* (Penguin Books, 1960).

[8] *The Autocar*, 15 March 1929.

French critics, with the Italian contribution clearly to the fore, coined the description *style nouille* or 'noodle' style for the contribution. Despite these problems of presentation, which were not of his doing, Carlo received a silver medal for his work. His style had evolved into a more discernible one with an increasing use of velum and a more assured employment of geometrical forms.

By this time Carlo was creating entire rooms, in the manner of William Morris and his associates in Britain, in which to display his furniture. Indeed in about 1900, he produced such a commission for the newly ennobled British barrister and rising politician, Lord Battersea, for a bedroom in his London residence at Surrey House, Marble Arch. However, one commentator who viewed the room soon after its completion remarked that the only merit of the visit was that it saved him the cost of a train fare to Granada to see the Alhambra![9]

It is an interesting historical aside that Lord Battersea's wife was the daughter of Sir Anthony de Rothschild, and when the latter's great nephew became the third Lord Rothschild in 1937, he had just taken a delivery of a rare and visually delectable Type 57S Bugatti Atlantic coupe. The Bugattis, father and son, were appealing to a similar clientele.

Carlo's increasing international stature was also reflected by a Turkish commission for the Constantinople residence of the Khedive's mother. Soon came another exhibition. Despite the unfavourable reaction to its display at the 1900 Paris event, the Italian government persisted in propagating an international artistic profile. In 1902, they staged an exhibition in Italy itself and the city of Turin was chosen as the venue. There, the Italian architect, Raimondo d'Aronco, created yet another Byzantine extravaganza. *The Studio* was unimpressed and maintained that 'the whole effect of it was a huge bazaar, rather than an exhibition of artistic work'.[10] Entries were attracted from America, Britain, Germany and France. Italy, naturally enough, was the largest exhibitor, being represented by no less than 250 contributors.

There Carlo Bugatti, from nearby Milan, displayed no less than four furnished rooms, with specially woven carpets, which constituted the *Maison d'Escargots* (The House of Snails), together with a separate display of furniture. The suite had in fact been executed by Carlo for a client at around the turn of the century. Once again, velum was much in evidence, while his so-called 'snail chairs' displayed Art Nouveau influences at a time when the style was enjoying great popularity. For some commentators they contain the elements of Ettore Bugatti's curvilinear radiator design which first appeared in 1910.

These exhibits apart, it has proved extremely difficult to date some examples of Carlo's furniture where they share common features with, in particular, Bugatti cars. This is because Carlo Bugatti continued to occasionally make pieces, particularly tables incorporating tantalizing, radiator-shaped forms, throughout his life. It is therefore difficult to decide which came first: the car, or the item of furniture? A further problem is that there is little contemporary documentary evidence of his portfolio and Carlo, apart from scrawling his signature, may have been illiterate. So it is only through surviving examples of his work that the progression and maturing of his style can be charted.

Throughout the Turin display, the rendering is bizarre and utterly individual but, for all its extravagance, there is an overall impression of total control. This is due to Carlo's assured interplay of decoration and form and, above all, his magnificent eye for proportion. For although some of his earlier chairs are inspired by Gothic arches, his work can only have been sharpened by the inspired, understated geometry of the classical architecture of the Renaissance, which was in abundant presence in his native Milan.

However, it would be unfair to suggest that Carlo's furniture and interiors were, in their day, universally acclaimed. In Paris, his work had been dismissed as representing 'the yachting style'. At Turin, his contribution was either ignored or cri-

---

[9] Duncan, *Art Nouveau Furniture*.

[10] Ian Bennett, 'Italy and Spain' in *The Encyclopedia of Decorative Arts*, ed. Philippe Garner (Chartwell Books, 1988).

ticized. With great irony, the German critic Fritz Minkus, writing in *Kunst und Kunsthandwerk*, described Bugatti's interior as producing 'the effect of a newly invented automobile, which cannot be quite comprehended at first glance, of outstanding technical refinement, together with an extremely ugly or ponderous appearance'.[11]

Nevertheless the exhibition underlined Carlo Bugatti's position as the most important manufacturer of Italian furniture of his day. He was, above all, the only one of his contemporaries to create an original, totally individual style. But for Carlo, Turin represented the climax of his career as a designer and manufacturer of furniture and he was awarded a gold medal there. About two years later, in around 1904, he sold his business to the De Vecchi company of Milan with the subsequent, and inevitable, decline in quality.

The Move to Paris

Carlo and his wife, Thérèse, left Milan and the family home at 13, Via Marcona and settled in Paris. He knew the French capital from his days at the Beaux Arts and, for many years, had made an annual pilgrimage there. The move, though, was prompted to some extent by his wife's ill health. Carlo was soon at work at his studio on the left bank at 13, rue Jeanne d'Arc, where he continued to produce some furniture though the designs were executed by local craftsmen.

Carlo Bugatti was not the only member of his family to have left Italy. By 1904 his eldest son, Ettore, was established in the German city of Strasburg and was already making his name as an engineer and designer of motor cars. Yet, as will have been apparent, the father's influence on his eldest son was profound. L'Ébé was to recognize that 'Carlo Bugatti and his environment had a great influence on . . . Ettore, who learned from him to regard art as a flowering of one's personality and not as a means of making money.'

Ettore himself was later to acknowledge his debt to his father, who also taught him to work with his hands. He records that Carlo attached 'great importance' to developing this skill though he was, by all accounts, a hard taskmaster and occasionally resorted to beatings. Also, Ettore later recognized that 'a cabinet-maker's work is the best grounding for mechanics'. It was an ability that was to prove its worth when he was able to make the wooden patterns for his first car.

Ettore's Early Life

When Ettore was four, the third and last of Carlo and Thérèse's children, a boy, was born on 16 October 1885. He had a particularly large head which was considered to be a sign of greatness and, as a result, he was christened Rembrandt at the suggestion of his godfather, Ercole Rosa. (Rosa was a sculptor friend of the family, who was perhaps best known in the city for the creation of the statue of King Victor Emmanuel II in the Piazza del Duomo in Milan.)

Thérèse Bugatti had a difficult confinement, followed by a bout of pleurisy which weakened her and necessitated a long period of recuperation. Consequently, Rembrandt never received the same affection that she lavished on Ettore, who was her favourite. This may have been a factor in Rembrandt's withdrawn and melancholic personality although, as will emerge, he was also found to possess an extraordinary and precocious artistic talent.

As Ettore grew up, he later remembered that: 'My first ambition was to be a great artist and so earn the right to bear a name distinguished by my father.' Ettore had developed a manual dexterity in his father's workshop and Carlo decided that his eldest son should follow him to Milan's Brera Academy where he studied painting, sculpture and architecture. Ettore recalled having 'the best possible teachers in painting and sculpture, especially our family friend, Prince [Paul] Troubetskoy'. An Italian-born expatriate Russian, he is best known for his bronze sculpture which included beautiful women and busts of the Russian writer Leo

[11] Bennett, *op. cit.*

Ettore's younger brother Rembrandt (1885-1916), pictured in about 1910 at a café at the Antwerp Zoo. It was there that he executed many of his finest animal bronzes.

Tolstoy, Carlo Bugatti and the composer Giacomo Puccini. He was another distinguished member of the Bugatti coterie and Carlo had been responsible for furnishing the dining room of his villa at Torre del Lago.

The Bugatti's middle class, but distinctly Bohemian circle, in both Milan and Paris also included the composer Leoncavello, poet and Puccini's librettist, Illica, music publisher Tito Ricordi, the artists Arturo Rietti and Giovanni Segantini. The latter, it will be recalled, had married Carlo's sister. Then, there was the art critic Rossi Sacchetti, who published a monograph on Carlo and Rembrandt in 1907. Just how much intellectual influence this group had on the young, extrovert Ettore is open to question. However, such a formative social grounding was to serve him well so that, throughout his life, he could move with ease across the social divide. His circle included members of European royalty and aristocracy as well as the inevitable associates from the world of motoring and racing.

Not long after Ettore had started at the Brera, 'My brother suddenly took to drawing. I saw at once, and confided to my dear mother, that he was the true Bugatti and would soon be far better than I, in spite of my studies.' This was an ironic development for initially 'Rembrandt wanted to be an engineer and build locomotives. I wanted to be an artist, but I was no more gifted to art than he was for mechanics.'

Rembrandt's talent was finally, and dramatically, recognized when he was fifteen and Ettore

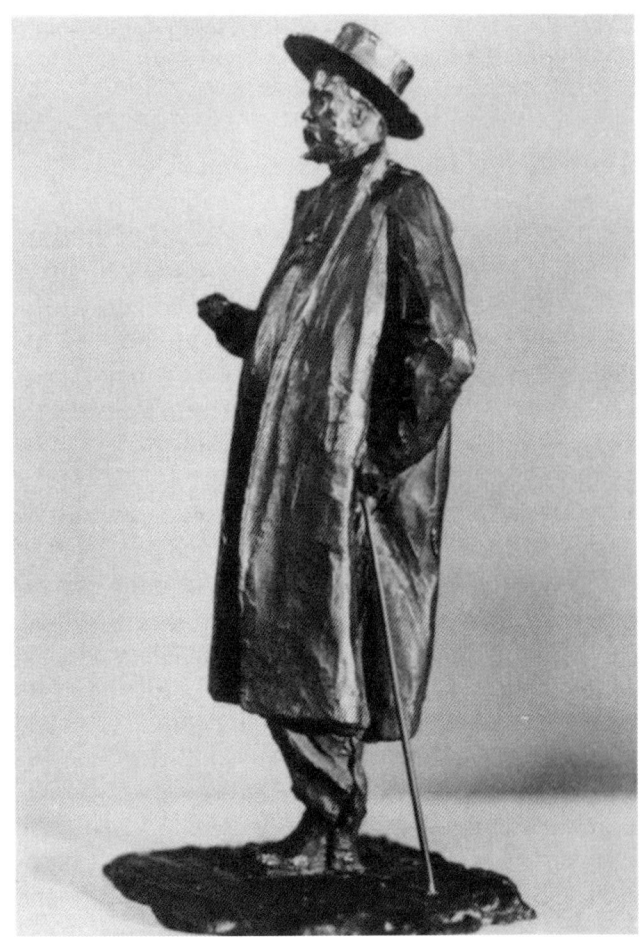

In 1899 Paul Troubetskoy, who was to teach Ettore painting and sculpture at Milan's Brera Academy, executed this fine bronze of his friend, Carlo Bugatti.

An example of Rembrandt Bugatti's sculpture, a team of horses pause from the toil of hauling a large stone slab. Entitled 'The Big Dray' or 'Ten Minutes' Rest', the wax of this sculpture was exhibited at the Salon National in 1905 and the bronze in 1907. It won the Grand Prix de Sculpture at the Milan Exhibition of 1906 and is pictured here at the Bugatti villa at Molsheim.

nineteen. A connoisseur visiting Carlo's Via Castelfidaro workshop 'noticed a bucket in which something was hidden under damp cloths. Lifting them, he discovered a beautiful clay group representing an old peasant leading three cows. Exclaiming his delight and admiration, the visitor asked Carlo about it and he was just as surprised and captivated as his guest. Young Rembrandt had sculpted the group after a scene he had observed out in the country.'

Rembrandt thereafter usually concentrated on animal sculpture, which he first moulded in clay and then cast in bronze. His earliest study dates from 1901 and he continued to produce work of exceptionally high quality throughout his tragically short life – he committed suicide in 1916 at the age of thirty.

The Young Engineer

By the time Rembrandt had begun his career as a sculptor, his brother Ettore had left the Brera in a similarly dramatic change of course. This followed an event which probably took place early in 1898, when Ettore was sixteen: 'One day, some friends of my father asked me to try out a motor tricycle which had been built a year after the appearance of the De Dion tricycle. It had been made by the firm of Prinetti and Stucchi.' The machine produced an instantaneous, intuitive response in the young Bugatti. 'In a short while, just by looking at the machine, I had grasped all the intricacies of its mechanism', he remembered.

Ettore could also ride the tricycle: 'Without knowing anything about its construction. Signor Stucchi, seeing that I was interested in this means of locomotion, asked my father to allow me to go to his factory. This was granted.' But once at the works at 11, Via Tortana, Milan, Ettore knew what he wanted to do and that was to cease his artistic education at the Brera Academy and join Prinetti and Stucchi as 'an unpaid apprentice'. Not surprisingly he 'had some difficulty in making my father understand' but, eventually, Carlo did agree to his eldest son's request.

The basic Prinetti and Stucchi tricycle. Although it superficially resembles a De Dion Bouton machine, it lacks the bridge-type support to which that version's engine was attached. The P and S's frame would appear to be by Rochet of Paris, or a copy of the same. It is fitted with a De Dion Bouton surface carburettor, complete with badge, which can be seen just below the saddle.

This impulsive decision by Ettore Bugatti resulted in him becoming technically self-taught and not receiving any real theoretical education. In retrospect, this was to have a major influence on the way in which he designed his cars when he began producing them under his own name. His lack of technical training permitted the artist in him to flower, which is superbly expressed in the Type 35 racing car but conversely, although extremely successful, the car was mechanically outdated when it appeared in 1924. In the longer term, however, Bugatti's ultra-cautious approach to engineering design would be the contributory factor in the firm's demise.

The irony was that, by the 1890s, what technical education did exist in Italy was accessible to Ettore – and of high quality. Its provision was to play a key role in the Italian motor industry's progressive, innovative designs, which were already apparent in the years before World War I.

The origin of much of Fiat's great technical strength is to be found rooted in the instruction rooms of the Turin Polytechnic and Alfa Romeo was equally well served by its Milan equivalent. The superiority of these two establishments was noted by Sir Philip Magnus, director of Finsbury

A young man in a hurry: a formal portrait of Ettore Bugatti in Milan at around the turn of the century...

...and a more informal picture taken at around the same time.

Technical College in London, after a visit to Italy in the 1880s. Writing in 1888, and all too aware of the backward nature of Britain's technical facilities, he noted that: 'Education was given on the same lines as in German polytechnic schools. Their professors enjoyed the rank and privileges of university professors, and were supported and controlled by the Ministry of Agriculture, Commerce and Industry.'[12]

Ettore bypassed these provisions when he joined Prinetti and Stucchi's workshop in 1898. It was a time, he was to write, 'when people were awakening to the possibilities of self-propelled vehicles'. This then is the moment to briefly examine the origins of the car and, in particular, its impact on the new kingdom of Italy.

It had been in 1886 that Carl Benz and Gottlieb Daimler, each working independently of the other, created the world's first practical motor cars. But, above all, replicas of these German designs were offered for public sale, which triggered the birth of the world's motor industry. Both vehicles were powered by rear-mounted single-cylinder power units, developed from stationary gas engines, which employed the four-stroke cycle patented by Nicholas Otto in 1876.

French Initiatives

However, it was the French who took up the invention with even greater alacrity, and its industry was to be Europe's largest until it was overhauled by Britain's in 1932.

One of the most significant French cars of the 1890s was the product of the Panhard Levassor concern, a Parisian company which specialized in the manufacture of woodworking machinery. It began selling cars in 1889 but one of the firm's partners, Émile Levassor, a graduate of the prestigious École Centrale des Arts et Manufactures, made a lasting contribution to motor car design in 1891 when he placed Daimler's new 3.5hp V twin engine at the front of the car, instead of the rear position which had hitherto been the norm. From its forward location, drive was conveyed to the chain-driven rear wheels, via a clutch, gearbox and differential. Soon to be identified as *Système Panhard*, this new layout proved itself to be capable of almost infinite development, while the earlier horizontal-engined, rear-mounted layout gradually fell into disuse. Levassor's innovative configuration was to dominate motor car design for the next eighty or so years.

Although Panhard was to be a major force in the early days of the industry – and its machines provided France with some of its foremost racing cars – by the turn of the nineteenth century, one of the country's most successful car companies was the De Dion Bouton concern, which also hailed from the French capital. In 1882, the worldly Count Albert de Dion had teamed up with Georges Bouton, an inventive, jobbing engineer, to build light, steam-powered vehicles. However, in 1895 Bouton created a 0.5hp single-cylinder air-cooled engine which ran at the hitherto 1,500 dizzy revolutions per minute, compared with about 700 to 900rpm of the contemporary Daimler. This figure was partly made possible by fine machining tolerances of the unit's pioneering steel cylinder and partly by the fitment of Bouton's make-and-brake electric ignition system which survives, in essence, to this day. Bouton also opted for an innovative aluminium crankcase, with the result that the engine weighed 18kg (40lb). At this time, France was the world's leading producer of the metal, following Paul-Louis-Toussaint Heroult's invention of the process in 1886, sparked by the arrival of cheap and readily available electricity, which made the production of abundant quantities of aluminium a practical reality.

The process could not have appeared at a better time for the nascent French motor industry. Later in 1895 the De Dion single was experimentally fitted to the rear of a tricycle. The resulting model was immensely successful and was the Parisian company's best-selling product for the next five years. It is estimated that some 15,000 examples were produced between 1895 and 1901, making it by far the world's most popular motor vehicle.

[12] George S Emmerson, *Engineering Education* (David and Charles, 1973).

Not only did the De Dion tricycle make plenty of friends among the general public, but it also proved its worth as a race-winning machine, following Viet's third place in the Paris–Marseilles–Paris race of 1896. It says much for the reliability of the little trike that it completed the 1,710km (1,063-mile) course. In the Marseilles–Nice–La Turbie event of the following year, Chesnay won the tricycle class and no less than six of the seven De Dion three-wheelers entered finished the event. It was to set the pattern for countless racing successes over the next four years.

The Popular De Dion

In addition to producing complete tricycles, De Dion Bouton supplied other manufacturers with its engines and, by November 1901, the company claimed to have built no less than 25,000 of them. They were also manufactured by licensees. Between around 1898 and 1908, about 100 different makes the world over were powered by the De Dion's engine and its two-cylinder successor. In France, over twenty-four firms alone used the little one-lunger. (The De Dion Bouton single grew in capacity from 0.5 to 1.25 and, ultimately, 2.75hp in its original air-cooled form.) They were also employed by such companies as Argyll and Humber in Britain and by the German Adler and Opel concerns. Across the Atlantic, such respected American companies as Packard and Pierce-Arrow began their sorties into automobile production by manufacturing De Dion-powered machines.

In addition, countless firms did not trouble to employ the genuine article but, to save money, directly or indirectly simply cribbed the overall design of the De Dion Bouton single. One such firm was the Prinetti and Stucchi company of Milan, which Ettore Bugatti had joined in 1898.

Compared with France and, to a lesser extent Germany, Italy was a relatively poor and backward country. Invariably, the first cars to cross the Alps hailed from its richer neighbours, though the market tended to be dominated by French imports. The first all-Italian motor car was not built until 1894, which was eight years after Benz and Daimler horseless carriages had appeared. It was the work of Count Enrico Bernardi, a former professor of physics at Padua University.

Bernardi had earlier produced two experimental engines, a 144cc 'Pia' liquid-fueled unit in 1882, which had won a silver medal at the 1884 Turin Exhibition, while the 265cc 'Lauro' dated from 1893 and was used in a motor cycle. Bernardi subsequently developed a single-cylinder 624cc engine, progressively fitted with overhead valves in a detachable cylinder head. This was horizontally mounted in a tricycle, adjacent to the chain-driven single rear wheel. It became Italy's first production car when, in 1896, it entered production in Padua as the Miari e Giusti.

The Birth of Fiat

One of Bernardi's disciples, who visited him at Padua and inspected this engine while it was under development, was a young cavalry officer named Giovanni Agnelli. The latter was to become one of the principal architects of *Fabbrica Italiana Automobili Torino*, what today we know as Fiat, established in Turin in 1899. By the outbreak of World War I in 1914, Fiat had not only become Italy's largest car maker, with 4,644 examples delivered, but was also one of Europe's mainstream motor manufacturers. Only the Ford factory in Manchester, England, which produced 8,352 Model Ts, made more cars than the Italian manufacturers.

When Fiat had been created, it purchased the modest Turin concern of Ceirano and Company which had just produced one car, the 3.5hp Welleyes. It was so called because Ceirano had originally built bicycles under the same curious English-sounding name, there being no W in the Italian alphabet. The design was basically an Italian copy of the English Rudge, which Ceirano had originally sold. Bicycles were becoming increasingly popular and Italy was following France and Germany in establishing its own industry.

Inevitably, this was concentrated in the industrialized north and one of the country's largest bicycle manufacturers was the Bianchi company, established in the city of Milan in 1875. The firm prospered and, in 1888, it had the distinction of offering the first bicycle in Italy to be equipped with pneumatic tyres. Bianchi subsequently shared the approach adopted by many European bicycle makers (Peugeot and Opel were the largest) by diversifying into car production.

In 1897 Bianchi had experimented with motor tricycles and in 1899, the same year as Fiat's creation, it unveiled its first car. This had a tubular chassis, which reflected its bicycle associations, and was powered by the ubiquitous single-cylinder De Dion Bouton engine. In 1902 Bianchi expanded in a new, larger factory and has been part of the Italian motoring scene ever since. It remained independent until 1955 when, under the Autobianchi name, it was effectively absorbed by Fiat.

Whereas Bianchi had from the outset built bicycles, the smaller Milan company of Prinetti and Stucchi, established in 1874, had initially produced sewing machines. However, in the manner of its English contemporaries, Prinetti and Stucchi later switched to bicycle manufacture and was to choose a spoked wheel as its corporate badge. According to Ettore Bugatti, the firm's 'unpaid apprentice' of 1898, 'Prinetti was an engineer and Stucchi a very good industrialist'.

The firm's further diversification into the manufacture of the De Dion-type tricycles a year – ahead of both rivals Bianchi and Fiat – shows it to be one of the handful of companies which constituted Italy's embryonic motor industry. Ettore Bugatti was a young man in a hurry and he could not have been better placed to benefit from the opportunities presented by the arrival of the motor car in Italy.

Although the Prinetti and Stucchi tricycle closely resembled its De Dion progenitor, closer inspection reveals some detail differences. These show that it was not a direct copy of the De Dion but was almost identical to the French Rochet tricycle which was, in turn, De Dion-inspired.[13]

There are also detail differences in the design of its engine. While the heavily finned cylinder and head of the Italian version appears a virtual carbon copy of the French unit, the Milan-built one seems to feature a ferrous crankcase, divided horizontally across its centre line, rather than being cast in aluminium and split diagonally, to permit the fitment of the crankshaft.

Two Instead of One

It seems likely that for all his initial enthusiasm for the tricycle, young Bugatti found it a trifle pedestrian for his tastes and began thinking of ways to improve its performance. His ingenious solution, which could be simply and speedily executed, was to add a second engine to the machine.

On the single-engined P and S three-wheeler, the power unit was offset to the right-hand side. The drive was transferred via a pair of spur gears – the smaller one was driven off the crankshaft, which meshed with its larger opposite number and drove the rear wheels via the differential. It should also be remembered that the tricycle's pedals remained in place for starting and helping the rider up inclines.

If what contemporary illustrations we have for Bugatti's machine are to be believed, the conversion apparently entailed retaining the original engine position, with the second unit added on the other side of the large, centrally located drive gear. The crankcase on this second engine was mounted back to back with its opposite number, so that its crankshaft extension could be coupled to work in tandem with it.

Ettore was proud of what may have been his first mechanically innovative idea. He was to recall that 'working in this factory enabled me to get permission to build *the first two-engined tricycle.*' [His italics.] Significantly, Bugatti was to adopt this modular method of increasing the power of a car throughout his career.

Having, by 1910, established his four-cylinder theme, he was to subsequently double, treble and even quadruple his original configuration. Consequently, no six-cylinder Bugatti engine was ever built.

---

[13] I am indebted to Malcolm Jeal for this information.

When Bugatti came to produce his twin-engined version of the Prinetti and Stucchi trike in 1898, he had to reinforce the front forks and also introduce what appears to be an external contracting brake on the rear wheels. The presence of the second engine also involved him in moving the original 4 volt De Dion coil from its original position to the right of the single engine and relocating it, together with a second, above the engines and just below the small saddlebag.

Birth of a Racing Driver

Sixteen-year-old Ettore had his 4CV, two-engined tricycle ready for the Turin–Asti–Alessandria–Turin race held on Sunday 17 July 1898.[14] This was the second pioneering competition to have started from Turin. The first 109km (68-mile) 'trial race', from the city to Asti and back, had been held three years previously, on 29 May 1895, only ten months after France's Paris to Rouen trial of 1894 – generally considered to have been the world's first motor competition. As there was not yet an Italian motor industry, all the five entries were foreign and the event was won by a Daimler from Germany.

[14] Cyril Posthumus, 'Early Italian Races, Part I' in *Veteran and Vintage Magazine*, vol. 7 no. 2 (October 1962).

The 1898 race was more ambitious and was held to coincide with Turin's Italian Exhibition, staged to celebrate the 50th anniversary of the start of heroic struggles which culminated in the creation of the Italian state. The Bernardi-designed Miari e Giusti three-wheeler car was on display there, where it was judged to be the most significant invention present.

The race itself attracted thirteen starters, of which Bugatti was one. The report of the event in the Exhibition's own magazine does not mention him by name but refers to two Prinetti and Stucchi tricycles entered, one of which had two engines. We have the authority of Carli's *Stettant'Anni di Gare Automobilistiche in Italia*, that one of the trikes was ridden by Ettore, so it is fair to assume he piloted his *bimotore*. The other P and S three-wheeler was, incidentally, ridden by Giovanni

Ceirano of Welleyes bicycles and car fame. But neither rider completed the course. Indeed, a total of nine entrants dropped out, leaving four survivors. The first entrant to return to Turin was Luigi Storeo, on a De Dion tricycle of the type he sold in the city. Storeo averaged a speed of 33.99kph (21.12mph) over the 193km (120-mile) course. Italian morale was upheld by the winning car (driven by one Ehrenfreud), since it was a Miari e Giusti.

Ettore had better luck in 1899, in what was to prove the busiest (in terms of competition) of his three years with Prinetti and Stucchi. He was to run in nine Italian speed contests during 1899, of which the following were the most important. The by then seventeen-year-old Ettore, and his tricycle, were featured on the front cover of the Italian motoring magazine, *L'Automobili*, of 15 January. In March, he rode one of eighteen tricycles entered, and had the distinction of winning the 160km (100-mile) Verona–Brescia–Mantua–Verona race, ahead of Count Carlo Biscaretti di Ruffia, who would become one of Fiat's founding fathers four months later, on a French-built Phénix tricycle. He would recall his second place 'just behind Ettore Bugatti in his powerful twin-engined Prinetti Stucchi. [The race] was an unforgettable experience for all of us.'[15] The four-wheel class was won by Giovanni Agnelli on a Phénix quadricycle.

Ettore entered his tricycle (or at least is said to have been listed as entering) for what was probably his first foreign event: the French 120km (75-mile) Nice to Castelanne race on 21 March, though it is not known whether he participated. What we do know is that, early in May, as one of forty-two tricycles entered, he won the Turin–Pinerolo–Turin event, run over a distance of 90km (56 miles). This was, he said: 'The best of these races, the one which gave me most satisfaction ... I beat Gaste and Rigal, who had come from Paris to beat me. Even before the start, I was sure of winning.'

Later in the month Padua staged an ambitious 175km (108-mile) road race between the two towns of Padua and Treviso and back. For this competition, Ettore Bugatti forsook his twin-engined tricycle and opted instead for a more powerful four-wheeler.

In addition to producing tricycles, Prinetti and Stucchi also built quadricycles, which either took the form of an additional wheel and seat mounted behind a tricycle, resulting in a diamond-shaped configuration, or a more conventional four-wheeled rendering with an additional seat mounted at the front of the machine. Bugatti may have first ridden such a machine in the Padua–Treviso event where, according to *L'Automobili*, he won the quadricycle class, covering the distance in a time of 4 hours 35 minutes.

In May 1899 Ettore undertook his most ambitious race to date by entering his tricycle in the Paris to Bordeaux race. This event had its origins in the celebrated Paris–Bordeaux–Paris contest of 1895, which had been so convincingly won by Émile Levassor in a Panhard Levassor. It had then lapsed but had been revived by *Le Velo* magazine in 1898 – though the return journey to Paris was dispensed with. In 1899 when Bugatti entered his tricycle, it had been reduced to one day, 24 May, rather than the two of the previous year, though the race was, to some extent, overshadowed by the nine-day *Tour de France* staged later in the year. Nevertheless the Paris–Bordeaux attracted sixty-five starters and Ettore was one of thirty-seven 'motor cyclists'.

Bugatti Abroad

His racing successes may well have spread beyond his own country for *The Autocar* recorded that, in the motor cycle section: 'All the professionals were there, and the foreign element was composed of S F Edge and C Jarrott [from Britain] and Bugatti, of Milan, who was riding an Italian-made tricycle.'[16] In the event, Ettore recalled '... running second [in his class], twenty minutes behind Osmont, when I ran out of petrol and had to give

---

[15] Carlo Biscaretti di Ruffia, 'Origin, rise and early development of the Fiat', in *Fiat – a Fifty Years' Record* (Fiat, 1951).

[16] *The Autocar*, June 1899.

up (damaged fuel tank)'. The race was won by Charron at the wheel of a 12hp Panhard, a make which occupied the first five positions.

Back in Italy, Bugatti won the Guastalla–Reggio, and Emilia races. He had a busy September, entering the Brescia and Verona contests. He was second in class in the latter, while he was placed third in the Trevise 3km (1.8-mile) race, though he was a non-starter in the 80km (50-mile) event due to a petrol blockage.

In May 1900 he rode his trike to a class victory at the Bologna 'Riunione' in a 75km (47-mile) run to Corticella and back. A Prinetti and Stucchi quadricycle also won its class, driven by a friend of Ettore's: Count Gian Oberto Gulinelli from Ferrara. Bugatti repeated his previous year's success at Padua: once again he entered his quadricycle and won his class. The event marked successes for the new Fiat marque, for it was won by Vincenzo Lancia, ahead of a Panhard, with another Fiat in third place. Vincenzo Lancia, like Ettore, was to later found his own car company. Held on 1 July, this 220km (136-mile) event between the towns of Padua, Venice, Treviso and back, was Bugatti's last recorded race with a Prinetti and Stucchi machine.

While Ettore undoubtedly enjoyed what was a highly satisfactory competitive record, which could only have brought his name and that of Prinetti and Stucchi to a far wider public, he '. . . regarded the tricar as a test machine to be constantly altered and developed, and racing it to be the means of judging modifications, of deciding whether they should be retained or discarded'.

Despite his hectic racing programme, Ettore continued to develop Prinetti and Stucchi products, though with no doubt more than half an eye on the competition:

> I was studying the different types of engines being used at the time, examining their qualities and discussing their defects. And I determined to build a car of my own. As soon as I obtained permission, I designed it and got it built. The car was quite small but had four [rear-mounted] engines – two in front and two behind the rear axle.

Four-Engine Excursion

With this vehicle, which was probably based on a P and S quadricycle, Bugatti took what had proved to be the successful concept of his twin-engined tricycle to an illogical conclusion by doubling up the number of its single-cylinder units. Even the usually optimistic Ettore later conceded that he was 'not happy with it', which suggests that the results of this experiment must have been fairly calamitous. He did, however, get the Pirelli company to produce pneumatic tyres, of 650 x 120mm dimensions, for it.

This Milan company had been established in 1872 by Giovanni Battista Pirelli, one of the Milan Polytechnic's first graduates, to manufacture rubber and elastic products. Bicycle tyres followed in 1890 and, maybe prompted by Bugatti's experiments, Pirelli introduced the Ercole in 1901, which was Italy's first tyre specifically designed for use on cars.

Recognizing the limitations of his four-engined concept, Ettore '. . . decided to build a second car. But Prinetti and Stucchi refused to have it made, saying that they intended to concentrate on the manufacture of cycle-cars.' Nevertheless, Bugatti's ideas may have influenced them to put a twin-engined four-wheeled car on the market in 1899 and there is a surviving example of such a machine in the Turin Motor Museum.

In this model, which is quite unrelated to a quadricycle in construction, the engines are at the front rather than at the rear of the car. They are coupled together and their total capacity is given as 516cc, making 258cc per engine, developing 4bhp at 1,500rpm. As this does not coincide with any known De Dion Bouton engine capacity (the nearest is 269cc), it can be assumed that these were purpose-designed P and S units. Drive was somewhat unsatisfactorily transferred to the rear-mounted gearbox by a single belt. Pirelli tyres also featured.

Just how many of these twin-engined cars were built is unknown and, in any event, by 1902 Prinetti and Stucchi was no more. Ludovico Prinetti had in the meantime decided to join with lawyer Cesare Isotta and the brothers Vincenzo and Oreste Fraschini, in the creation of the Isotta Fraschini company, established in Milan in 1900. For reasons which are unclear, Prinetti did not take up the shares to which he was committed and was eventually, in 1913, sued by the company for defaulting on his allocation.

Isotta Fraschini had its origins in an agency which sold French Mors and Renault cars and Aster engines. Its first car, a 6.5hp single-cylinder model, arrived in 1901. It was not until the following year that what was to become one of Italy's most famous makes began to feel its feet with the arrival of a four-cylinder 12hp car. This was the work of the company's talented consulting designer, Giuseppe Stefanini.

By this time Bugatti had not only left Milan but also Italy. He does not indicate the circumstances of his departure from Prinetti and Stucchi in 1900 but, as he participated in his last reported race on a P and S machine at Padua on 1 July, he may have left the firm at about then. All he says was that he had 'time to reflect and think over my projects, and *I suddenly found that I was left to my own devices*. [My italics.] (The same thing happened several times in later life.)'

On His Own

Ettore's three years with the firm had been extremely valuable ones. In addition to his racing activities, he was becoming well versed in the practical aspects of motor vehicle construction, while he was also 'taught how to use materials to the best advantage by a kindly man who saw everything very clearly, and his advice was the best teaching I could have had . . . I often thought to myself that he was full of years and experience, and how happy I should be if I could ever know as much as he did . . .' By the time Bugatti left the firm: 'I had realized that by then I was completely taken by mechanics, in which I could clearly see such imperfections.'

Before moving on from this episode in Ettore's life, reference should be made to the Type numbers he allocated to his car and engine designs which will continue to pepper this text. In about 1907 he set down tantalizingly brief details of his first eight Type numbers and Type 1 was simply described as 'P.S.'. This can only refer to Prinetti and Stucchi and to one of two projects: Ettore's twin-engined tricycle of 1898 or his four-engined car, which may have been built in 1899.

Of these vehicles, the four-engined four-wheeler is far more likely to be the Type 1 because it was closer to a car than the tricycle and Bugatti had no hand in the overall conception of the latter, as the design was completed before he joined the firm. Even more significantly, in *Confidences de Monsieur E. Bugatti*, dictated by Ettore in 1926, 'E.B. says that his first car was the four-engined quadricycle . . .'[17] This allocation also excludes the front-engined Prinetti and Stucchi from any such candidacy because Type 2 is listed as 'C. Guli' which clearly refers to the car sponsored by the Counts Gulinelli – a car that Bugatti designed after leaving P and S.

Having severed his connection with Prinetti and Stucchi, Ettore 'decided to get down to designing my car first, and to plan the assembly of it later.' His efforts were witnessed by a correspondent, 'E.V.', who wrote an account of the creation of this car for the 10 May 1901 issue of the Milan-based newspaper, *Gazzetta dello Sport*. He recalled, in the previous year, witnessing the sight of Carlo Bugatti's studio 'full of vast rolls of paper – Bugatti spent whole days designing the vehicle in pencil and crayon in all details. How often the worthy Bugatti took me to his studio, showing me and explaining on a score of sheets of paper the function of his car! His enthusiasm in describing it was such that it seemed to be already on the road at 60kph (37mph): indeed I feared it might turn his mind.'[18]

Ettore had probably mentioned his project to one or both of two brothers, the Counts Gulinelli. (At least one of the brothers had competed in many of the same races as Ettore Bugatti.) L'Ébé,

[17] Hugh Conway, 'Those missing type numbers' in *Bugantics*, vol. 41, no. 1 (spring 1978).

[18] HG Conway, *Bugatti Le pur-sang des automobiles* (Foulis 1987).

Ettore in his first car, completed in 1901. It won a gold medal in the Milan International Exhibition held at the city in May of that year and followed Panhard practice in having a front-mounted engine, a 3-litre four, with chain drive.

tells us that her father '. . . was often invited to stay with them on their huge estate outside Ferrara, and it was there that he became keen on horses and riding'. Ettore remembered the 'pleasant surprise' he had 'towards the end of the summer of 1900, when the Gulinelli brothers asked me to join them'. Perhaps it was this endorsement that persuaded Carlo Bugatti 'to participate in a small way' though Ettore ruefully notes 'although he had little faith in the success of my undertaking'.

Bugatti says that: 'IQ October I began making the wooden patterns which were needed for the casting of the various parts at the foundry.' Work must have proceeded at a good pace for, in his May article, E.V. mentions that 'a month ago', dating his visit to early April 1900:

> On my return from the Circuit of Italy trials I went to visit Bugatti in his workshop in company with Chevalier Ricordi and my friend Georges Berteaux. The car was under construction and Berteaux was most enthusiastic and told Bugatti it would be an undoubted success even abroad. So sure was he of its success that he undertook to buy one of the first type.
>
> The day before Bugatti fitted the wheels, installed the provisional ignition system and drove out of the workshop for the first test. The car went superbly and can at present do 60km/h [37mph] but the power of the engine is such that it can without any effort pull a higher gear ratio.

This might suggest that the car was slightly undergeared.

It was ready in time for Milan's International Exhibition, which opened in May 1901. Ettore's car was displayed on a stand belonging to Giuseppe Ricordi, the younger son of music publisher Tito Ricordi and a member of Carlo Bugatti's circle.

Giuseppe had cornered the market in all the significant Italian makes of the day; namely Bianchi, Fiat, Isotta Fraschini and, significantly, in the light of future events, the products of the De Dietrich company which was building cars in both France and Germany. The locally published *Gazzetta dello Sport* was in little doubt of the most impressive motor car present, when it described as 'the sensational novelty of the Exhibition: The Bugatti-Gulinelli carriage. The engaging daredevil of Italian motor cycling conceived the idea of producing a kind of light car which was not at the time on the market.'

Triumph at Milan

There was a competition, organized by the Association of Lombardy Journalists, for the best Italian car present. The young, confident Ettore decided to enter his car in the trials held to evaluate the respective vehicles. He remembered that: 'The competition was very severe. The members of the organizing committee and the Automobile Club included some highly qualified motorists.'

He recalled Vincenzo Lancia, entering a car on behalf of his employer, the two-year-old Fiat company. This 10hp two-cylinder model of 1,082cc won the principal pize donated by King Victor Emmanual. There were nine further awards and the one which Bugatti won was 'For the car of private construction proving to be the fastest among those exhibited and which joins to speed the requirement of solidity, optimum functioning, and of clever construction.' For this achievement, Ettore received a diploma and a handsome silver cup bearing the arms of the city of Milan and decorated with enamelled panels. It was a considerable accolade for the nineteen-year-old self-taught engineer.

So what made Bugatti's car such a worthy winner? Although we lack specific details of its design, it seems highly probable that its overall concept formed the basis of the models he built up until 1903 and, with that knowledge, we can piece together its overall specification.

The fact that the car which Ettore created during his time with Prinetti and Stucchi was unhappily powered by four single-cylinder engines, suggests that he had been converted to the need to build a single four-cylinder unit. Although of 3,054cc capacity, this was considered to be a relatively small car

A subsequent photograph of Bugatti in his first, much garlanded car. By this time it had acquired a single headlamp and third gear lever, perhaps for engaging reverse.

Ettore in the same car at a seaside location after it had covered some miles. The front headlight stirrup has lost its lamp.

in its day and, Bugatti thought it, 'a real triumph. I think it must have been the first *voiture légère* [light car].' This is something of an exaggeration because this term not only referred to a type of car but was also applied to a racing class. The latter was formalized in 1901 and would relate, for the next six years, to vehicles of 250 to 400kg (4.0 to 7.9cwt) in weight.

What Bugatti, in all probability, did was to echo the theme of the most successful racing car of 1899, namely the 90 x 130mm, 3,308cc four-cylinder 12hp Panhard which fell within the *voiture légère* category. This car had swept the board at that year's Paris–Bordeaux race, in which Ettore had unsuccessfully competed, and later in the year it had gone on to win the Paris–Ostend and Paris–Boulogne contests.

As far as the power unit of his car was concerned, Bugatti himself tells us that: 'The four-cylinder engine was water-cooled and had a bore and stroke of 90mm x 120mm . . . [It] had overhead valves and I believe that I was the first to build engines with this feature . . . An important detail for that time, was that each pair of valves could be instantly dismantled by a single nut.' Although cars had been fitted with mechanically operated overhead valves prior to this date, the Lanchester of 1896 is a case in point, they were extremely rare. The conventional variety of valve layout, typified by the De Dion Bouton engine with which Ettore was so familiar, was a mechanically operated, side-mounted exhaust valve while the spring-loaded overhead inlet valve was of the automatic type which opened by atmospheric pressure as a result of the vacuum created when the engine's piston descended on its inlet stroke. This was the type of inlet valve fitted to the prize-winning 10hp Fiat.

Bugatti can be seen to have anticipated European trends, for Daimler is considered to have introduced a pushrod-operated overhead inlet valve to the motor industry in its 60hp Mercedes of 1903. Griffith Borgeson[19] had convincingly suggested that the inspiration for the overhead valves of Bugatti's car was the by then obsolete Bernardi-designed Miari e Giusti tri car of 1896.

Local Influences

Ettore would have had plenty of opportunity to study these advanced little cars as they competed

[19] Griffith Borgeson, *Bugatti by Borgeson* (Osprey, 1980).

in many of the race meetings in which he himself had participated. An example had won the Turin–Alessandro–Turin race of 1898, which was the first event in which he had taken part. They continued to put up a good showing and in Bugatti's last race, at Padua in July 1900, Bernardi's son, Lauro, was second in the *voiturette* (small car) class. From 1899 the Miari e Giusti was renamed the Bernardi, though only lasted until June 1900.

While the Bernardi single-cylinder engine featured advanced pushrod actuation of his overhead valves, Bugatti perversely opted instead for *pull*rods, what E.V. in his description of the car refers to as a 'completely new system for closing the inlet valves'. This is perhaps the first notable example of Bugatti's own, utterly idiosyncratic approach to car design. His engine was fitted with twin camshafts either side of the blocks in the manner of the newly introduced T head Mercedes. But the lower ends of the rods were circular and surrounded the cams, the motion being transferred via rollers. These pullrods were to be a distinctive feature of Bugatti's engine designs until 1907.

Each pair of cylinders was contained within a circular water jacket, in the manner of the Bernardi, with the low-mounted carburettor on the right-hand side of the engine and exhaust on the left. This resulted in a very neat configuration.

Bugatti tells us that 'Ignition was by electric tube or magneto'. He adds that the car 'had chain drive,

There are no photographs of the engine of Ettore's Type 2 but it probably looked like this. The distinctive circular water jackets of Bugatti's later De Dietrich cars are revealed on his first racing car. He has changed the seating position to the rear of the car with luggage space positioned between the driver and engine. This car was probably completed in Italy, prior to Bugatti going to Germany in 1902. But who is in the driving seat?

four forward gears and one reverse'. We have it on the authority of E.V. that 'the engine, gear change and all the rest of the machinery are fixed on a double frame of rectangular steel of very great rigidity.' This probably refers to the subframe on which the mechanical components were mounted.

Most cars of the day – the racing Panhard and the award-winning 10hp Fiat were typical examples – had wooden chassis, though Daimler's Mercedes model of 1900, which boasted a steel frame, was soon to render it obsolete. Certainly, the cars which Bugatti designed for De Dietrich were described as having a 'main chassis . . . built of stamped steel, lined with wood' and this advanced specification applied to the Milan one, as indicated in the hitherto unpublished photograph shown below. Ettore's 12hp car, to which he alloted the Type 2 designation, employed half-elliptic springs all round and a simple open two-seater body was fitted.

At some stage during the exhibition, which was held during May and June 1901, Ettore left Milan for the picturesque Lake Maggiore, about 50km (31 miles) to the north-west of the city, with eighteen-year-old Barbara Mascherpa Bolzoni. Barbara was a childhood friend of Ettore's and a Roman Catholic like himself, who would eventually become his wife. Both families were well known to one another and, on occasions, the respective Bolzoni and Bugatti mothers had looked after the other's offspring.

L'Ébé Bugatti has left us with this picture of nineteen-year-old Ettore, an 'elegant young man, with a touch of dandy about him' who was poised on the brink of an extraordinary career. He was:

A well-built, healthy-looking young man, with clear-cut features and a more than average height. He had a rather long nose and deep-blue eyes which were full of intelligence and kindness. His shapely mouth had a ready smile, revealing fine, white teeth; his rather full lips indicated a love of the good things of life, yet with the high, wide forehead and chestnut hair – soon to start thinning – he gave the impression of being a thoughtful, clear-minded man. A strong-willed, dimpled chin complete the portrait of this lively, fine-drawn face. His hands were a distinctive part of him, vigorous yet controlled in their gestures, like all the movements of his refined person.

By the end of the following year, Ettore would have left his native Milan and become resident in Alsace, which thirty years previously had been annexed by Germany. This area would be home for Bugatti and his family for close on forty years, though, as will become apparent in the next chapter, it was a move in which the element of chance was to play a crucial part.

Ettore Bugatti demonstrating that the steel chassis of his Type 2 was strong enough to carry his and additional weights and taking the precaution of having the feat photographed.

2
Opportunity in Germany

'I built a new model which went into series production for a few years at the oldest metallurgic factory in Alsace–Lorraine, which thus became one of the first automobile factories in the world.'

Ettore Bugatti

In the eight years between 1902 and 1909, Bugatti would gain valuable experience by working for three German car makers: De Dietrich, Mathis and Deutz. During this period, his heady optimism knew few bounds and his designs would progressively evolve so that, by 1907, they would be at the very forefront of European thinking. Also at this time, Ettore would perpetuate the practice, established in his native Italy, of regularly campaigning his cars in competition, which provided excellent publicity for his products. All these activities would pave the way for him to start building cars under his own name in 1910.

Back in 1901, the finance necessary to manufacture Bugatti's car had been readily forthcoming. As E.V. reported in his article of 10 May 1901: 'The car now built is the welcome result of the anxieties, the difficulties and efforts of the past. Now Bugatti, teamed up with the brothers, Counts Gulinelli, will create a vast factory in Milan for the building of motorcars.'

Societa Bugatti e Gulinelli had been established in 1901 to these ends and the resulting cars would have been called by the Bugatti-Gulinelli name. But destiny was to intervene for, 'this fine project had to be shelved, because of the death of one of the Gulinellis.'[1] We do not know when this was. It could have been in 1901 or even 1902 because, by the middle of the latter year, Ettore had sold a licence to manufacture his cars to the De Dietrich company.

It will be recalled that a De Dietrich was one of the other vehicles displayed on Ricordi's stand. There, Bugatti's car was seen, either by a De Dietrich engineer, or director; accounts differ. Bugatti himself says that the individual was 'very interested in my ideas and construction. He said that Baron de Dietrich would like to see the car, and he asked me if I felt inclined to go to Niederbronn. The idea and the journey appealed to me – to travel through Switzerland and see new regions. I set off without any passports or papers – travel was free of all formalities in those happy days.'

This account would suggest that Ettore made his journey to Niederbronn during, or just after the 1901 Milan Show, and this may well be the case but, if this was so, no agreement was made between the two parties during that year. This was probably because Ettore was still committed to the Gulinellis, though this agreement was subsequently unscrambled, following the death of one of the brothers. Ettore then renewed his contact with De Dietrich. This scenario of events which, I must stress, is purely speculative would explain the gap of a year between the Milan exhibition and the signing of the contract between De Dietrich and Bugatti. This is dated 26 June 1902, which confirms 'the agreement we made . . . this morning',

[1] L'Ébé Bugatti, *The Bugatti Story*.

suggesting that the firm was most anxious to clinch the deal.

The De Dietrich company was France's oldest industrial enterprise. Shortly after the French had annexed Strasburg in 1681, the firm's founder had purchased an iron works 41km (26 miles) to the north at the city of Niederbronn, Alsace being rich in iron ore deposits. The business flourished and, by the mid-nineteenth century, had responded to the growth of the railways by making rails and wheels for rolling stock. But Germany's victory in the Franco–Prussian War saw Alsace–Lorraine ceded to that country in 1871. This left Niederbronn isolated from the French railway system, so in 1879 De Dietrich established a new works 24km (15 miles) inside the new French border at the town of Luneville, on the Paris to Strasburg railway line.

Once again rolling stock was the firm's principal product but, during the slump of the 1890s, this branch of the company, managed by Baron Édouard de Turckheim (who had married a de Dietrich daughter), decided to diversify into car production. In this venture de Turckheim was greatly encouraged by his son Baron Adrien, who in 1897 purchased, for 500,000fr. (£125,000), the manufacturing rights of the two-cylinder horizontal-engined 2.3-litre Léon Bollée, which it built for the German market.

The cars enjoyed moderate success, though Amédée Bollée also maintained production at his Le Mans factory. De Dietrich sold 72 cars in 1898 but this figure had risen to 164 by 1901. This was a year in which Panhard, as the market leader, built 723 cars. By then, however, the Bollée concept was becoming outdated, vertical engines quickly establishing themselves as the norm, and in 1902 Luneville took up a Turcat Méry design which was to form the basis of the French-built De Dietrichs until 1914.

The German headquarters of the company, run by Baron Eugène de Dietrich, was based at the Reichshoffen district of Niederbronn. There, the Léon Bollée was also briefly made, though this was abandoned in 1899 and, instead, the lightweight Vivinus *voiturette* from Belgium was manufactured. This was only a short-term expedient, so Bugatti's arrival could not have been better timed.

A Fine Contract

Because Ettore was not yet twenty-one (he would have been three months later, on 15 September 1902), the letter from *De Dietrich et Cie*, of 26 June setting down details of its offer, was addressed to Carlo Bugatti at the family home, 13, Via Marcona, Milan. Its terms were extremely favourable to Ettore and may have been the result of negotiation by his father.

The letter, which fortunately survives, makes no mention of Bugatti's 12hp Milan car but does refer to what appears to be two new designs, 'the production of motor cars of the type that Mr Ettore Bugatti has designed and built.'[2] These are listed as 10 and 15hp models and Ettore would receive a royalty of 400fr. (£16) and 500fr. (£20) respectively on each.

Even more intriguingly, Bugatti was obviously pursuing his competitive activities because there is reference to '... the racing car which he has shown us.' This may well be the machine Bugatti campaigned in Germany and Austria later in the year. There was obviously an intention to manufacture this vehicle because a royalty of 2,000fr. (£80) was payable to Ettore 'on each racing car, when the sale price is above 20,000fr. [£797]. If the price of these racing cars is below this amount, we will pay him 10 per cent of the difference.'

In addition, De Dietrich offered to pay Bugatti the sum of 50,000fr. (£1,992) by instalments. This was made up of 10,000fr. (£398) for 'his car' which is presumably the one from the Milan show. On arrival at Niederbronn, he would receive a further 11,000fr. (£438). There would be two additional sums of 6,666fr. (£265) payable on completion of the 15 and 10hp design respectively and he would receive the slightly untidy 9,002fr. (£358) balance on the design of a singularly uncharacteristic Bugatti machine: an omnibus.

[2] L'Ébé Bugatti, *op. cit.*

All these models would be sold under the *De Dietrich–Bugatti* name and the firm would have exclusive rights to sell them in all countries, other than Italy. In that instance, Bugatti was awarded the Italian rights '. . . and we agree to supply him with cars at the usual trade price, less the above mentioned royalty and usual commission allowed to all our sales representatives.' The agreement bound Ettore for seven years, until 1909. There was also a recognition that he could engage in other designs but De Dietrich would have first refusal on them. The letter concluded in requesting that Carlo 'be good enough to have your son put his signature on your reply, and he will agree to keep to contract after he attains his majority.'

It would appear that Ettore did 'put his signature' to the contract. He was later to perhaps optimistically recall receiving '25,000 to 30,000fr. [£996 to £1,195] when I handed over my designs . . . I drew satisfaction through being able to support myself without anyone's help'. He clearly relished in being a free agent and '. . . without being on the staff of some firms and paid for my works as I did it, and I found satisfaction in receiving a sum of money for a completed job of work which had given me pleasure to think out and do – which had even been fun to do – and which left me free at the end of it.'

We know from his June contract that Bugatti had a racing car in his possession and it may have been this machine he entered for the Frankfurt Automobile Club's race meeting at the city's Oberforsthaus course on 31 August. This was reported in the 15 September issue of the German motoring magazine *Der Motorwagen*, and refers to Ettore being placed second in a 20hp De Dietrich. The race was an 13km (8-mile) handicap event for fourseaters which was won by one J Goebel in an 8hp Bergmann. Émile Mathis, the De Dietrich distributor in Strasburg, and a friend of Bugatti's, was third in a 24hp De Dietrich. At least one authority refers to this as a Luneville car but matters are complicated by the fact that both arms of the De Dietrich empire produced 16 and 24hp models!

Ettore also did well in a 16km (10-mile) scratch race when he was placed second at 63.83kph (39.66mph) behind C G Densmore's 40hp Mercedes, which won at 68.48kph (42.55mph). The magazine was greatly impressed and opined: 'The real winner is the new Dietrich–Bugatti car, with a 20hp engine, which achieved almost the same performance as the Mercedes–Simplex with 40hp!' The crowd was clearly delighted with Ettore's performance and: 'They greeted Mr Bugatti, who proved himself to be a driver of the very first rank, every time he drove past the Grand Stand. Even

Bugatti enjoying himself in his racing car at the Frankfurt meeting of 31 August 1902. He was placed second in both the handicap and scratch races.

the officials . . . left their places in order to get as close as possible to the bends, where one could see exactly in what masterly fashion Mr Bugatti took them every time.' It was a good start.

Bugatti was similarly successful in the following week at the Automobile Club of Austria's Semmering hill climb held on 7 September. There, he is listed as having entered a 24hp De Dietrich, and whether this is the same car he raced at Frankfurt, when it was mistakenly described as being 20hp, is not known. However, we can be certain what it looked like because there are two pictures of it in the 20 September 1902 issue of *The Autocar*. It is the car, often pictured with Ettore at the wheel and Émile Mathis in the passenger seat at the extreme end of the vehicle, with its engine exposed and its distinctive circular banks of cylinders clearly visible.

Driving this car, Ettore succeeded in putting up the fourth fastest time of the day, which was equivalent to 42.65kph (26.5mph). He came just behind Ferdinand Porsche in his ingenious locally built petrol/electric Lohner Porsche. Eight days after Semmering, on 15 September, Ettore celebrated his twenty-first birthday. He had financial security, was associated with one of the most powerful industrial establishments of its day and was also making his name as an accomplished racing driver. The future must have looked bright indeed.

Ettore's De Dietrich

This racing car may have been an early version of the 24/28hp De Dietrich–Bugatti, which entered production in 1903. It had a 114 x 130mm bore and stroke, giving a capacity of 5,307cc.[3] Ettore allotted Types 3, 4 and 5 to De Dietrich, which is simply listed as 'D.D.' so it could be assumed that this model can be assigned the Type 3 designation.

From contemporary photographs of the engine, we know that it followed Bugatti's familiar circular water jackets which were initially made of copper

[3] Hans-Heinrich von Fersen, *Autos in Deutschland 1885–1920* (Motorbuch Verlag, 1976).

though aluminium is also mentioned. The crankshaft drove two camshafts, positioned either side of the engine, via exposed front-mounted gearwheels. Overhead valves were actuated by angled, offset brackets from Bugatti's distinctive pullrods. The neat appearance of the power unit was accentuated by the fact that the carburettor was low mounted, while the exhaust manifolds were integral with the water jackets and also emerged below the level of the chassis frame. Whether this curious arrangement had any practical advantages is not known, but it seems more likely that Ettore was more preoccupied with the visual appearance of the unit. Ignition was by four dashboard-mounted trembler coils, one per cylinder, the spark being generated by a commutator, driven off the right-hand camshaft. There was a total loss lubrication system.

Drive was taken to the gearbox, which incorporated at its other extreme the differential unit, via the customary drive shaft while the clutch is a good example of Bugatti's perverse approach to design. Hugh Conway has pointed out that the male internal cone is on the engine shaft with the heavier outer one on the gearbox, instead of the other way round!

With the layout of the gearbox the limitations of Bugatti's knowledge, and his youth, become apparent because the all-indirect gears he employed would have been very noisy in operation and the change itself heavy. The input shaft is at a lower level than the drive shafts which resulted in a rather high unit, the chassis side members being raised to clear the top of the box. Consequently, when the body was fitted, the driver and his passengers were perched *on*, rather than *in* the car. The chassis itself was a simple rectangle of steel reinforced with wood, and half-elliptic springs were fitted all round.

In addition to this 24hp machine, there was also a 30/35hp version, with a 130 x 140mm, 7,433cc engine. This may have been Ettore's Type 4. Both models were destined for a limited production life, maybe the 24hp for longer than the 30, of a little more than eighteen months, from early in 1903 to the middle of 1904. Interestingly, they seem to

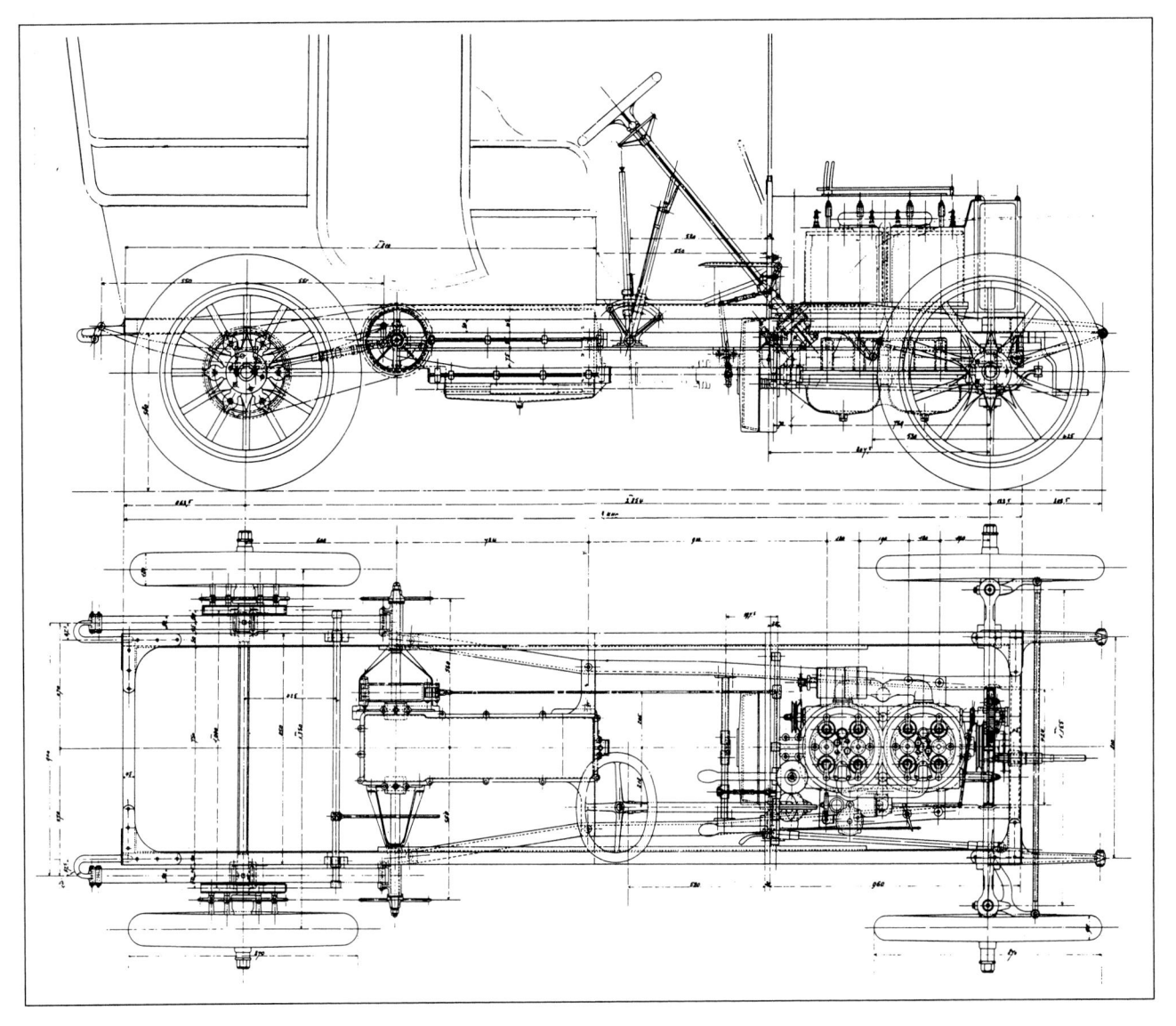

A drawing of a Bugatti-designed chain-driven De Dietrich of 1903. It was available with two wheelbases, one of 2,395mm (7ft 10in) and this longer one of 2,854mm (9ft 4in).

have been sold under the De Dietrich name, with the Bugatti half of the title omitted. This can only have added to the confusion between the products of the De Dietrich company's French and German divisions.

Examples of both types of De Dietrich reached the lucrative British market through the efforts of Luke Terence Delaney, who had begun his career in 1900 with the International Benz Company and subsequently sold the Birmingham-built Allard car in London. These were marketed by the Burlington Carriage Company of 315–317, Oxford Street, London, which combined its traditional coachbuilding activities with that of selling cars.

In 1902, Baron de Turckheim, head of De Dietrich's French company, brought two Luneville-built cars to London and Delaney arranged for Burlington to take up the agency for the make. Its manager at this time was Alfred Mays-Smith, later Sir Alfred, who became president of the Society of

A Neiderbronn-built De Dietrich of 1903. These earlier cars had gilled tube radiators . . .

. . . though later examples were fitted with honeycomb ones. Here Ettore, well protected from the elements in a fur coat, is at the wheel of one of his cars.

Vienna 1903 and two Bugatti-designed De Dietrichs on display. The radiator and fan on the chassis in the foreground, revealing Mercedes influence, can be clearly seen.

Motor Manufacturers and Traders in the 1919–1921 period. Burlington's concession for the French De Dietrich was short-lived because, early in 1903, it was taken over by Charles Jarrott, one of Britain's best-known racing drivers, who had just joined forces with William Letts to establish Jarrott and Letts Ltd, at 45, Great Marlborough Street. Jarrott accordingly forsook the Panhards he had previously campaigned and raced De Dietrichs in two important events in 1903.

Paris to Madrid Ban

Delaney's associations with De Dietrich were to reap benefits because he was offered a place as a driver in the firm's multiple entry for the celebrated Paris to Madrid race held on 24 May 1903. No less than ten De Dietrichs were entered, which represented the firm's serious attempt at motor racing. The front line trio of 45hp cars was headed by Jarrott, and a pair of French-domiciled Englishmen, Stead and Lorraine Barrow. On arriving at the start of the event, Delaney, in another 45hp car (race number 125), observed no less than Ettore Bugatti in a car with 'a centre of gravity as low as in the modern racing car [he was writing in 1945] and the driving seat as near the ground as the modern racer'.[4]

[4] LT Delaney, 'A 19th Century motorist looking back' in *Bugantics*, vol. 43, no. 3 (summer 1980).

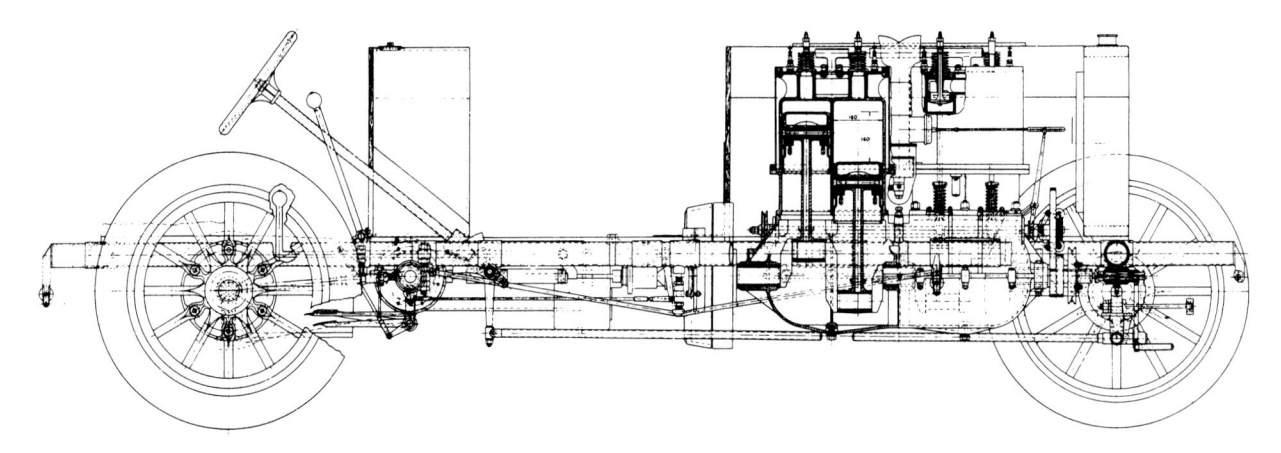

A drawing of Ettore's massive 12.8-litre Paris to Madrid racing car. Note that the engine has a three bearing crankshaft. In its issue of 18 April 1903, The Automotor Journal *reported its wheelbase was 'about 2,997mm [9ft 10in] and the track about 1,219mm [4ft]'. The circular chassis members carried cooling water.*

This purpose-designed, massive 60hp racing car, which may be the Type 5, shared a similar layout to that of the production De Dietrichs. It therefore employed the layout of its pullrod engine, though with 'square' dimensions of 160 x 160mm which gave a capacity of 12,867cc. This was mounted in a tubular chassis which carried, in the best Peugeot traditions, its cooling water. Ettore perpetuated the layout he had adopted for his first racing car, in that the driver was positioned at the extreme rear of the vehicle. According to Delaney: 'Although Bugatti went to the starting point, [the car] was not allowed to run, as the driver's line of vision was such that he could not see the road immediately in front of the bonnet.'

This decision probably reflected the anxiety of the Minister of Mines, who had responsibility for the event, and had only reluctantly sanctioned the race in the wake of the uproar which had followed the death of Count Eliot Zborowski at the La Turbie hill climb during Nice Week that month. These forebodings were vindicated, following a spate of accidents (some of them fatal) amongst drivers and spectators alike. The race claimed the life of Louis Renault's brother, Marcel, while the De Dietrichs of Stead and Barrow crashed, the latter suffering fatal injuries. Delaney was also unlucky but fortunately survived an accident in his car. The race was accordingly stopped at Bordeaux. Jarrott, it should be noted, was in fourth place at this time.

In these circumstances, Bugatti in his 60hp monster, was fortunate to be prevented from participating in what had been dubbed 'The Race of Death', particularly as he later confided to his son, Roland, that 'he was eternally grateful . . . because the car's brakes were merely symbolic and, had he participated in the race, he would have come to no good end.'[5] Ettore subsequently modified his Paris to Madrid entry with more conventional seating and, as far as is known, only ran the car in one event during 1904.

The De Dietrichs entered in the fateful Paris to Madrid race were, of course, Turcat-Méry designed, French-built cars but, in Britain, Delaney was also familiar with the Niederbronn-built models. 'I arranged with E.E.C. Mathis to purchase a number of German De Dietrich cars . . . I had no difficulty in disposing of these to clients who wanted early delivery and, in addition, the . . . Bugatti–De Dietrich gave a better performance than the French model'.[6] These were unveiled at

[5] Borgeson, *Bugatti by Borgeson.*

[6] Delaney, *op. cit.*

When Ettore created this car he perpetuated the curious rear seating position of his original racer. But the authorities refused his entry on grounds of safety and . . .

. . . it was rebuilt in this form with a conventionally located seat, so that the driver could see ahead of the engine. It has also been considerably tidied up with the bonnet lined, the colour scheme being echoed on the new seat.

The Automobile Show of February 1904, when the Burlington Carriage Company displayed the Burlington car which was, in truth, a 24hp Bugatti-designed De Dietrich. The unusual appearance of its engine came in for contemporary comment, for *The Autocar* made mention of its 'most remarkable feature [of] . . . the cylinders . . . placed within copper drums'.[7]

Photographs of the engine, which accompanied the account, show the unit little changed, apart from the replacement of the trembler coils by a high tension magneto. As by this time the Niederbronn-built De Dietrichs boasted honeycomb radiators and fans in the Mercedes manner and, as will emerge, the company's German arm had already decided to discontinue car manufacture, it can only be assumed that these 24hp chassis were earlier unsold examples which had been unloaded on the British market by the wily Mathis. Just how many of these Burlingtons were sold is unknown, probably only a few, and none survive.

The year 1903 was marked by the arrival in Milan of Ettore and Barbara's first child, a girl, born on 21 November. She was christened L'Ébé Maria Teresa, Ettore having concocted her unusual first name from his own initials! The Bugattis were not married at this time, a state of affairs which probably did not concern Ettore's parents in the slightest – they themselves led a distinctly Bohemian lifestyle – and the couple would remain in this unwedded state for a further three years. They finally married in Milan on the 25 February 1907 and their second child, another girl, was born later that year, on 14 July. Once again Ettore's influence in the choice of her name was evident. It was Lidia Germania Ettorina Maria . . .

A Parting of the Ways

Matters had come to a head between Bugatti and his employers on 22 January 1904 when De Dietrich informed Ettore that it was dispensing with his services. Ettore responded with a letter (25 January), but Niederbronn countered on 3 February, informing him that:

> Mr Bugatti constantly and in spite of repeated warnings on our side and our declaring him in default has not discharged his obligations to comply with the contract into which he had entered, and in view of the fact that as a consequence of this default of Mr Bugatti, manufacture and sales have become onerous for our house, our board of directors has decided not only to stop the fabrication of automobile car, system Bugatti, but also to drop the construction of automobiles in Reichshofen altogether.[8]

It was just eighteen months since the two had signed a contract and we can only speculate for the reasons of Ettore's dismissal. The most likely one is that he spent too much time pursuing his passion for racing and, as recently as 18 October 1903, he had been placed second in the 30hp and over class in the Berlin races. The creation of the abortive 60hp Paris to Madrid car must have absorbed much of Ettore's energies and such an involvement maybe flew in the face of the clause in his contract which stated that: 'Mr Bugatti agrees to supervise the execution of his designs and models, and the sound construction of the cars in our workshop.'

Thus the production of Niederbronn-built De Dietrichs ceased in 1904 though the announcement of the ending of their manufacture was not made until the month of October of that year. We can only guess the number of these Bugatti-designed cars built, but there is documentary evidence which refers to the construction of sixty-three engines.

This decision did not affect the production of the French-made De Dietrichs and, maybe to underline the fact that the cars were solely being built at Luneville, from 1905 the marque was renamed Lorraine–Dietrich and endured until 1934.

Bugatti makes no reference to his rift with De

---

[7] *The Autocar*, 27 February 1904.

[8] Griffith Borgeson, 'Portrait of the Artisan Engineer' in *Automobile Quarterly*, vol. 19, no. 4 (1981).

Dietrich in his reminiscences but simply says that: 'I changed tactics and dropped light cars in order to make big ones.' This followed Ettore having consolidated his association with Émile Mathis, whom he had known since 1902.

The Move to Strasburg

By this time Émile Mathis was on his way to establishing what he claimed, in 1906, to be Germany's largest car agency and the third biggest in the world. Like Bugatti, he was a keen and competent racing driver but was less scrupulous as a salesman and, for instance, when the manufacture of the German-built De Dietrich ceased, he imported genuine Turcat-Mérys from Marseilles and simply fitted them with De Dietrich badges! Mathis also handled such makes as Panhard and Rochet-Schneider and became the representative for both Fiat and Minerva for a large part of Germany and Austria.

He clearly had aspirations as a car maker and, as early as 1900, the prototype of a massive 11.3-litre four-cylinder model was produced. While *La Vie Automobile* maintained that it was being built in limited numbers,[9] this statement smacks more of Mathis the salesman than the aspiring motor manufacturer.

With the severance of Bugatti's involvement with De Dietrich, Mathis no doubt saw his opportunity to fulfil his ambition of becoming a car maker in his own right. The resulting automobile could then be marketed through his sales organization. Therefore, on 31 March 1904, he wrote to Ettore Bugatti setting down the terms of his commitment. He agreed to pay for the cost of the prototype car, which should not exceed DM12,000–15,000 (£587–734). If this proved to be satisfactory, Émile Mathis would be granted the world manufacturing rights. However production, for which he would be responsible, would be modest and would amount to no more than twelve to fifteen cars per annum. This suggests very limited manufacturing facilities.

Mathis agreed to pay Bugatti a royalty of 2,000fr. (£80) per car on a projected annual production of twelve cars though, in the unlikely event of output reaching 500 cars per annum, the royalty would then be cut to 600fr. (£23). The vehicles would not cost more than DM10,000 (£400) to make 'with some sharing of benefits' if less. Although the name of the car was not specified, Mathis stated that the publicity relating to it would credit the design as 'Licence Bugatti'. Ettore wasted little time and signed this agreement on the following day of 1 April.

This meant a move from Niederbronn, where Bugatti and his wife-to-be Barbara and baby daughter, L'Ébé, had moved from Italy. They established a new household in a pleasant rented villa in the Strasburg suburb of Graffenstaden.

The car was conceived by Ettore in a room on the top floor of the Hôtel de Paris on the rue de la Nuée-Bleue, which was owned by Mathis's father and this new generation of Bugatti-essayed cars reflect a clear maturing of his style. The chassis was of steel and, instead of perpetuating its simple oblong, the new chassis was waisted beyond the scuttle to increase its turning circle, which may have proved to be a limitation of the De Dietrich. Bugatti also dispensed with the subframe that he had previously employed to mount the engine and gearbox, which saved weight. Half-elliptic springs were perpetuated, though the track rod was more fashionably relegated behind the front axle.

When it came to the engine and gearbox, Bugatti clearly recognized the eccentricities of his 1901–1904 concepts and adopted a more orthodox approach. In his design for Mathis, Bugatti allowed himself to be influenced by the most successful racing car of its day, the Daimler Motor Company's 60 and 90hp Mercedes models, built only 100km (60 miles) or so to the east on the outskirts of the city of Stuttgart. It should be said that he was in good company, with such reputable firms as Fiat and Isotta Fraschini in his native Italy, Darracq in France and Napier in Britain, all following in the wheel tracks of the car from Cannstatt.

---

[9] Lord Montagu of Beaulieu, 'The Alsatian that lost its bite: Mathis of Strasburg' in *Lost Causes of Europe*, vol. 2 (Cassell, 1971).

A Car for Mathis

With the Mercedes, engineer William Maybach retained the twin camshafts of the pioneering T head side-valve Mercedes of 1901 but the left-hand one was relegated to operating, via pushrods, the low tension ignition points. The eight-lobed right one was used for actuating the new valve configuration, with the new pushrod and rocker arm operated inlet valves moved from their previous right side position and relocated on the left of the engine above the exhaust ones which remained in their existing position. This engine first appeared in 60 and 90hp forms in 1903 and the smaller capacity car subsequently won the principal race of the year, the Gordon Bennett event, which gave Mercedes its first major win in international motor racing.

We do not know just how much influence Mathis had on the specification of the car that Bugatti designed for him. He might have simply asked Ettore to 'do a Mercedes' and the outcome was almost exactly that, being predictably offered in 60 and 90hp forms while there was also a 50hp version. Ettore perpetuated his customary four-cylinder configuration, with two blocks each containing two cylinders apiece, but dispensed with the distinctive circular water jackets and associated manifolding he had previously employed.

The outcome was an outwardly less ordered but inwardly more efficient power unit. The concept of the aluminium crankcase, with two camshafts either side, was perpetuated though the front-mounted drive gears were now enclosed. Ettore followed Maybach's example by relegating the right-hand one for operating the points of a low tension ignition system. As on the Mercedes, the left-hand camshaft operated both the exhaust and inlet valves. For the latter, Bugatti permitted himself the indulgence of perpetuating his pullrods, though they would prove a limitation as far as engine speed was concerned. He had, it should be recalled, consistently positioned his valves in the overhead position, ahead of Mercedes. However, he relegated the exhausts below the inlets and relocated them, Cannstatt-wise, in the side position where they were conventionally tappet operated.

The carburettor remained on the right-hand side of the engine with the nearby magneto driven, by bevel gearing, from the rear of the crankshaft and balanced by a similarly located water pump on the opposite side of the engine. A heavily chamfered

The engine of a Hermes which Bugatti designed for Mathis featuring his curious pull rods. The capacity of this sole survivor is not known though the stroke has been measured at 160mm. The car is at Mulhouse and bears the chassis number 30.

radiator of great elegance and proportion – the Bugatti hand is at once evident – and a development of that tentatively employed on the De Dietrich, completed the power unit.

Transmission was also an improvement on previous practice. The clutch was of more conventional design and accordingly lighter to use, as the driven disc was located on the gearbox shaft. Bugatti also radically re-designed the gearbox so that the input shaft was below that of the output one, which made for a lower unit and resulted in an improved seating position. Yet another bonus, attributed to the *Système Mercedes*, was Bugatti's employment of sliding selector rods, which meant that it was no longer necessary to pass through every cog each time a gear change was effected. Chain drive was perpetuated while the distinctive water-cooled brakes were also standard Daimler practice.

As already indicated, this model was offered in three capacities, a 7,433cc 50hp car, with a 136 x 150mm bore and stroke, a bigger 8,261cc 140mm bored 60hp[10] and a larger capacity 12.8-litre 90hp, which shared the same 160 x 160mm dimensions of Ettore's Paris to Madrid De Dietrich.

What's in a Name?

When it came to naming the car, it only appears to have been known as a Mathis when Mathis raced one, or a Bugatti when Ettore raced one! Production models were named Hermes, after the fleet-footed Greek messenger of the gods. It was also variously known as the Hermes-Simplex, once again a Mercedes plagiarism, and as the Hermes–Bugatti in Britain. That country was the recipient of the first car (chassis no. 351), which was delivered, on 15 April 1905, to the Burlington Carriage Company – the association no doubt being a perpetuation of Delaney's established contacts with Mathis.

Delaney himself was later to remember that he 'drove the first car that Bugatti built as an independent . . . and raced with the same at Blackpool in 1905'. The event was the city's speed trials, held at

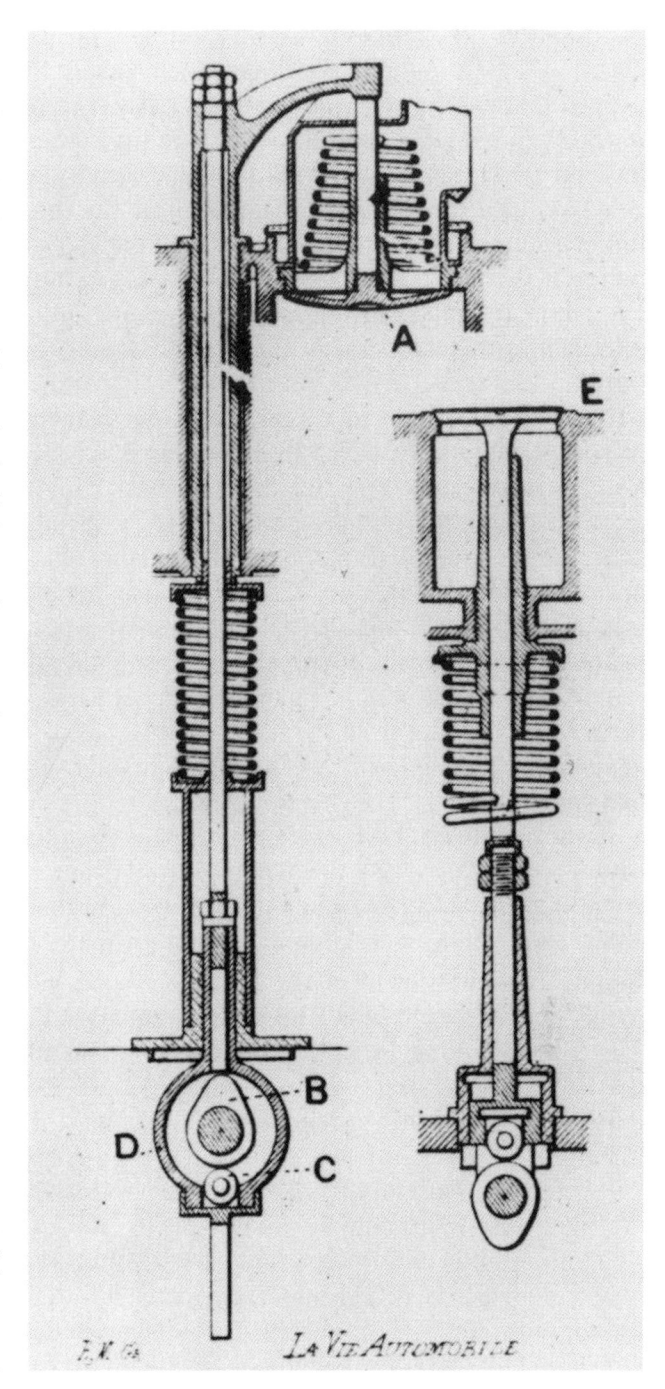

How the pull rod mechanism worked. Ettore was clearly an early disciple of desmodromic actuation as the overhead valve is both opened and closed by the cam. It was employed on all Bugatti's designs from 1901 until 1906.

---

[10] Variations are 136 x 146mm and 130 x 140mm.

Southport Sands, when he was placed fifth in his class, at a not particularly impressive speed of 97kph (60.5mph), in what was described as a 45/60 PS model. By contrast, Daimler's managing director Percy Martin recorded 117kph (73mph) in a side-valve Coventry-built Daimler. But by this time Delaney had become enamoured of the splendiferous Delaunay-Belleville from France and became fully committed to the marque, prior to him establishing his own Gallay Radiator Company in 1914.

Hermes car production – based at the SACM works at Illkirch–Graffenstaden, which is where Mathis was located – was relatively modest. Much work was undertaken by young Ernest Friderich who, at the time, was employed by Mathis. Friderich would recall that to get to the suburb, which was about 10km (5 miles) from the Strasburg city centre: 'I went by tramcar and Monsieur Bugatti on horseback.'[11] This was as characteristic as Ettore racing his cars and it was not long before he was campaigning the 'Mathis' in German competitions and speed trials.

Both Bugatti and Mathis were active, though with a conspicuous lack of success. Ettore drove a 60hp Mathis in the Herkomer Cup event in August 1905, while there is reference to Bugatti with 'a Bugatti' and Mathis in a car that also bore his name, in the Bleichroden Cup in the same month. The following year Bugatti entered a 45hp car in the Heidelberg Kaiserstuhl race. Although Mathis participated in the first eliminating trials for the inaugural Kaiserpreis event of June 1907, the 60hp car was not fast enough, having only averaged 58.10kph (36.1mph), and he did not qualify. But by this time Mathis and Bugatti had parted company, affably enough it would appear, as a celebrated photograph taken at the event of Bugatti's party, which includes Mathis, testifies.

In March 1906, two years after signing their original agreement, Mathis had once again written to Ettore, asking that the contract be renewed and requesting that he guarantee the manufacture of twenty cars a year. But Bugatti declined. Such limited production was hardly in his financial interests and he was no doubt anxious to design a new car which would reflect his formidable creative talents. Also the two men were temperamentally at odds with each other's approach to car manufacture. As French front-wheel drive pioneer, Jean-Albert Grégoire, who knew both of them, has pointed out: 'Bugatti loved beauty, quality of finish and the proper preparation of his cars ... Against this, Mathis's preoccupation was always to try and produce more cheaply.'[12] Grégoire also mentions Bugatti's concern at the nature of his contract, and so the pair went their separate ways.

Total Hermes production amounted to a mere fifteen examples. After Delaney's car of 1905, only thirteen were sold in 1906 and 1907. These were 40/50hp models though the last, of April 1907, was a 90hp car. One example of the Hermes survives; a 60hp car can today be seen at the *Musée national de l'Automobile* assembled by the Schlumpf brothers at Mulhouse.

Just how satisfactory were these cars? In January 1908 Mathis was to complain to Ettore that, 'You know the great problems I have with the Hermes vehicles. In fact no owner of these cars is satisfied with his and every sale is followed by legal action.'[13] This perhaps should not be taken too literally as Mathis was trying to get Ettore 'to buy the remainder of Hermes chassis which I still have' in view of their similarity to the Deutz frame. Mathis was to pursue his dream for producing a car under his own name. This finally appeared in 1910, though the 2.8-litre model was, in reality, a German Stoewer with a different radiator and hub caps...

On His Own Again

Back in 1906 Ettore had set up on his own account as a consulting engineer and had succeeded in getting Friderich to join him. Friderich remembered

[11] Ernest Friderich, 'How the firm of Bugatti was born' in *Bugantics*, vol. 12, nos. 2 and 3 (1949).

[12] Jean-Albert Grégoire, *Best Wheel Forward* (Thames and Hudson, 1954).

[13] HG Conway, *Bugatti* (Octopus Books, 1984).

that: 'We took up our quarters one kilometre from Graffenstaden in a shed in the centre of a large garden.' This Ettore equipped with machinery to produce yet another car, which he was to design. Finance for the venture was supplied through the good offices of the Darmstadt Bank, an enlightened establishment which, along with 'the Discount Bank and the Berlin Commercial Company played a key role in fostering German industry.'[14]

Friderich says that 'the shed and workshop adjoined the drawing office in which three draughtsmen started getting out the rough drawings to the Chief's instructions. There were also three of us in the workshop, one turner, one fitter and myself.' As speed was of the essence, Bugatti did not produce a completely new car. What he did was to take the existing Mathis chassis, as is echoed in the above letter, and designed an advanced single overhead camshaft engine for it.

However, there were a few problems that were not satisfactorily resolved. One was the presence of the existing steering box and, as will become apparent, Bugatti was forced to untidily squew the engine's water-pump drive to clear it. We understand that work began in March 1907. Friderich recalls that it was ready for the road in July and that 'its tests were highly satisfactory. After a few trips to Cologne Deutz, to the Deutz Gas Engine Works, the latter firm acquired a manufacturing licence.'

Gustav Langen, one of the firm's directors, had driven Hermes cars in the 1905 and 1906 Herkomer Trials, so was familiar with Bugatti and his work. Ettore moved to the city in late August or early September and was to take a house in the small town of Mulheim-am-Rhein just to the north of Cologne. He says: 'I took with me, to this factory, the draughtsmen who were in my employ and the mechanics who had assembled the model.' Significantly, instead of being a consultant, he joined the Deutz payroll. 'For the first time .. I agreed to take an appointment with a firm, while being remunerated by the royalties I received for the production of my car under licence.' But he also 'retained the right to work independently on other projects in which I might become interested'.

Despite what Ettore himself says, and other writers have subsequently asserted, Bugatti authority Norbert Steinhauser has discovered that Ettore was *not* empowered to undertake outside work and, when he did come to design what was to become his Type 10 in 1909, it proved to be a very difficult exercise.

Friderich, in the meantime, remained at the Bugatti workshops near Strasburg, where his 'last job was to dismantle the car, part by part', which he then delivered to the Deutz drawing office so that plans could be made from it. In October, Friderich went off to Luneville for two years of military service in the horse artillery.

Deutz Makes an Offer

The agreement between Bugatti and Deutz, which is dated 1 September 1907, is interesting in the way that it differs from previous arrangements. Perhaps Ettore's reputation of his time at De Dietrich had gone before him because the document refers to him giving his 'full-time attention' to the job and 'to work at the factory to supervise production.'[15] He was to be paid the sum of DM100,000 (£4,895), half on signing the contract and the balance within eighteen months. In addition he would receive DM1,000 (£48.95) a month. The equivalent of an annual salary of £600 was a handsome income for a twenty-six-year-old.

The car that Bugatti had sold to Deutz was an extremely advanced design for its day and both mechanically and visually displays characteristics that are to be found when he began producing cars under his own name. With this power unit, Ettore effectively started with a clean sheet of paper and accordingly dispensed with the curious pullrods which had been a feature of his approach since 1901. Once again all the valves were overhead ones though, for the first time, they were actuated by a single overhead camshaft. As Bugatti was to

[14] WO Henderson, *The Industrial Revolution on the Continent* (Frank Cass, 1967).

[15] HG Conway and M. Sauzay, *Bugatti Magnum* (Foulis, 1989).

thereafter feature one, and later two, such located camshafts on all his subsequent engines, the contemporary designs which may have influenced him are therefore of crucial importance to our story.

The first car in the world to feature a single overhead camshaft engine is said to be the British Maudslay of 1902, though this is of purely historical interest. It was Isotta Fraschini, from Bugatti's home city of Milan, that was responsible for what was, in all probability, the first competition car to feature a single overhead camshaft engine. The work of the firm's talented consulting engineer, Giuseppe Stefanini who, after a technical education at the Scuole Tecniche Operaie di San Carlo in Turin, worked for the Martina brothers. The latter were involved in the creation of the car that resulted from the pioneering tinkerings of candle maker Michele Lanza.

Later Stefanini moved to Milan and began his consultancy for Isotta Fraschini. His Type D 100hp monster was powered by a 17.2-litre four-cylinder engine, its in-line overhead valves being actuated by an overhead camshaft. For starting, the shaft was located in the first of two positions which released the compression of the cylinders. Once the engine had started, a centrifugal governor on the rear-mounted drive shaft moved the camshaft into position.

Two such cars were produced in time for the Coppa Florio race of September 1905 but both failed to finish, due to poor preparation. Despite this inauspicuous start, Isotta Fraschini displayed this pioneering machine at the 1906 Paris Motor Show. With his interest in design and racing, plus the Type D's geographical origins, it is inconceivable that Ettore Bugatti would have been unaware of this car's existence.

Perhaps the first company on continental Europe to apply the overhead camshaft concept to a road car was the small Gaggenau concern, named after the town of the same name and a mere 50km (30 miles) north-east of Bugatti's Stuttgart home. Theodor Bergman's Industriewerke was a firm which was originally known for its enamel signs and chocolate slot machine, and produced the notorious Benz-inspired Orient Express car of the 1895–1903 era. The motor car division was established in 1905 under the name of Süddeutsche Automobile-Fabrik. Two cars were announced at the 1906 Berlin Motor Show in February. The 35 and 60 models were significant because their respective 4.7 and 8.8-litre four-

The world's first overhead camshaft racing car? The 17-litre 100hp Type D Isotta Fraschini, designed by Giuseppe Stefanini, which ran in the Coppa Florio of 10 September 1905. Two examples entered and both cars made little impression as they had only been completed four days prior to the event. One car, driven by Trucco, retired after practice. This is Hubert Le Bon in the other Type D, conveniently revealing its engine, which retired during the first lap. Later, one of these cars was displayed at the 1906 Paris Motor Show.

cylinder engines employed single overhead camshafts. In-line valves in the Isotta Fraschini manner featured, though the drive for the camshaft was at the front rather than at the rear of the unit, and incorporated an engine-driven fan.

The Gaggenau car only remained in production until 1911, the German firm having been taken over by Benz in the previous year. These fast, potent cars were also active in sporting events and although their participation in the qualifying rounds for the 1907 Kaiserpreis proved unsuccessful, the cars did well in the Prince Henry and Alpine Trials of 1910.

A Mercedes Pioneer

The Gaggenau was a road car but, in September 1906, a six-cylinder Mercedes competition car broke the record at the Semmering hill climb in Austria. Driven by Hermann Baur for Theodor Dreher, its brewer owner, this car represented Daimler's excursions into the potential of the overhead camshaft engine. Back in 1902 William Maybach had produced a one-off overhead camshaft six for a Russian customer but work on this later project began during the winter of 1905/1906. The resulting oversquare 140 x 120mm power unit of 11,080cc capacity represented a combination of the archaic and advanced design features. The separate cylinders resembled an inverted T head design though the shaft drive from the camshaft was at the front, rather than at the rear of the unit. The drive shaft turned a fan, in the Gaggenau manner, while the engine's twin magnetos were driven, via bevel gears, from the front of the crankshaft.

Also, Maybach's design differed from both the Isotta Fraschini and the Gaggenau in that the overhead valves were located transversely across the cylinder rather than in line with the camshaft. This required the use of roller tipped rockers though these were later eliminated in some way. But the car's appearance at Semmering effectively represented the end of the project.

Maybach departed early in 1907, to be replaced by Paul Daimler and although Mercedes reverted to a single overhead camshaft for its celebrated 1914 French Grand Prix winning cars, their power unit derived from the firm's unrelated six-cylinder aero engines, introduced in 1913.

We do not know whether Bugatti knew of this car, though as will have been apparent, he was already well acquainted with Mercedes design practice. The point is an academic one. What is important is that during the years of 1905 to 1907, there was a climate of change in the automotive sector, as the overhead camshaft, with its greater efficiency at high revolutions, began to make inroads into the ascendency of the pushrod and rocker for operating the overhead valves of racing and sporting motor cars. And Ettore Bugatti was in the vanguard of this movement.

The other significant aspect of Bugatti's engine was that its cylinder block was cast in one piece. As Anthony Bird has pointed out 'cylinders were at first generally cast singly or in pairs; blocks of four cylinders or more were rare before 1907 and uncommon before 1912'.[16] Bugatti was therefore well in advance of contemporary thought when he began the design of his engine in March 1907 and he may well have been correct when he claimed that this power unit was 'the first, I believe, to have an overhead camshaft ... with the cylinders cast in the single block'.

It had been Amédée Bollée, back in 1899, who could probably claim to be the pioneer of casting a four-cylinder engine in one block, while the small Rolland–Pilain company from Tours was also an early advocate of that arrangement. Another 1906 contemporary was the extraordinarily advanced Aquila Italiana from Turin, the work of Giulio Cappa who, in 1914, would join Fiat as the head of its design office. In addition to a one-piece cylinder block, and a type of unit construction gearbox, it was also the first engine to be specified with aluminium pistons. By 1907, monobloc cylinder castings had reached Germany with NAG, for one, featuring such engines.

[16] Anthony Bird, *Early Motor Cars* (George Allen & Unwin, 1967).

Bugatti's first overhead camshaft engine, an advanced unit for its day because the block was also cast in one piece. This is the exhaust side of the engine. The trumpet-like device is the carburettor air intake which was warmed by the proximity of the exhaust manifold.

The Right Lines

Although the one-piece block was a more complex casting than the single and two-cylinder construction that it gradually replaced, its presence greatly strengthened the customary aluminium crankcase and resulted in a far more rigid engine. It also had the advantage of simplifying the process of assembly and it is easy to see Bugatti's artistic eye responding to the element of symmetry it bequeathed the engine.

Indeed, the Deutz power unit is imbued with a sense of proportion and order that was to feature in Ettore's subsequent engines. It is outwardly a less cluttered design than his earlier efforts, with its one-piece cylinder block and the inlet and exhaust manifolds echoing the circular cross-section of the camshaft casing. The latter was secured above the block by two triangulated bridge pieces, and also doubled as a housing for what we now recognize as Bugatti's distinctive banana-shaped tappets, with rollers at either end, which appeared for the first time on this engine. The shape was to be a feature of Bugatti's eight and later 16-valve fours until 1926.

The camshaft drive was by shaft and bevel gears from the rear of the engine, where its ancillaries were also located. These presented the only visually jarring note of the car's under-bonnet prospect and was the result of Ettore fitting his new power unit in a chassis for which it had not been designed. So that the water pump would clear the steering box of what was essentially the Hermes frame, Bugatti had to rather untidily squew it at a 45-degree angle. In truth, the magneto should have been located there, adjacent to the nearby sparking plugs, but there was no room for it between that intrusive box and carburettor, so Bugatti was forced to fit it on the left-hand side of the engine and the angle of its drive shaft echoes that of the pump.

The clutch was what proved to be Bugatti's long running multiplate unit and was subject of a German patent (no. 203453), which Ettore had registered on 22 March 1907. It consisted of a number of steel and iron plates running in oil, with pressure formidably and effectively applied by a spring and a pair of toggle levers. He would employ a similar arrangement on his engines up to 1935. The clutch release mechanism was particularly ingenious and unique to the Deutz. As the illustration on page 49 (bottom right) shows, it took the form of a flexible

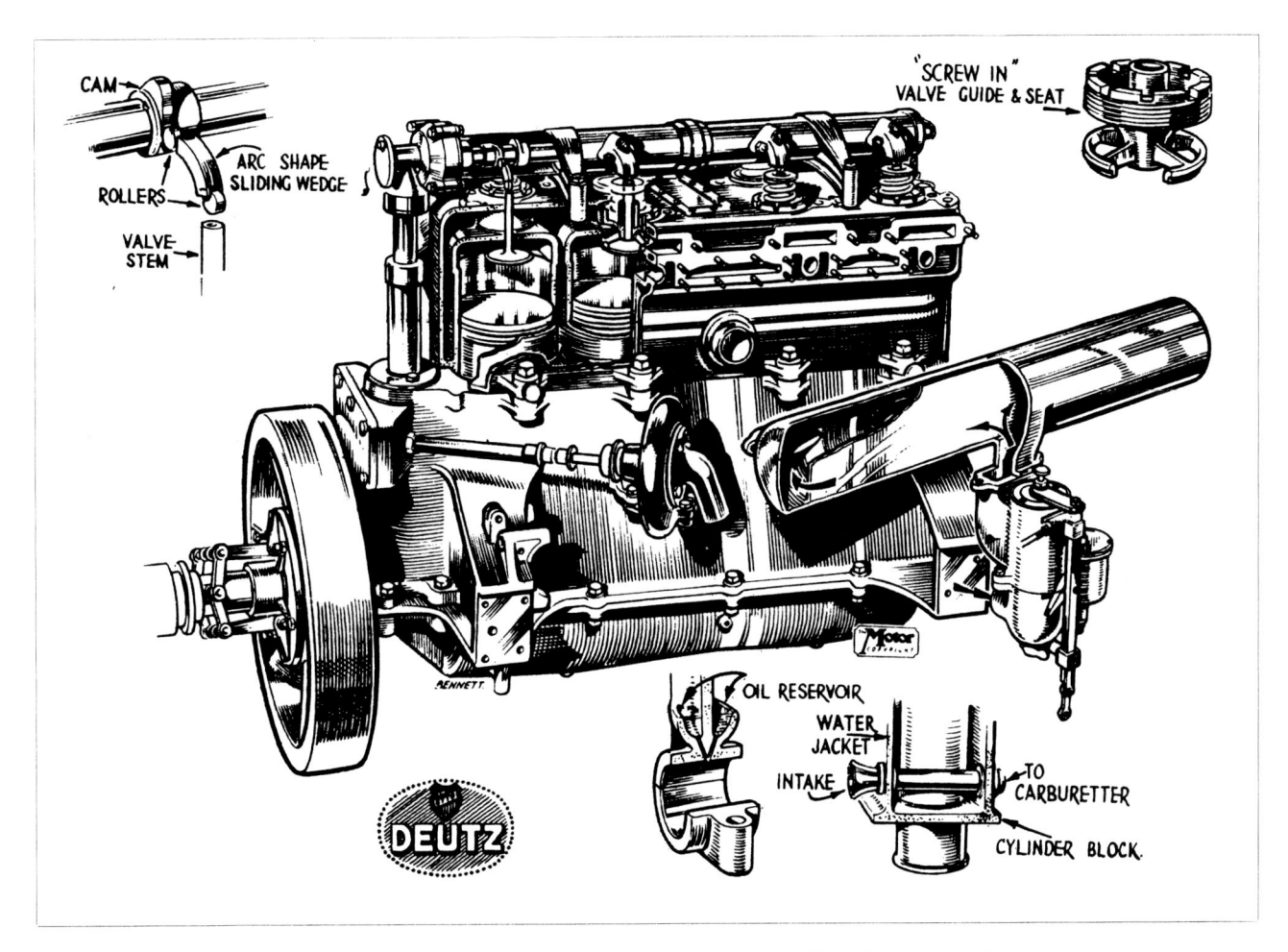

A drawing of the inlet side of the same engine. Noteworthy features are the detachable valve guides and seats, quadrant and roller interposed between the camshaft and valve stems and the angled drive for the water pump.

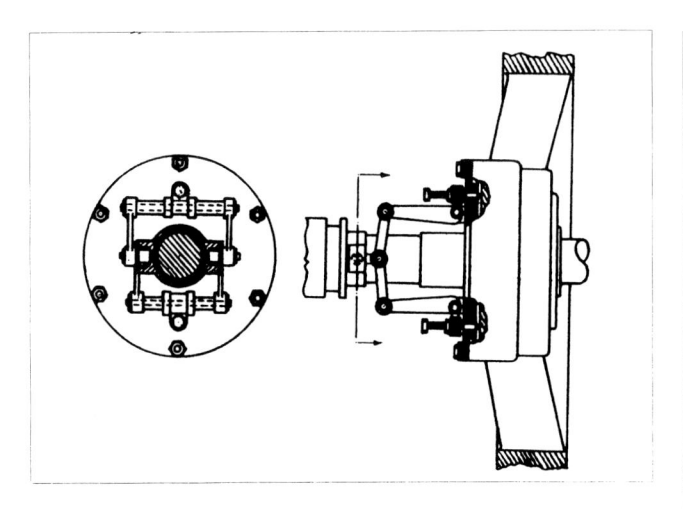

The famous Bugatti clutch is instantly identifiable.

(Right) Another example of Bugatti's ingenuity, the Deutz's clutch was actuated by this device, which flexibly transmitted force from the pedal to the unit.

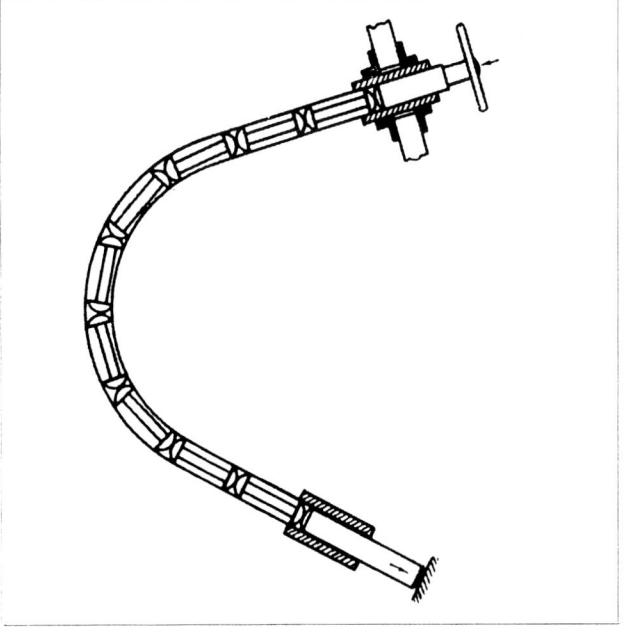

The overhead camshaft four mounted in a Mathis chassis, pictured at the works of the Société Alsacienne de Constructions Mécaniques at Graffenstaden. A large combine which produced such products as railway engines and looms, it also built Mathis's cars and the Type 8 which Ettore designed on his own account. It is the car featured in the . . .

. . . familiar but important picture of Bugatti's Type 8 taken at the first Kaiserpreis at Tanus on 13 to 14 June 1907. Note that the radiator is not fitted with a radiator badge though originally the Bugatti name was painted on the radiator grille. This was removed from the negative, which was held by Fiat, in the 1920s. At the wheel is Bugatti, photographed with some key members of Fiat and its drivers whom he would have known since his racing days in Italy. From left to right: Lodovico Scarfiotti, chairman of Fiat, Ernest Friderich, EB, Pierre Marchal of Berliet, racing driver Felice Nazzaro in the front passenger seat, who won the event, and Vincenzo Lancia, Rembrandt Bugatti, Louis Wagner from France, another Fiat driver, Émile Mathis and, secretary of Fiat and soon to be at its helm, Giovanni Agnelli.

tube containing a string of dumb-bell-shaped sections, which conveyed the downward pressure from the pedal and, at the same time, compensated for the curvature of the outer casing.

This engine was available in three sizes of 65, 45 and 35hp. The larger capacity form was a 145 x 150mm, 9,920cc unit, while the 45hp, of 6,400cc, had a 124 x 130mm bore and stroke. The 35hp car was powered by a 4,960cc engine, with dimensions of 110 x 130mm.

When it came to allotting the various models with their designations, interestingly, for the first time Bugatti's hitherto personal-type numbers, which he had assigned to his various projects since his first car, were allocated to the Deutzs. Therefore, the Type 8A was the 65hp engine mounted in the 3,098mm (10ft 2in) wheelbase chain-driven chassis, while the 8B was the 45hp version in the same frame, though it was also available with shaft drive in the form of an open propeller shaft and live rear axle. The 35hp car went under the Type 8C name.

The shaft drive chassis represented a new departure for Bugatti. Since it required that the gearbox and differential now be separated, Ettore took the opportunity of re-designing the former and it was fitted with a 'high speed' layshaft, which resulted in a lighter gear change than hitherto. The gearbox was attached to the chassis with a three-point mounting while the torque arm, to absorb both drive and braking forces, is also to be found on Bugatti cars. The Type 8 duly made its appearance at the 1907 Berlin Motor Show, which opened on 5 December, though it did not enter production until late in 1908. This may have been an intentional hiatus on Deutz's part in view of an arctic economic climate . . .

Soon after his arrival at Cologne, Bugatti began work on a second model, designated the Type 9,

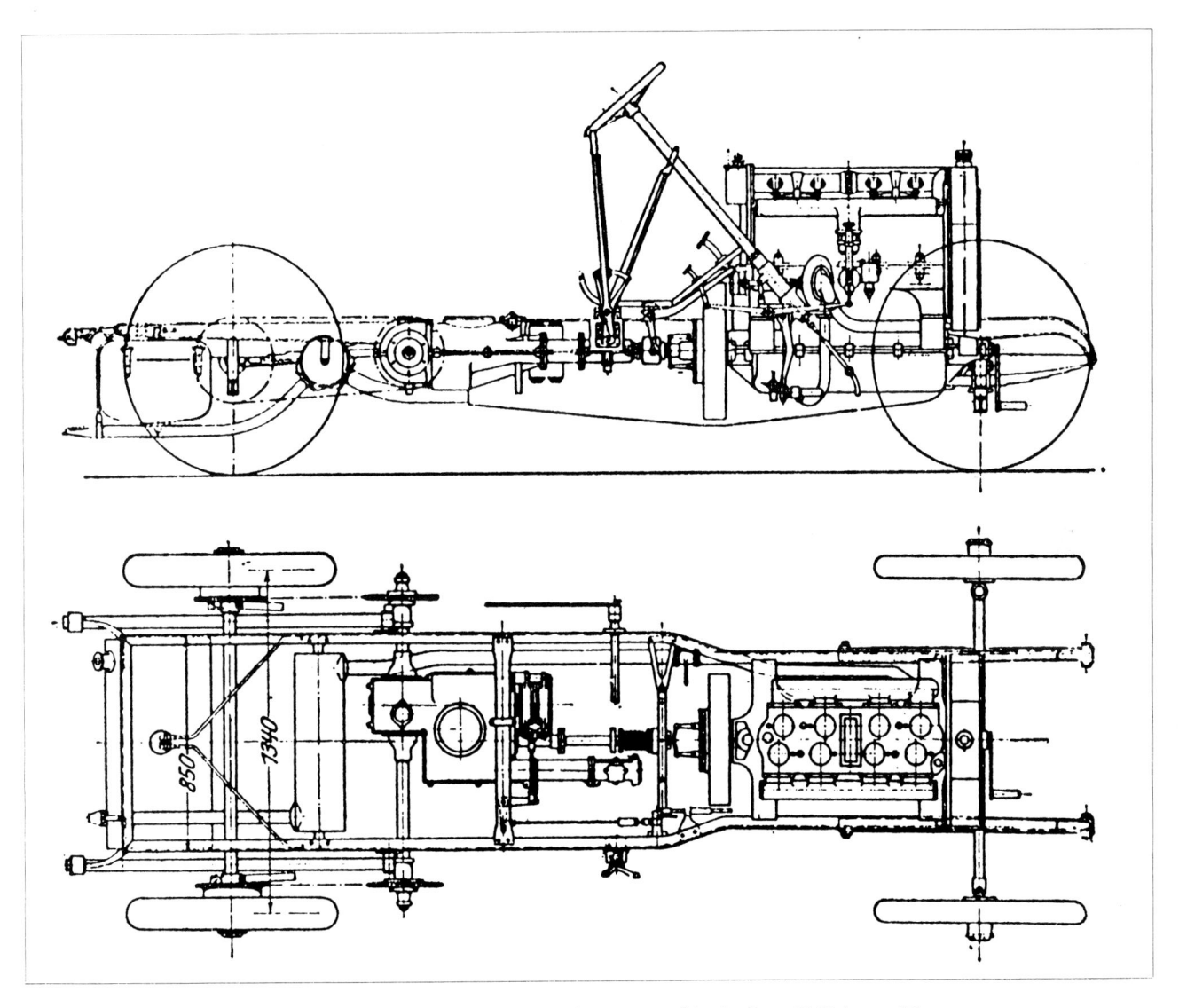

The Bugatti-designed Type 8 Deutz with chain drive. This drawing appeared in the June 1908 issue of the German magazine, Z.V.D.I. Note how the magneto just clears the nearby steering box.

The Type 8 Bugatti becomes the chain-driven Type 8 Deutz though it has yet to acquire a badge. It is pictured at the Deutz works at Cologne and is identifiable by its straight-sided radiator.

which was only available with shaft drive, and had a 3,180mm (10ft 5in) wheelbase. It was specified in type 9HA and 9B forms, with the 45hp engine and as the Type 9C with a 35hp unit.

In 1910 Deutz announced yet another model, the Type 21.[17] It was smaller than its predecessors, also shaft driven and with two new engine capacities, a 40hp of 3,565cc, with 90 x 140mm bore and stroke, and a 2,612cc 30hp car of 80 x 130mm cylinder dimensions. Intriguingly this is one of the 'missing' Bugatti type numbers and, if the precedent of the Types 8 and 9 were perpetuated in that Ettore's personal model numbers were also used by Deutz, then 21 fills a long-running gap in the register.

[17] von Fersen, *Autos in Deutschland 1885–1920.*

The basic layout of the Type 21's engine was essentially the same as the Type 8 and 9, but one difference was that Bugatti modified the camshaft drive, which was now located at the front of the engine, in the manner of his Type 10. The magneto and water pump changed sides; they were at last in their correct positions and driven, via bevel gears, at the front of the engine. Moving the water pump from the right hand back to the left front of the engine allowed Bugatti to reposition the steering box, now located above the engine bearer. The pump's new position meant that the exhaust pipe ran from the rear, rather than the front, of the manifold. Ettore also took the opportunity of tidying up the top of the engine by enclosing the camshaft, its casing and associated tappets.

The radiator of its shaft-driven Type 9 contemporary, by contrast, was enhanced by a gentle curvature, a feature which was also bequeathed to the Type 10, which led to the creation of the Bugatti marque. Ettore also borrowed the general appearance of the Deutz badge.

A chain-driven Type 8 Deutz in 1908, apparently being evaluated by a military customer.

The Type 21 Deutz which appeared in Z.V.D.I. of June 1910. Its engine differed from the Types 8 and 9 in that the overhead camshaft drive was at the front rather than the rear of the block, with the water pump and magneto driven from its base, as on the Type 10. But were many of these cars built?

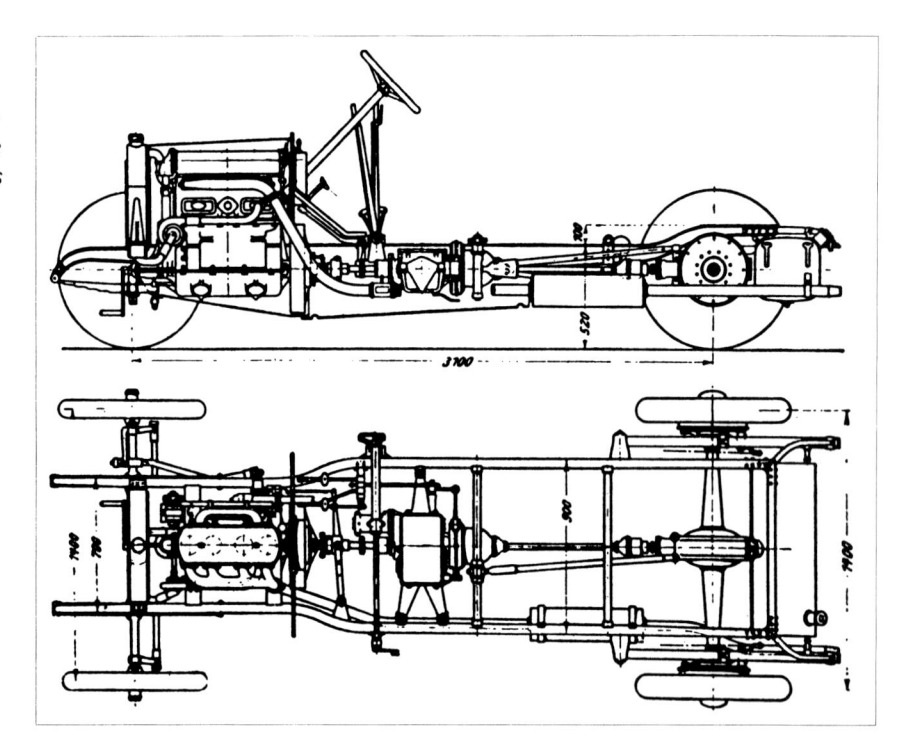

It Depends on the Competition

Inevitably perhaps, Ettore was soon to be found entering his latest product in competition. The first Prince Henry Trial was held between 9 and 17 June 1908 and Bugatti drove the single Deutz though he did not complete the course. He had the misfortune to doze off while at the wheel of his car, though fortunately without injury to himself, and had to leave it parked against a tree. Ettore was to be involved in an almost identical incident over twenty years later when he fell asleep at the wheel of the prototype Royale – though in that instance it was said to be the aftermath of a good lunch . . . We can only speculate as to whether that was the cause of the 1908 misadventure!

In the following year of 1909, Deutz was present in earnest and entered a team of no less than three cars, no doubt to reflect the fact that Bugatti's Type 8 was by then in production. What appeared to be the standard Deutz tourers carried race numbers 610 and 539, though the third of the trio, number 612, was clearly something special and carried a distinctive open two-seater racing body. The engines of all three were experimental. Their dimensions did not coincide with those of the production Deutz and are quoted as having bores and strokes of 124 x 140mm, 145 x 160mm and 95 x 155mm.[18] All the cars completed the course though did not figure amongst the winners. The victor (it was a handicap event) was William Opel, in one of his diminutive 8hp cars. However, by the time the third, and last, Prince Henry Trial took place in 1910, Bugatti had left Deutz and returned to Alsace and the small town of Molsheim, where he had begun producing cars under his own name.

To put the circumstances of his departure into context and, above all, to see how the Bugatti marque came into existence, we must first retrace our steps to the September of 1907, which was when Ettore joined the Cologne company. In retrospect, it is difficult to think of a worse time for any firm to embark on the manufacture and sale of large and expensive motor cars, let alone one which had never done so before. The fluctuations of the trade

[18] Jerrold Sloniger and Hans-Henrich von Fersen, *German High-Performance Cars 1894–1965* (Batsford, 1965).

cycle, which had been on an 'up' since 1902, resulted in a downturn in the European economy from mid-1907. It was first felt in Italy, where Fiat recorded its first large loss, of L.7m (£39,649), while Isotta Fraschini, which catered for the international carriage trade, also suffered and, as a result, the French Lorraine–Dietrich company we encountered earlier, took a substantial holding in the firm.

The following year of 1908 saw the French motor industry, which was Europe's largest, in the doldrums with many of its car makers in trouble. The established and respected Panhard and Levassor company, for instance, which catered for the more expensive end of the market, saw its car sales plummet by 40 per cent during the twelvemonth. But this chilly financial climate represented an opportunity for the manufacturer of small cars and Renault was one of the few companies to actually increase production during the year. Kent Karslake quotes a contemporary source which declared: 'Everyone was saying that the industry could only be saved by the *voiturette* which, if produced at a low enough figure, would sell in such considerable numbers as to fully compensate the trade for the sudden drop in demand for more expensive vehicles.'[19]

A Car for the Times

This fast deteriorating state of affairs was reflected late in 1907 when *L'Auto* magazine decided for the first time to open the *Coupe des Voiturettes* to four-cylinder cars. (*L'Auto* magazine had, in 1905, initiated its *Coupe des Voiturettes* for smaller racing cars. The event had hitherto only been open to single and twin-cylinder vehicles.) 'The promoters apparently thought that the *voiturette* of the future was hardly likely to be equipped with a four-cylinder engine but it appears that there were quite a number of firms who believed that the small vehicle should possess all the refinements of a big car.'[20] The bore limit was set at 65mm, which was criticized at the time on the grounds that 75mm was about the smallest bore that could be used for practical purposes ... However, this figure was subsequently revised by the Automobile Club de France (ACF) to an even more demanding 62mm.

In January 1908 the Club announced that the French Grand Prix would be preceded by the *Grand Prix des Voiturettes*, when it was prophesied that the event would 'certainly do as much for the cheap and light car as racing has done for the big vehicle'.[21] It was scheduled for 6 July 1908, to be held at Dieppe.

There were no less than 64 entries for the July event, of which 47 started and, of these, six firms entered four-cylinder models of which two, the Swiss Martini and Isotta Fraschini, were powered by overhead camshaft engines.

The Isotta Fraschini is particularly important because it is more than likely that Ettore was inspired by its roadgoing derivative when he came to design the first car to carry the Bugatti name. The racing car was designated the FE model and designed by Giuseppe Stefanini. Stefanini, it will be recalled, was also responsible for the pioneering overhead camshaft Type D of 1905, of which Bugatti would have been aware. In the small Isotta, he excelled himself with a beautifully executed, scaled-down version of a large car and even though such an Isotta Fraschini did not exist, there was a family similarity to the earlier Type D.

Under the bonnet was a 1,208cc four, with dimensions of 62 x 100mm, which developed an estimated 18bhp at 2,300/2,500rpm. It progressively consisted of a diminutive cast iron monobloc, which was mounted on an aluminium crankcase. An overhead camshaft was employed with a rocker shaft and rockers positioned above because, unlike his earlier overhead camshaft design, Stefanini opted for transversely located valves, instead of in-line ones. The camshaft was driven by a shaft from the front rather than the rear of the engine, which incorporated, like its 17.2-litre predecessor, the

---

[19] Kent Karslake, *Racing Voiturettes* (Motor Racing Publications, 1950).

[20] Karslake, *op. cit.*

[21] *Ibid.*

The car that probably inspired Bugatti's Type 10, Stefanini's 1.2-litre FE four-cylinder overhead camshaft Isotta Fraschini. A team of three was entered in the Grand Prix des Voiturettes *held at Dieppe on 6 July 1908. This is Alfieri Maserati who was placed 14th. In the following year came an FENC roadgoing version though its engine differed somewhat from the racer's and the overall concept of its engine resembles Ettore's Type 10 of 1909.*

vertically located water pump. The magneto was driven, via an intermediate gear, from the rear of the camshaft and was thus sensibly accessible. At the other end of the engine, the crankshaft ran on two ball bearings.

For the six-lap 460km (286-mile) race itself, a trio of these small Isotta Fraschinis was entered, driven by Buzio, Maserati and Trucco. Although a single-cylinder car won the event, Guyot in a De Dion Bouton-engined Delage gave the marque its first racing success; the winner of the four-cylinder class was Buzio, in an FE, who finished in eighth place.

The Grand Prix proper, which took place on 7 July, resulted in a clean sweep for Germany, with Mercedes winning and Benz in second and third places. But undoubtedly the faster and the most spectacular car present, which came fourth, was Rigal's Clément-Bayard. Despite a calamitous nineteen stops for tyre changes, it achieved an average speed of 102.3kph (63.6mph). As the fastest French car present, it could hardly have been overlooked by Ettore Bugatti, particularly when we find that it was powered by a new overhead camshaft engine, one of two Grand Prix cars in the Dieppe race to be so equipped. (The other, the British Weigel, had a power unit cribbed from the Clément-Bayard's.)

The 13,963cc four-cylinder engine, the work of M. Sabatier, was significantly powered by a single overhead camshaft four-cylinder engine, with valves inclined at 45 degrees, actuated by rockers, to reap the benefits of a hemispherical combustion chamber. Like the FE Isotta Fraschini, the drive for the overhead camshaft was at the front of the engine but, unlike the car from Milan, the magneto was driven by an adjacent cross shaft and bevel gears, balanced by the water pump on the left-hand side. This is precisely the layout adopted by Bugatti on his Type 10 though he may have been influenced by the roadgoing version of the FE Isotta Fraschini, which could have also copied the layout.

The Milan Effect

When Isotta Fraschini designed its production version of the FE, designated the FENC, its engine differed in some radical respects from the racing unit and incorporated modifications that may have been gleaned by examining the Clément-Bayard at Dieppe. As Stefanini was no longer constrained by the 62mm bore limit, it employed a 65mm one, which was subsequently agreed for the 1909 *voiturette* formula. Capacity therefore rose from 1,208cc to 1,327cc. The main difference related to the camshaft drive. This now drove the shaft via a bevel wheel, instead of a tapered pinioi, while at the crankcase end, it followed in M. Sabatier's wheel tracks and drove a stubby cross shaft, to which the magneto on the right and the water pump on the left were attached.

A detailed article on the FENC Isotta Fraschini appeared in the 1 September 1908 issue of the Italian monthly magazine, *Motori, Cicli & Sports*. However, the engine appears to have been an experimental one because it is shown in mirror image to the layout subsequently adopted, with the exhaust manifold and water pumps on the right and the induction and magneto on the left. The car itself did not enter production until the first half of 1909 when 100 examples were sold, making it one of the best selling Isotta Fraschini of the pre-World War I years.

It seems highly likely that Ettore Bugatti, probably recognizing that the large cars he had designed for Deutz were unsuitable for the depressed economic climate of 1908 and perhaps inspired by this article, decided to design his own *voiturette*. Probably the best indication that Ettore was inspired by the Isotta Fraschini, is that it was the only four-cylinder car in the 1908 race with engine dimensions of 62 x 100mm, which Ettore subsequently followed when he came to design his own small car. When the FE's bore was enlarged to 65mm, Bugatti also followed suit for the production version of his design.

According to Ernest Friderich, who was to rejoin Ettore during the latter part of the following year of 1909, Ettore had 'built the first Pur Sang Bugatti in the cellar of his villa'. Like the Type 8 and 9 Deutz, this was a single overhead camshaft unit though it may have followed the Clément-Bayard inspired Isotta Fraschini layout, in that the camshaft was driven from the front of the engine, the magneto moved to the right side, and the water pump positioned on the left. Ettore was to perpetuate this revised layout on his Type 21 Deutz of 1911.

Small is Beautiful

The diminutive cast iron monobloc was dominated by its overhead camshaft which, on the prototype, was located by two pieces of bent steel! Ettore once again employed the banana tappets that were such

The Type 10, 'Le Petit Pur Sang', which Ettore built while at Deutz. Ettore's wife, Barbara, is at the wheel and Ernest Friderich is about to crank the engine. The wheelbase of this car is not known.

A study in extremes, the 1.2-litre Type 10 of 1909 and the 12.8-litre Kellner-bodied Royale, pictured in the early 1930s. The exposed tappets of the small car are readily apparent as is the Bugatti-designed carburettor. Both cars thankfully survive, the Type 10 in the collection of General William Lyon in America and the Royale in Japan.

The exhaust side of a Type 10 engine. It is slightly taller than the body and the bonnet had to be modified to accomodate it. A carburettor inlet tract running through the block, similar to that employed on the earlier Deutz cars, can be seen between the pairs of exhaust pipes.

The rear view of the Type 10. The rear axle of this car was not perpetuated on the Type 13 Bugatti.

a feature of his earlier Deutz designs. The cams and tappets were lubricated by simple felt pads, with access provided via four small trapdoor-like covers in the top of the cam housing. The engine had a capacity of 1,200cc, while its 62 x 100mm bore and stroke was the same at the FE Isotta Fraschini's.

There was a not inconsiderable amount of space between the top of the block and the camshaft to permit the intrusion of the inlet manifold which reached its ducts through the top, rather than the side of the block. The inlet valves were carried in separate cages though the exhausts were not. A Bugatti-designed carburettor also featured. The crankshaft was carried, *à la* Isotta, in two ball bearings. Once again the familiar multiplate clutch was employed. There was a separate gearbox, while drive was via an open propeller shaft and torque arm in the manner of the Type 8/9 Deutz.

The car's chassis was a simple channel section unit, with half-elliptic springs all round, and there were wooden wheels. It had an utilitarian open two-seater, doorless aluminium body, which was initially left unpainted and Friderich recalls that it 'was known as the "bath tub" because of its peculiar carosserie'.

What Ettore described as his Type 10 (a plate on the top of the block reads 'L' ℬ 10, 1908–9') was probably built over the winter of 1908/9. We know that it was completed by April 1909 because he mentions, in a letter he wrote to Émile Mathis, that 'the little car is finished and is marvellous!'[22]

Perhaps to gauge public response to his car, Ettore ran it, on 22 August 1909, in a speed test organized by the Frankfurt auto club. It was entered in the 6hp or 1,500cc class as a Deutz, though the space on the radiator for a badge was left blank in the manner of Bugatti's Type 8 prototype of 1907. With its creator at the wheel, it was placed sixth, where he recorded a speed of 88kph (55mph).

It is difficult to be precise about when Ettore decided to leave Deutz and put his Type 10 in production under his own name. His biographer, W F Bradley, says that 'as it was not the type of vehicle which appealed to the Deutz company he began to entertain the idea of making it himself.'[23] In September 1909 Ettore played host to the famous French aviator Louis Blériot, who in July had made his celebrated crossing of the English Channel in his own monoplane. Friderich confirms the fact, stating: 'He came every day to the aviation meeting at Cologne on board the famous Pur Sang.'

Bugatti's daughter, L'Ébé, credits Blériot with a catalytic role in Ettore deciding to leave Deutz and set up in business on his own accord, though perhaps we should be cautious in accepting this speculative version of events. She wrote: 'Blériot's enthusiasm for this new machine *must have* [my italics] convinced my father that the important decision he had made was the right one – to set up in business for himself.'

[22] Conway, *Bugatti*.

[23] WF Bradley, *Ettore Bugatti, Portrait of a Man of Genius* (Motor Racing Publications, 1948).

Friderich at the wheel of the Type 10. It has been fitted with a new rear axle and has been painted, probably red.

Soon afterwards Ernest Friderich, accompanied 'by the General Manager of the Gas Engine Works', went on a trip in a Deutz car to eastern Prussia and Russia; though 'on my return the Chief told me that he had resigned from his position with the firm and that he intended returning to Alsace.' This was probably in November 1909 because, on the 16th of that month, Deutz had written to Bugatti, cancelling their licensing agreement of 1907, to take effect from 15 December. However, by the time of his departure, Ettore had completed the design of his Type 21 car for Deutz, effectively a scaled-up Type 10. The 21 was announced in June 1910, and briefly entered production in 1911.

As for Friderich, he decided to leave Cologne with Bugatti and he later remembered that in December, 'we left . . . in the small Bugatti car . . . We stopped at the Bank of Darmstadt, where an appointment had been made with Mr de Vizcaya, senior.' It was the Darmstadt Bank, it will be recalled, which had assisted Ettore back in 1906, after he broke with Émile Mathis.

Friderich's 'Mr de Vizcaya' was, in fact, a Spanish aristocrat, Baron D Agustin de Vizcaya[24] who was to play a key role in Bugatti choosing Molsheim for the location of his car factory. He had been introduced to Ettore by a Milanese Marquis in 1905 and was also a shareholder in the Darmstadt Bank, a position which must have been

[24] JMR de la Vina, 'de Vizcaya', in *Bugantics*, vol. 33, no. 2 (summer 1970).

of considerable assistance to Bugatti as he was striving to establish his car company. Born in the Spanish city of Bilbao in 1870, de Vizcaya soon excelled at sport. He enjoyed riding and hunting, as did Ettore, and also became fencing champion of Spain, despite the fact that he is said to have smoked 120 Russian cigarettes a day! The Spaniard also liked fast cars, a passion which he also shared with Bugatti.

At the time of his marriage to a French lady, the Baron bought a substantial property, *Jaegerhof*, in Alsace, which was around 3.5km (2.5 miles) from the small town of Molsheim, itself about 20km (13 miles) to the west of the city of Strasburg. The estate, which de Vizcaya had purchased from Cardinal Rohan, consisted of 9,000 hectares (16,000 acres) of fertile land and forest, and the Baron is said to have hosted some of the finest hunting parties in Europe there. Unfortunately Ettore fell out with him in 1914. Two of de Vizcaya's sons, however, Fernando (better known as Ferdinand) and, in particular, Pedro (Pierre), were to become members of the Bugatti racing team of the early 1920s.

Back on that winter day of 1909, de Vizcaya took Ettore and Friderich to visit a derelict dye works on the outskirts of Molsheim, which he thought might be suitable for Bugatti's purposes. Ettore clearly liked what he saw. The old buildings would form the nucleus of his car factory and there was also a château attached to the property which would become home for him and his family. This had grown to five with the birth on 15 January 1909 of his third child, a son, who was christened Gianoberto Carlo Rembrandt, though he would become universally known as Jean.

From thereafter, the Bugatti name would become inexorably linked with that of Molsheim, for the cars which Ettore produced there were destined to be some of the most successful ever seen on the racing circuits of Europe and would make him famous the world over.

Ettore at the wheel of his Type 10 in 1909, still lacking a radiator badge but with E. Bugatti painted on the side of the bonnet.

3
The Move to Molsheim

'My small factory acquired greater importance. Production was increasing. The nights were often short but orders were flowing in.'

Ettore Bugatti

Bugatti's small car entered production in 1910 and was built – apart from when its manufacture was interrupted by World War I – for the next sixteen years. Updated in 16-valve form, it endured until 1926, by which time over 2,000 examples had been built, making it Ettore's best selling model.

Before taking a closer look at the model, its evolution and its derivatives, we should briefly pause to consider the context of this 1,327cc Bugatti in Germany's pioneering role in the creation of the small four-cylinder car. For 'from about 1908 onwards, when English and French cars in the 8–10hp range were almost invariably twin cylinder (some indeed still had single-cylinder engines) . . . nearly all the leading German firms brought out light cars with excellent four-cylinder engines, some with overhead valves, in the range of 850–1,200cc.'[1] Such cars offered their customers the refinement and, in the case of the Bugatti, the type of performance previously reserved for the big car owner. Born as we have seen from racing experience, the light car was a peculiarly European concept. The theme was not perpetuated across the Atlantic by the Americans.

Prior to the 1908 *voiturette* formula, which specified a 62mm bore for a four-cylinder engine, 75mm was considered to be the smallest acceptable size for such a power unit. An exception was the NSU company, which can be credited with pioneering the small four-cylinder car in Germany. This successful bicycle and motor cycle manufacturer of Neckarslum had briefly, in 1905, built the Belgian Pipe car under licence. The following year, however, it had introduced its own Pfaender-designed 68 x 90mm, 1,420cc 6/10 model. It therefore comes as no surprise to find that this firm, no doubt in response to the economically bleak year of 1908, produced in 1909 its 5/12 model. This model, powered by a 60 x 100mm four-cylinder, 1,132cc engine, remained in production until 1913. It was one of the most popular German small cars of the immediate pre-war years.

NSU was not alone in its advocacy of the German small four. In 1906 the Loreley company was offering a 1.5-litre car, though in 1909 marketed a model with identical dimensions and 1,132cc capacity as the NSU. Yet another popular contribution came from another bicycle manufacturer in the shape of Opel, whose 65 x 90mm 1200 model appeared in 1910, the same year as Bugatti's car. Wanderer had its 1,145cc Puppchen model in 1912, and in the following year Émile Mathis, never one to ignore a good idea (particularly if it belonged to someone else) marketed a 5/14 under his own name, with similar dimensions to Ettore's car.

Of all these offerings which employed side and overhead valves, the overhead camshaft car from Molsheim was alone in its refinement, performance and general well-being. It was, in short, the progenitor of the small sports car at a time when

[1] Bird, *Early Motor Cars*.

performance on the road invariably meant a large car with an appropriately powerful engine.

The eight-valve Bugatti was not cheap. It was marketed as a 10/12 in Britain before World War I, where it sold for £350, and £275 in chassis form. This was about the same figure as a vehicle of the quality of a 12hp Talbot and was about twice the price of many of the light cars then on the market.

The Sporting Thoroughbred

Ettore was all too aware that he was not only producing a new car but also a new type of vehicle and he wasted little time in informing his potential customers of the fact. The following is the introductory paragraph of an open letter published in his first catalogue, which appeared in late 1910.

> Gentlemen:
>
> Considering the enormous expense given rise to, until now, by fast and powerful cars, I have decided to create a new *breed* of light car, able to render the same services, enjoying the same qualities, the same comfort, but freed forever from that great source of expense: WEIGHT.

One of the most distinctive features of the eight-valve Bugatti is its visual delicacy coupled with the strength of its component parts and, as a result, it turned the scales at a mere 349kg (6.8cwt). Such a car, born as it was of both economic depression and *voiturette* racing, might also have represented something of a tacit recognition by Bugatti of the commercial failure of the earlier large cars he had designed which, fortuitously, he had not had to underwrite. There were therefore compelling reasons for him to produce a small car, which was cheaper to build than a large one though he would have to sell more of them. Fortunately, the production version of his Type 10 was a great sales success. But first the Molsheim factory had to be transformed from its former function as a dye works into a car factory.

Following Bugatti and Friderich's visit to Molsheim in December 1909, Ettore departed for Paris and then Cologne. Subsequently Ernest Friderich was 'left alone ... with the job of whitewashing the walls and having a general clear up. About 25 January 1910, the first machine tools arrived, and continued arriving at the rate of two to three per week. It was also my job to find a nucleus of workers, turners, millers, fitters, smiths, etc.'[2] The workforce, which grew to sixty-five by 1911, was completed by the arrival of a draughtsman, who had previously been employed by Ettore at Deutz.

The car entered production at Molsheim in 1910 and Friderich says that five were produced in that year. This first Bugatti model was a mildly modified version of the Type 10 which Ettore had built at Deutz. Indeed, if the Type numbers are any guide, all his designs up until the Type 21 Deutz were completed at Cologne, *prior* to his arrival at Molsheim at the beginning of 1910.

The identities of the Types 11 and 12 are unknown but the Type 13 was the first production Bugatti. As already indicated, it had a larger capacity than the prototype, with the bore size upped from 62 to 65mm, giving 1,327cc. This modification was hardly apparent but the principal visual difference was that the overhead camshaft was now enclosed. The cambox was initially of plain aluminium, with a total of twelve brass inspection plugs on either side and at the top of the cover. But soon after the first few cars had been built, the box was re-designed and greatly enhanced by the presence of a rendering of Ettore's signature, very effectively set off against a criss-cross background.

While a Bugatti-designed carburettor had been employed on the Type 10 and on the first few production cars, Ettore was fortunate when, in 1909, the Zenith carburettor subsidiary of the Rochet-Schneider company began producing units at Lyon and the Type 13 benefited from such an updraught component. The engine's inlet tracts had, by then, been re-designed so that they now emerged from the side, rather than the top, of the block. A Bosch magneto was employed. On the other side of the engine the so-called, and by then obscured banana-

---

[2] Friderich, 'How the firm of Bugatti was born'.

shaped tappets, were echoed by Bugatti's famous 'bunch of bananas' exhaust manifold which was to remain a feature of the marque until 1935.

Splash Instead of Pressure

The Type 13 inherited the two ball bearing crankshaft from the Type 10, though these were soon replaced with three bronze plain bearings. With the initial layout, a pressurized lubrication system was not crucial and, indeed, the Type 13 employed a total loss arrangement, with lubricant supplied by a dash-mounted reservoir. The oil found its way to the crankshaft, whereupon the rotation haphazardly distributed it to the bearings, cylinders and pistons. This arrangement was perpetuated, despite the ball bearings being replaced by plain ones, and the archaic system persisted until 1928, a good ten years or so after the rest of the industry!

The Deutz-style radiator was retained, though was fitted with Bugatti's famous red oval radiator badge, a feature he had borrowed from the Cologne company which had used a similarly shaped emblem since 1895. Another minor modification, which is to be found on the Cologne-built Type 10 and the larger Deutzes is that the rear axle's large hub was reduced in size.

We do not know the wheelbase of the Type 10, though the Type 13 was so called to indicate it was 2,000mm (6ft 6in) – following the Deutz example where the 8 and 9 Type numbers were used to differentiate between the varying lengths of the cars' chassis frames and not the capacities of their respective engines. Initially, the type numbers were only used for the convenience of the Molsheim drawing office. It was not until his 1912 catalogue that Bugatti publicized type numbers – a practice which, once again, followed Deutz precedent.

A shortcoming of the Type 10 and indeed the Type 13, was its relatively short wheelbase which meant that it was only suitable for two-seater bodywork, which would have limited its appeal. Therefore, to permit the fitment of larger bodywork, the eight-valve engine was offered in two additional chassis lengths. The Type 15 had a 2,400mm (7ft 10in) wheelbase, and was mechanically very similar to the 13 but with double, rather than single, rear springs to cope with the heavier bodywork. The third variation was the Type 17, which used a chassis with a 2,550mm (8ft 4in) wheelbase. In all variants, the track remained a constant 1,150mm (3ft 9in).

Although Bugatti did not formally break with Deutz until November 1909, at least two further cars had been built at Cologne: factory records indicate that the first two production chassis (numbers 361 and 362) were delivered in September, though their respective invoices were not issued until the following year. 'Vizcaya' is listed as the recipient of the next chassis (363) and this was Ettore's friend, Baron Agustin, who had suggested the move to Molsheim. Interestingly, the next but one chassis (365) may well have been the first Type 15. It went to Prince Hohenlohe in Bohemia and survives today in the technical museum in Prague as the oldest survivor of the marque. We know it has a 2,400mm (7ft 10in) wheelbase, is still fitted with its Bugatti carburettor and its cambox is bereft of Ettore's signature.

These cars were produced in chassis form only because there was no bodyshop at Molsheim until 1923, so coachwork was supplied by such local firms as Forrier at nearby Strasburg and Durr of Colmar, about 50km (30 miles) to the south of Molsheim.

Bugatti did design and offer his own factory specification bodywork, which initially consisted of three body styles on the Type 13 and 15 chassis. There was a two-seater 'Carrosserie Torpedo' on the shorter wheelbase frame; a longer four-seater variant which, progressively, offered the coachbuilder some elements of standardization because the lines of the rear half of the body were effectively a duplicate of the front.

The final and most expensive body was a slightly bizarre three-seater saloon. Its height was such that it gave the appearance of having been designed for a larger car. It is of particular interest because its *fiacre* (coach) lines, inspired by horse-drawn vehicles of the previous century, were to

The world's oldest surviving Bugatti, the 1910 Type 15, chassis no. 365, is owned by the National Museum of Technology in Prague, Czechoslovakia. The car has been in what was Austro-Hungary since new. Its first owner was Prince Hohenlohe and it later passed to Robert Patocka, manager of a sugar factory in northern Bohemia.

reappear on Bugatti's bodywork in the late 1920s. This body which survives, though not with its original chassis, was built to Ettore's design by Widerkehr, an old established coachbuilder, also of Colmar. Widerkehr was subsequently taken over by Gangloff of Berne, a firm which was to thereafter be closely associated with Bugatti throughout his years as a car maker.

At the Paris Show

Mounted on Type 15 chassis 366, this body was delivered in December 1910 to the Paris Salon, where Bugatti exhibited with a small stand for the first time. The same car was later displayed by the company at the 1911 Berlin Motor Show and the chassis, in re-bodied form, survives in Britain, having been owned for many years by the well-known *Bugattiste*, Peter Hampton.

A visitor to the Bugatti display at the Paris Salon was Dr Gabriel Espanet. Espanet had abandoned surgery in 1907 for aviation (an industry in which France then led the world), and was destined to play a peripheral role in the affairs of the Bugatti company. As soon as he entered Paris's Grand Palais, he noticed the *Bugatti* name on a stand on the left, because he had recently been to the Louvre to view an exhibition of sculpture by Ettore's brother, Rembrandt. He was fascinated to find the same name though in different surroundings. 'It was like seeing Nieuport's first aeroplane,' he remembered.[3] The parallel is an appropriate one because Édouard Nieuport had introduced his Blériot-inspired monoplane in the previous year, which was notable for its pioneering fully enclosed, streamlined fuselage.

With Bugatti's first car, Espanet found himself looking at a vehicle '. . . completely different from the usual form of construction. I was so taken by

[3] L'Ébé Bugatti, *The Bugatti Story*.

This car, which also appeared in Bugatti's 1910 catalogue, is thought to be Type 15 chassis number 366 of 1910 and may have been used by Madame Bugatti. Note Ettore's signature. When 366 was bought by Colonel C P Dawson of Norwich from Ettore in about 1912, he recalled that 'the chassis had been humping a four-seater limousine which Ettore built for himself'. Bugatti retained the saloon body and 366 was fitted with an open two-seater body of a type which it retains to this day. What happened to the original body is not known but that great Bugatti collector, Uwe Hucke, has a similar one fitted to a Type 17.

English Bugattiste Peter Hampton bought 366 from specialist Jack Lemon Burton in 1938 and owned it until his death in 1991. He is seen here, on 1 July 1939, winning an Edwardian Race at the Crystal Palace circuit at 68.28kph (39.32mph) from Clutton's Itala. Immediately behind the 1.3-litre Bugatti is the 15.5-litre 1912 Lorraine-Dietrich Vieux Charles Trois.

An eight-valve car pictured on the Bugatti stand at the 1910 Paris Show. The engine's cambox has now acquired Ettore's signature, although it had originally been left blank. The two-seater coachwork, designed by Bugatti, featured in his 1910 catalogue.

the simplicity and harmony of its styling, as well as by the impression of power it gave, that even without a trial-run I ordered one there and then.' Bugatti was not present on his stand at the time of Espanet's first visit though, when he returned on the following day, he asked the doctor why he had decided to buy one of his cars. ' "Did someone tell you about it?" "No one at all", I replied. "Well, you're quite right!", he said with assurance.'

One of the first journalists to drive a Bugatti car was an Englishman, W F Bradley, who was to be associated with its creator for the rest of his life and would be responsible for writing his posthumous biography. Bradley's preferential treatment may have been a reflection of the importance that Bugatti attached to the English market, which at this time was responsible for importing more continental cars than any other European country. The account appeared in the 1 November issue of *The Motor* magazine. Ettore clearly approved of its sentiments because he quotes from it in his 1910 catalogue, which had been prepared for the 1910 Paris Show. Although much of the piece is taken up with a technical description of what is headed 'A New Light Car', one paragraph is worth reproducing because, in many respects, it sums up the distinctive quality of the recently introduced Bugatti:

The claims made by the makers of this little car are that it can hold its own in the matter of speed with any pure touring car on the road. Frankly, we did not think it could ... Fully equipped, with mudguards, lamps, horn and all touring accessories, we made a demonstration run in the suburbs of Paris. After being warned twice that speeding was not allowed in the straight avenues of the Bois de Boulogne ... fate sent us a big Benz Prince Henry car of 105 x 165mm bore and stroke, four passengers, and a pure touring body to act as pacemaker. The Prince Henry car was driven by a hot-blooded sportsman in his teens, who went over the most abominable roads between Paris and Saint-Germain at a speed only limited by the ability of his car to hold the road. Yet it was only possible with the little Bugatti to keep within 50 yards of his pointed stern all the way, and with less discomfort than can be experienced in many cars three times the size and weight. The makers guarantee that the little Bugatti can maintain 60 miles an hour, and although no opportunity was given of definitely proving this, the claim seems to be well founded.

From the five cars built in 1910, manufacture gradually built up and Friderich has recorded that it 'amounted to three chassis in February ... and then four chassis in each of March and April; May

With the exception of the Type 13 that Friderich ran in the Grand Prix de France in 1911, most pre-1913 eight-valve Bugattis had wooden wheels. This then is a photograph of the wheel rather than the car at Molsheim.

A pre-1913 two-seater Bugatti with imitation canework and rare wire wheels.

The angular radiator which Bugatti had fitted to his eight-valve cars from 1910 began to be replaced by this new and distinctive oval-shaped one from 1912. For the 1914 season the Types 15 and 17 were redesignated the Types 22 and 23 respectively and fitted with reversed rear quarter elliptic springs in place of the original half elliptics.

and June six chassis each month, whilst for the third quarter our production had risen to eight chassis per month and I remember that in December the Chief wanted nine chassis in order to obtain a total of 75 for the year 1911. By good will, hard work and perseverance we succeeded.'

In 1912 the Types 13 and 15 had been joined by the longer wheelbase Type 17 which contributed to a further rise in output, while, in 1913, Friderich says that monthly production at the beginning of the year was twelve and progressively increased, so that by December, nineteen chassis were built. All in all, a total of 175 Bugattis left Molsheim in 1913.

A New Look

Meanwhile the line was continuing to evolve. Perhaps the most obvious modification was to the radiator – the shape of which, it will be recalled, was a scaled-down version of one which Bugatti had used on the cars he designed for Deutz. Ettore no doubt wanted to imbue his car with a distinctive new radiator to underline the arrival of the new Bugatti marque. He already had a suitable shape, in the form of a graceful oval which could not be possibly confused with any other car. As for the inspiration for its contours, Ettore's youngest son, Roland, has maintained that his grandfather Carlo

A pre-war eight-valve Type 22 or 23, pictured in Britain in the early 1920s.

was the source of its memorable lines. He had 'taught his sons that the most perfect shape in nature . . is that of the egg. That was the whole, sole reason behind the original choice of that shape.'[4]

The new radiator had in fact already appeared, in enlarged form, on a big 5-litre car that Bugatti had run in the 1910 Prince Henry Trials. It was fitted, in scaled-down form, to the 1.3-litre cars from 1912 and survived, in essence, until 1926.

Yet further mechanical refinement came in 1913 when the rear half-elliptic springs were replaced with reversed quarter-elliptic ones on the longer wheelbase Type 15 and 17 cars, which were re-designated the Types 22 and 23 respectively. The Type 13 briefly continued with half-elliptic rear suspension.

The eight-valve Bugattis were not the first recipients of the new rear suspension. It had first appeared, maybe as a cost conscious exercise, on a new design which was to emerge as the Bébé Peugeot for 1913. The layout had been the subject of a Bugatti patent (21160/1911), which shows a chassis with such springs positioned both front and rear, though they were only employed in the latter position on production cars. Bugatti claimed that, for a given axle motion, the chassis motion was reduced by the outboard hanging of the springs, when compared with conventionally mounted quarter elliptics. However, Hugh Conway[5] has said that the advantages of the system were that the spring was in tension during traction and the axle motion was forward when the springs deflected, which would promote a slight degree of understeer. This is an instance perhaps of Bugatti arriving at the correct solution for the wrong reasons! This unmistakable rear suspension is to be found on every Bugatti car from 1913 onwards.

Bugatti took the opportunity to make some modifications to the engine's rather primitive lubrication system in 1913, though it still remained unpressurized. These changes show the extraordinary lengths to which Ettore would go to make what was rapidly becoming an outdated arrangement, work more efficiently. It highlights a conservatism in his approach to design which would become more apparent as he became older. What he did was to introduce an oil pump, driven from a skew gear at the top of the drive shaft, to the lubrication system. Thereafter, the oil reached the sump via external copper tubing. The pump (containing two independent pistons) and its distribution mechanism were contained within an oblong box, located at the front of the cambox. For an account of just how the device worked, I can do no better than to quote from a contemporary description of the arrangement and the writer's opinion of it!

> A peculiarity of the oil pump is that, while it is of the plunger variety, the moving element not only rises and falls, but oscillates to the extent of about one third of a revolution in the pump cylinder. This oscillating action serves in one position of the plunger to bring into register the holes through which the oil is drawn from the supply tank, and in another – namely, when the plunger is at the top of its stroke – to register the holes through which the lubricant is forced out to its destination. Admirably as this pump does its work, we fancy that practically the same results could be obtained by a less complicated mechanism . . .[6]

[4] Borgeson, *Bugatti by Borgeson*.

[5] *The Automotive Inventions of Ettore Bugatti.* (Paper presented to The Newcomen Society, 11 February 1959).

[6] *The Autocar*, 19 May 1917.

THE MOVE TO MOLSHEIM

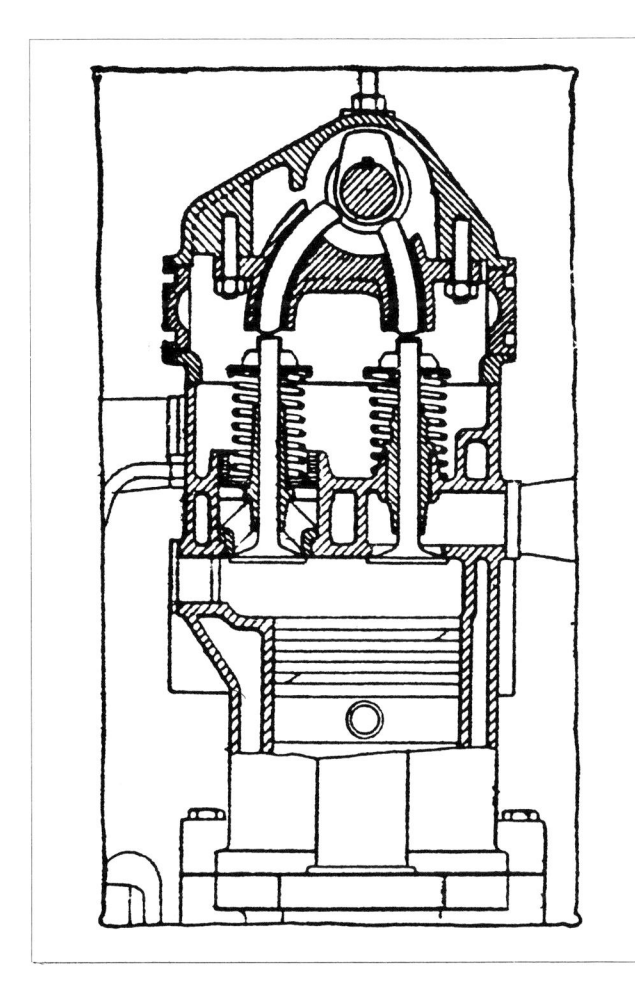

The eight-valve cambox removed which, for 1914, also incorporated the oil pump. The upper dog clutch engaged with the vertical camshaft drive.

(Left) A cross-section of the eight-valve cylinder head with banana-shaped tappets.

A contemporary photograph of an eight-valve engine which, interestingly, is painted. Today the aluminium is usually polished! This is a post-1913 unit as it is fitted with an oil pump mounted at the front of the cambox.

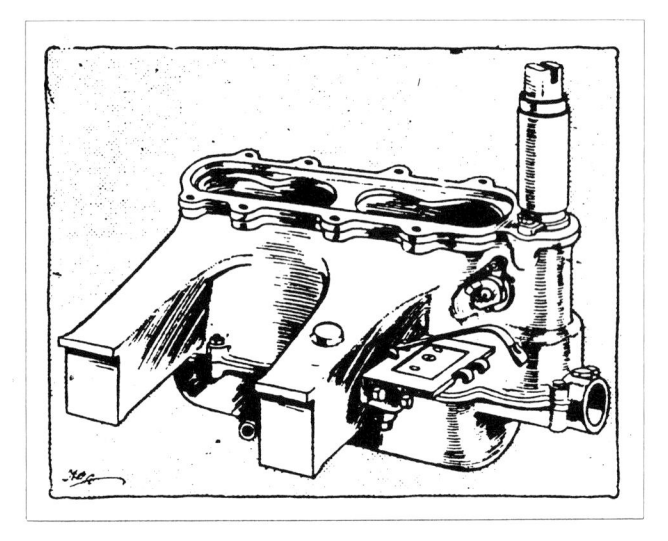

The eight-valve crankcase, complete with camshaft drive but with the magneto detached.

broke out. The last chassis to be delivered, in 1914, was number 706. As the first was 361, provided that the serials ran consecutively, this makes a total of 346 Bugatti cars built during the five years between 1909 and 1914. If this is the case, this means that the vast majority, no less than 339, were eight-valve ones, with the remaining seven being allocated to the enigmatic 'Type Garros' Bugatti.

Production of the eight-valve Bugatti restarted after the war and continued until 1920, by which time the model had been largely superceded by its 16-valve derivative. But before considering this unit, it is appropriate for us to retrace our steps to see how Bugatti's cars behaved in the field of competition. Such a pursuit was established practice for Ettore and there was the incentive of publicity that could only boost sales for the new make.

Friderich says that 'at the beginning of 1914, our work was increasing, and the rate of output reached twenty-seven chassis in March' and production continued until August, when World War I

A Racing Start

Perhaps the first public appearance of a Bugatti in a competitive event occurred on 25 September

An eight-valve Type 23 of about 1914 fitted with a handsome boat-tailed body.

An eight-valve car, bearing the 56 competition number, in a pre-First World War event.

1910 at the Gaillon hill climb in Normandy which was run on a sidespur on the main Paris to Rouen road. There M. Darritchon, in a standard bodied Type 13, took second place in the touring car class. In his write-up in *The Motor* of 4 October 1910, W F Bradley records a time of 1min 3⅘sec and his account suggests that this was his first acquaintance with the marque. He makes reference to 'a little car, known as the Bugatti . . . The dainty-looking vehicle, having more the appearance of a toy than a real motor car, is a product of Alsace.' There was time to include this success in Bugatti's 1911 catalogue, which probably appeared early in the year.

In January 1911, M. Huet, Bugatti's Paris agent, of 3, Saint Ferdinand Square entered a rally which ran from the French capital to Monte Carlo, and he completed the 700km (435-mile) run in 51¾ hours, so underlining the reliability of the little car. A Bugatti ran with success at Lorraine, while at Mont Ventoux Baron de Vizcaya in a bonnetless Type 13 finished first in the 65mm Touring class. In May a Bugatti took second place in the Limonest hill climb while cars, driven by de Vizcaya and Gilvert, achieved first places at a meeting at Sarthe in June 1911.

By far and away Bugatti's most celebrated performance of the pre-World War I years came in the Grand Prix de France, held at Le Mans on 23 July 1911. Friderich reveals that the idea of entering a Bugatti in the event came from the publicity-conscious M. Huet, though it should be made clear that this race was not the French Grand Prix proper. The last event of that name had been run in 1908 when German cars had swept the board with a celebrated victory by Mercedes, while six out of the first seven cars to finish also hailed from that country. The Automobile Club de France had been unable to convince the French manufacturers, already weakened by the 1908 economic depression, to produce costly racing cars for a Grand Prix in 1909 and 1910, though maybe they would have been less reluctant had a French car taken the chequered flag in 1908 . . . However, the Automobile Club de la Sarthe, of the *département* which included the town of Le Mans, was keen to hold a race. The Automobile Club de France therefore gave permission for the local club to include the celebrated *Grand Prix* name in the title of its event. Regrettably, the French press soon dubbed the race *le Grand Prix des vieux tacots* (the old crocks' Grand Prix)!

In the first instance it was decided that cars would have to be powered by engines of not greater than 110 x 120mm dimensions, which gave a four-cylinder unit the capacity of 7.6 litres. However, only one firm, that of Rolland-Pilain, fielded a purpose-designed team of 110 x 160mm

Ernest Friderich at the wheel of the most famous of all Type 13 Bugattis, thought to be chassis 415. The date is 23 July 1911, the race the Grand Prix de France and the venue Le Mans, where it won first prize in the 110 x 200mm class. The car is painted white for Germany though the radiator is black, a feature which Ettore sometimes used to differentiate between his road and competition cars in pre-First World War days.

cars. In order to attract more entries, it was decided to make the event a free-for-all, though there would be a *voitures légères* class for light cars to be run concurrently with the 'Grand Prix'.

The car that Bugatti entered was a Type 13 fitted with a purpose-designed doorless, two-seater body with exposed petrol tank. It was finished in white, which was Germany's racing colour as Alsace was so annexed at the time. The radiator shell was painted black. We do not know if any mechanical modifications were made to the car and although wire, rather than the usual wooden wheels were fitted, the production models of that year could be so equipped. Friderich says that his car 'was, of course, the 65mm bore and 100mm stroke eight-valve model', though W F Bradley quotes the engine as a stroke of 110mm, which would have meant a capacity of 1,457cc.

Kent Karslake describes the fourteen cars that assembled at Le Mans as 'a pretty motley collection'[7] and it is difficult to argue with that contention. The chain-driven giants were perhaps typified by the 10-litre S61 Fiat, driven by Victor Hémery, a rugged former seaman and a native of Le Mans. The Fiat had a sports rather than a racing chassis and had originally been rejected by its prospective

[7] Karslake, *Racing Voiturettes*.

owner, a Paris café proprietor, 'on grounds of late delivery'. In addition to the specially tailored Rolland–Pilains, the sole Lorraine–Dietrich was an 18-litre monster which had been built for the 1906 French Grand Prix, a Corre-la-Licorne for the 1907 race and the Porthus for the 1908 one. There was also a team of 3-litre Excelsiors entered for the *voitures légères* class but, eventually, only one competed. It would be imagined that Friderich's Bugatti would have been similarly classified but its weight, of a mere 304kg (6cwt), meant that it was too light and the Type 13 was put in with the heavy metal.

The cars were intended to make twelve circuits of the 54km (34-mile) triangle, which incorporated some of the 1906 Grand Prix course, making a total of 651km (405 miles). The event was due to start at 8 a.m. and had to be completed by 4 p.m. when the roads would be re-opened to the public. It was a sweltering day. W F Bradley, who witnessed the event, says that the weather was 'torrid, the temperature in the shade being over 100 . . . tyres just melted away' and during the race 'cylinders cracked, rear axles went to pieces, the drivers wilted under the strain, only Friderich, with his little Bugatti appeared to be in a happy mood.'[8]

There were casualties, both mechanical and human, but the race settled down to a tussle between two giants, Hémery's Fiat and Duray's Lorraine–Dietrich, though Friderich kept going. By the halfway mark, the Lorraine had taken 3hr 24min 24sec for the 325km (202 miles) while Friderich had taken 4hr 27min 35sec, which meant that he had averaged over 74kph (46mph). Then Duray's differential gave way and, soon after 3 o'clock in the afternoon, Hémery was declared the winner in 7hr 6min 30sec, having averaged 84.5kph (52.5mph). Incredibly the Bugatti was second and also won the light car class, having covered ten laps in 7hr 16min 50sec, which meant that Friderich had maintained his average of over 74kph (46mph). He was followed by a Rolland–Pilain and a two-stroke Cote. But the point had been made. According to Bradley: 'The contrast with the big brutes was so great that every time Friderich appeared, a rousing cheer went up from the spectators.'

An Experimental Eight

In about 1912 in his quest for more power, Bugatti created a rather crude straight-eight engine by mounting two eight-valve units in a common chassis frame. Friderich says that the two engines were joined by a 'rubberized leather flywheel', whatever that might have been, and the concept has echoes of the two-engined Prinetti and Stucchi tricycle which Ettore had created as an eighteen-year-old, back in 1898. The origins of the chassis are unknown but its large wooden wheels are unrelated to those of the small Bugattis, so it might have been a Deutz frame. Friderich says that, in this form, it was capable of 138kph (86mph), whereas the Type 13 he had run at Le Mans could only manage 106kph (66mph) flat out.

The 2,654cc 'eight' was clearly designed for competition and Friderich recounts that he 'took part in the hill climb at Gaillon in October 1912 in which I failed to complete the course, the third gear having broken in the middle of the hill'. If Bugatti was relying on the standard eight-valve gearbox, as seems to have been the case, this is hardly surprising. This was a time, it should be noted, when there were virtually no straight-eight racing cars, and the significance of this pioneering experiment will be considered in Chapter 6. The fate of this fascinating vehicle is unknown but was reported, in 1921, to then being 'in service in Paris'.[9]

By 1913 the Bugatti name, largely through Friderich's racing activities, was becoming known to a far wider public though, he says 'not so much as we desired. To enhance our reputation, therefore, I took part in numerous races: on the Côte du Val-Suzon, Limonest to Lyon, Ventoux to Avignon, Nancy, La Baraque, Toul, etc.' Although in many instances he was victorious, it was important that these activities did not interfere with his duties at the factory: 'All this travelling had to be undertaken

---

[8] Bradley, *Ettore Bugatti*.

[9] *The Autocar*, 7 May 1921.

between Saturday and Monday morning, as I had to be present at the Works during the normal working hours of the week, where the task of carrying out trials and final adjustments of chassis absorbed the whole of my time, often till very late in the evenings.'

The Type Garros Mystery

In 1914 came the firm's most ambitious competitive sortie when Bugatti entered a car for Friderich to drive in America, at that year's prestigious Indianapolis 500 Miles Race. This car was not a diminutive Type 13 but a much larger one which, on this occasion, had its usual 5-litre engine enlarged to 5.6 litres. It was an example of what we have come to know as the type Garros and its origins, date of design and inspiration have probably absorbed more speculation and correspondence amongst historians and Bugatti enthusiasts over the years than any other Molsheim model. What follows is my interpretation of the known facts which, inevitably, contains speculative elements of the sort that always arise when this model is discussed!

On the face of it the available facts are straightforward. In 1912 Bugatti introduced a new car, which has been identified as the Type 18. This was powered by a 5-litre engine, with three valves per cylinder, which Bugatti would standardize on his single overhead camshaft cars from 1922. This was, perhaps surprisingly, mounted in a chain-driven chassis of the type that was reminiscent of Bugatti's Type 8 Deutz of 1907. This then is a car of contradictions but let us begin by setting down some known facts about the model. In describing the 'Type Garros', W F Bradley – in his biography of Bugatti written in 1947, just after Ettore's death – says that 'There were negotiations for selling the design of this car to the new Peugeot racing organization.' This is confirmed in *50 Ans de Compétitions Automobiles*, the privately published Peugeot history of 1968, written by Philip Yvelin, who had full access to the company's archives.[10]

[10] 'Letter from Edward Eves' in *Bugantics*, vol. 43, no. 2 (summer 1980).

It was during 1911 that the trio of Boillot, Goux and Zuccarelli (waspishly branded *Les Charlatans* by Peugeot's resident engineers) and assisted by draughtsman, Ernest Henry, were engaged in the design of a revolutionary Grand Prix Peugeot to be powered by a relatively small 7.6-litre twin overhead camshaft engine. A team of three of these L76 cars made their competitive debut at the 1912 French Grand Prix, which Boillot won in one of the Peugeots, so once again upholding French competitive honours. This victory finally banished forever the era of the big chain-driven racers; the epochal Peugeot, in Laurence Pomeroy's words, providing 'the very foundation of the modern high-speed engine'.[11]

In 1911 the concept was far from proven and the hard-nosed Robert Peugeot, according to Yvelin, asked Bugatti to 'discreetly design a team of three large cars as a safety measure'. The reference to a large car is probably a relative one, either to the 855cc Type 16 Bugatti, which was soon to emerge as the Bébé Peugeot, or to the production 1.3-litre Bugatti. The only *large* car that this could possibly refer was the Type 18 5-litre because all Ettore's other designs depended on his 1,327cc four-cylinder engine. The fact that the Type 18 was originally produced for Peugeot is, to some extent, strengthened by the fact that a batch of four 5-litre chassis was originally laid down (471/2/3/4), which would represent the traditional team of three cars, plus a spare. At least one chassis was fitted with a racing body, while its black-painted radiator echoed that of Friderich's competition Type 13 of 1911.

According to Griffith Borgeson,[12] W F Bradley, 'recorded that, in the run off between the two rivals for Peugeot's patronage, the Bugatti's best speed was about 159kph (99mph), while that of the Charlatan's creation was superior by more than 24kph (15mph).' This figure approximates the top speed of the Type 18, which was about 160kph (100mph). Bradley himself says that 'the Peugeot

[11] Laurence Pomeroy, *The Grand Prix Car, Volume 1* (Motor Racing Publications, 1949).

[12] Borgeson, *Bugatti by Borgeson*.

One of the large Bugatti-designed Deutz cars which ran in the 1909 Prince Henry Trials.

directors cast their vote in favour of the trio of Boillot, Zuccarelli and Goux.'[13] It was a decision that left Bugatti with a batch of uncompetitive racing cars on his hands but, not wishing to run them in the Grand Prix, he decided to campaign at least one example himself in less demanding hill climbs. The remainder could be sold off to private owners.

Three-Valve Innovation

That is my scenario. But what was the Type 18 and when was it designed? We have it on the authority of Ettore himself that 'the first one was built in 1908 and the first sale was in 1912.'[14] This suggests the type Garros as a spiritual successor to the big-

[13] Bradley, *Ettore Bugatti*.

[14] Barry Eaglesfield and C. W. Peter Hampton, *The Bugatti Book* (Motor Racing Publications, 1954).

engined Deutz cars which Ettore ran in the Prince Henry Trials from 1908 onwards. As already mentioned, these were of varying capacities and in the 1910 Trial Bugatti ran a 5-litre car fitted, for the first time, with the distinctive oval-shaped radiator, which was subsequently scaled down for use on the eight-valve cars from 1912.

The 1910 car was chain driven, which suggests that it was probably a Deutz which Ettore brought with him from Cologne, and fitted with a new radiator at Molsheim. This particular car raised some eyebrows at the weigh-in for the Trial because the bathtub-type four-seater touring body was finished in a rather vivid sea-green paint, while its chassis was painted red. Although it did not complete the event, Ettore illustrated it in his 1910 catalogue as the 'Prince Henri' type. Incidentally, this third and final run of the Prince Henry series stopped for a champagne breakfast at Molsheim when, by all accounts, Ferdinand Porsche, the eventual winner,

The chain-driven car that was marketed as the Prince Henry model in Bugatti's 1910 catalogue. This marked the first appearance of the oval radiator, the shape of which was transferred to the eight-valve cars in 1912.

had to be prised away from the newly equipped works, before continuing in his chain-driven 22/80PS Austro–Daimler!

Apart from its 5-litre capacity, we do not have precise details of Bugatti's Prince Henry model. Bearing in mind its date, it was probably a single overhead camshaft unit, with two valves per cylinder, shaft driven from the front of the engine, with the associated magneto and water pump in the manner of the contemporary Type 13 Bugatti and Type 21 Deutz. If there had been anything particularly radical about the design, I feel that we would have heard about it! For instance, such a modest modification as the separate valve springs employed on Ettore's 1908 Prince Henry Trials car, was faithfully recorded in the contemporary press.

Working on the assumption that this car formed the basis of the Type 18 Bugatti, then at what stage did it acquire its distinctive three-valve layout? My contention is that, while the main portion of the Type Garros was, as Ettore stated, designed in 1908, its cylinder head came later and dates from 1910 or 1911.

My thinking in this instance is that the three-valve layout is a perfect example of Ettore's lateral thinking. His *pull*rod engine of 1901 was produced in response to the contemporary *push*rod, his clutch design on his De Dietrich was an inversion of current design and the same, I feel, applies to the three-valve layout, in that Bugatti would require a feature which he could then re-interpret in his own unorthodox way.

As it happened, the year of 1910 saw cars, with which Bugatti would have been familiar, undergoing experimentation in the field of valve design. Porsche had flown in the face of convention with his celebrated Prince Henry model, as the 22/80 had been renamed, in view of its success in the 1910 event. Its 5.7-litre single overhead camshaft engine boasted a total of five valves per cylinder – one inlet and four exhaust. However, the catalyst in this instance was, I believe, a Mercedes, a make with which Ettore was well acquainted.

Late in 1910 Mercedes introduced its 37/90 four-cylinder model and a notable feature of its pushrod engine was an innovative three-valve layout, though with two exhausts and one inlet. If previous examples of Bugatti's thinking are anything to go by, it would have been perfectly in order that he should reverse this configuration and opt, instead, for a two inlet and one exhaust layout. Although the arrangement clearly worked, the exhaust valve area was greater than that of the inlet valves; it

Bugatti introduced his distinctive three valves per cylinder on the 5-litre Type 18. This is Black Bess's engine.

'seriously reduced the power potential of his three-valve engines'.[15]

What may well be the first published photograph of Bugatti's new 5-litre engine appeared to illustrate an article by W F Bradley in the 1 August 1911 issue of *The Motor*. There, it is incorrectly captioned as the power unit of Friderich's triumphal Grand Prix de France Type 13, but even so there would have been little point in picturing an established engine. The chassis was clearly not designed for it, as distance pieces are shown above the frame. With this three-valve layout, Bugatti took the opportunity to dispense with the banana-shaped tappets he had previously employed and replaced them with a new layout. This is where, I think, a secondary design influence has come into play.

As long ago as 1955,[16] Laurence Pomeroy, famed technical editor of *The Motor*, drew attention to the similarity of Bugatti's Type Garros engine and that of the S61 Fiat. Pomeroy opined that Bugatti himself might have had a hand in the Fiat's design though I think that this is an instance of Ettore being retrospectively acclaimed from a reputation greatly enhanced by the Type 35 which did not appear until 1924.

We now know that the Fiat in question was designed, late in 1908, by Carlo Cavalli, who had joined Fiat's drawing office in 1905 and became the firm's technical director in 1919. The S61 was powered by a massive 16-valve, 130 x 190mm 10,087cc engine, with a single overhead camshaft, driven from the front of the unit, and the magneto

[15] Borgeson, *Bugatti by Borgeson*.

[16] 'The Missing Link, an Exercise in Design Detection' in *The Motor*, 24 August 1955.

and water pump located at right angles to it – in the manner of the FENC Isotta Fraschini and, of course, the eight-valve Bugatti engine. But the really significant aspect of the design, certainly as far as the Type Garros unit is concerned, was the method Cavalli employed to actuate the four valves per cylinder. The cams operated them via fingers containing rollers, located between them and the tops of the valve extensions.

The Fiat Influence?

The S61 (it is not certain whether it was simply a super sports or a racing car) was nevertheless celebrated for its initial appearance with a victory in the first American Grand Prix, held at Savannah in 1910. Accordingly, it was also described on occasions as a Savannah Fiat. It was produced until 1913, by which time around fifty examples had been built. An S61 Fiat, it will be recalled, won the Grand Prix de France in 1911, which was the model's only win in a European race. The Type Garros engine had been completed by then though, if he had needed to, Ettore may have been able to study an S61 through the good offices of Émile Mathis, who held the Fiat franchise. We know that he handled the model, because one of the most celebrated examples was sold by him as an '100–120hp Mathis' to Sir Frederick Richmond in 1913!

The Type Garros engine was of 5,027cc, with 100 x 160mm bore and stroke and three valves per cylinder. The camshaft, running in five ball races, was built up with individual cams. This was, perhaps, to permit the fitment of experimental lobes, with fingers, positioned in the Fiat S61 manner, between the lobes and the valves. At the rear, this shaft drove an oil pump, which looked after its lubrication, though that of the camshaft appears to have been overlooked! The front-mounted camshaft drive reflected that of the eight-valve engine, driving the magneto and water pump which were positioned on the right and left respectively.

The first engine featured a five bearing crankshaft, though subsequent units employed a three bearing one with an improved lubrication system which squirted oil on to the crankshaft and rods by jets. The original arrangement, by contrast, relied on oil being thrown about by the crankshaft and rods and it was then supposed to find its way on to little trays on the tops of the big end caps and above each main journal.

A Chain-Driven Chassis

As the eight-valve gearbox was unsuitable for a 5-litre engine, the only appropriate chassis, and no doubt time and expediency were also factors, was a chain-driven one of the type that had been used on Bugatti's Prince Henry cars of the 1908/1910 era. However, it is only the general layout of the Type Garros that is similar to that of the Deutz. The detail design appears different. The model shares the 2,550mm (8ft 4in) wheelbase of the Type 17 frame though has a wider track. The large engine is reflected by the use of twin half-elliptic springs at the front, another instance of Ettore 'doubling up', while reversed quarter-elliptic springs were used at the rear.

Chain drive would still have been appropriate for a racing car if, as I suspect, the Type Garros had been designed as a potential 1912 French Grand Prix entry. Although shaft drive had decisively ousted chains on smaller road cars, they were still employed on more powerful models, the Prince Henry Austro–Daimler being a case in point. Racing cars still retained the layout and it is also worth recording that the hotly tipped favourites for the Grand Prix were the large, chain-driven Fiats.

According to Ernest Friderich, Type Garros 'manufacture was pushed forward, and the car saw the light of day at the beginning of 1912; it was with one of these cars that Monsieur Heille, a great friend of the Chief, took part in the Herkomer Fahr (a race reserved for car owners) and I was his mechanic.'

The all-important French Grand Prix took place at Dieppe over two days, on 25 and 26 July, and, as already recounted, it witnessed a triumphant

A 5-litre fitted with a four-seater touring body.

vindication of the new Peugeot twin overhead camshaft engine. But if that company did not want his 5-litre car, then Bugatti would put it to good use. Taking what was probably the first car to be completed (471), which had been fitted with a stark two-seater body featuring a small and apparently unrelated pointed tail, Ettore decided to use it for hill climbs. This meant travelling to and from the appropriate venues and the only difficulty was that there was no room for any luggage. So a basic, but functional, rectangular rack was constructed at the front of the car, ahead of the radiator and just the right size for a leather trunk, which was then strapped in place, for personal effects.

A Hill Climb Car

With Friderich in attendance in 'a little Bugatti', Ettore drove the 5-litre in the Mont Ventoux hill climb near Avignon in the south of France on 11 August 1912. Described as 'driving the large chain-driven Bugatti car'[17] Ettore put up fourth-fastest time of the day, with a climb of 19min 16.4sec for the 21.3km (13.25-mile) run.

[17] Karslake, *Racing Voiturettes*.

Friderich was also entered for the second Grand Prix de France, on 9 September 1912, which was once again held at the Le Mans circuit. But Bugatti was unable to repeat his success of the previous year because the Type 13 was a non-starter. The event was won, against little opposition, by Zuccarelli in a 3-litre L3 version of the twin overhead camshaft Peugeot.

A further two 5-litre chassis (472 and 473) were later built, either in 1912 or 1913, though we do not know exactly when. What is certain is that, on 18 September 1913, the fourth chassis (474) was delivered to its most famous recipient, France's celebrated pioneer aviator, Roland Garros, and his name would thereafter be associated with the model. W F Bradley tells us that it was Dr Gabriel Espanet whom, it will be recalled, had forsaken medicine for aeroplanes, who introduced Garros to Bugatti and 'he placed an order for one of the 100 x 160mm racing cars.'[18] This was duly fitted with a handsome touring body, with its two seats distinctively staggered, by Labourdette. After World War I it was exported to Britain, later christened Black Bess, and is, thankfully, one of three 5-litre cars to have survived.

[18] Bradley, *Ettore Bugatti*.

A photograph which perhaps encapsulates Ettore Bugatti's twin passions, that of motor cars and thoroughbred horses. Taken in August 1912, the location is Molsheim and Ettore (left) is about to depart for the Mont Ventoux hill climb in a 5-litre, chassis 471. Luggage storage was clearly a problem, hence the leather case mounted ahead of the radiator. The car survives in Britain and the case is also extant!

Two 5-litre chassis at Molsheim; 471 (left) and 715 yet to acquire its coachwork. Two of Ettore's children, Jean and Lidia, look quite at home in 471. Note the black finished radiator. Also behind the chassis on the left is the box used for testing the chassis on the road. The date is about 1914.

The 5-litre chassis number 474, delivered to aviator Roland Garros on 18 September 1913. This lovely two-seater body by Labourdette was at least the second to be built on this chassis. We know that because, on 6 January 1914, Garros wrote to Ettore: 'I didn't use the car much in Paris . . . I left the car with Labourdette who has drawn a splendid design . . . the car should be ready in three weeks'. It was . . .

. . . subsequently brought to Britain and run by Louis Coatalen of the Sunbeam Talbot Darracq combine and, in 1923, bought for Miss Ivy Cummings, who christened the car Black Bess, on account of its black coachwork and radiator. In 1925, it was acquired by H L Preston and then passed to James Justice. He entrusted the Bugatti to Michael McEvoy to tune but the gearbox casing split. It was discovered in a sorry state by a Mr Aylward and was purchased, in 1935, by that dedicated Bugattiste Colonel Godfrey Giles, who restored it. Black Bess then went to Rodney Clarke and was bought by Peter Hampton in 1948. Here, veteran car expert Kent Karslake is at the wheel. Peter Hampton retained Black Bess until 1988.

The location of this 5-litre car, probably chassis 472, is something of a mystery. The registration number of LE 6169 or 8189 suggests a London County Council issue of 1912. But is the background the British capital or is it continental? Black Bess is thought to have been the only 5-litre to have come to Britain. Intriguing!

Garros was something of a national hero at the time he met Ettore and Bradley says that 'an immediate friendship, based on mutual admiration, sprung up between Bugatti and Garros, the latter frequently making visits to Alsace and, on two occasions, flying there in his Morane-Saulnier plane and landing in a field close to the Bugatti house.'

A further three 5-litre chassis were completed in 1914, one of which was converted from chain to shaft drive. Ambitiously Bugatti decided to enter it for the fourth Indianapolis 500 Miles Race, which was to be held on 30 May 1914. In the previous year, the Indianapolis had been won by a European car for the first time in the shape of a Grand Prix Peugeot. Ettore no doubt thought that a good showing in the event would help his sales there, as a small but regular stream of eight-valve Bugattis had been exported to New York since at least 1912.

For the prestigious Indianapolis race, the stroke of the Type 18's engine was increased from 160mm to 180mm, which upped capacity from 5,067 to 5,657cc. This was the equivalent of the American 390 cubic inches though this was well within the capacity limit for the race which was 450cid or 7,375cc. The Bugatti was fitted with a basic two-seater body, finished in Germany's racing white. But there is evidence to suggest that Bugatti had misgivings about the race because, on 25 April 1914, just five weeks before the event, he confided to his friend, Dr Espanet: 'I am taking part with my large car at the Indianapolis race, unhappily I decided at the last moment, but I hope that luck will lead my driver to a good result.'[19] The driver was, not surprisingly, the competent Ernest Friderich, though Ettore's forebodings were well justified.

Across the Atlantic

The Bugatti qualified for the event at 141.18kph (87.73mph) and, for many years it was thought that Friderich put up a good showing in the race itself, no doubt because he later claimed that he was in third place when he retired after covering 683km (425 miles). But the lap chart of the race, which set down the position of the cars every eight laps, or 30km (20 miles), reveals a rather different story. This shows that the car was last, or almost so, for the majority of the time, until Friderich's retirement with a broken ball bearing in the drive pinion after completing around 547km (340 miles). He was lying thirteenth out of a field of fifteen surviving cars at the time. The event was once again a vindication of the technical superiority of the European racing car over the home-designed product, for René Thomas won that year's 500 in a 6.2-litre Delage, the team of two cars being managed by no less a personage than W F Bradley. Not

[19] HG Conway, 'Ettore Bugatti: Correspondence 1913–16' in *Bugantics*, vol. 40, no. 1 (spring 1977).

A shaft-driven 5-litre with fiacre coachwork. This may well be chassis 714 which Ernest Friderich drove in America in 1914.

Friderich in the 5-litre which ran unsuccessfully at Indianapolis in 1914. The car would appear to have been painted white and the curious damper positioned between the front dumb irons will be noted, suggesting that this is probably the previous chassis.

The 5-litre campaigned by Johnny Marquis in America. The date is 17 May 1915 and the place Venice, near Los Angeles, California. The event was a 93-lap race held on . . .

. . . the first of the board circuits, opened in 1910, at Playa Del Ray. Behind Marquis is Barney Oldfield in a Maxwell who won the race in 4 hours, 24 minutes and 2 seconds. Marquis was placed fourth in 4 hours, 31 minutes and 39 seconds.

surprisingly, Friderich says that he 'returned to Molsheim without glory'. Another 5-litre was acquired by Charles W Fuller of Pawtucket, Rhode Island. This car was run in a number of events in 1915, driven by Johnny Marquis. (The latter was Louis Strang's former riding mechanic and had progressed to a six-cylinder Grand Prix Sunbeam in the Vanderbilt and Grand Prize races on the West Coast in 1914.)

Marquis ran the Bugatti in the US Grand Prize at San Francisco, on 27 February, but retired after five laps with ignition failure. Marquis also competed in the Vanderbilt Cup on 6 March though again did not finish. At the Venice Grand Prix in California eleven days later, on 17 March, he succeeded in finishing a race and came in fourth behind two Maxwells and a Mercer. By this time the car had been fitted with Miller carburettors and Marquis also finished eighth in a race at Providence, Rhode Island on 8 September. He also appeared in the same car at Santa Monica in 1916.

It was in 1915 that a second chain-driven Bugatti made its appearance in America. George Hill, who was the cigar-chomping Barney Oldfield's sponsor, had visited Europe[20] to try and find Oldfield a Mercedes, no doubt because the marque had so sensationally taken the first three places in the 1914 French Grand Prix. Oldfield's contract with Maxwell was due to expire in March 1915 and the fact that America was not yet at war with Germany made Hill's trip possible. We can only surmise that, at the time, the Daimler factory at Stuttgart was fully occupied with the demands of war-time aero engine manufacture to come up with a Grand Prix car. In the event, Hill returned to America, having secured a 5-litre Bugatti, which he had presumably obtained from Molsheim.

By then a new 4,916cc (300cu in) limit had been instituted for the 1915 Indianapolis 500. The Bugatti's stroke was therefore reduced to 150mm, giving a capacity of 4,712cc (287cu in), the work being undertaken by the White Motor Company of Cleveland, Ohio. Even so, Oldfield was not particularly enamoured with the car and complained that it was 'as slow as hell.' It threw a connecting rod during trials for the 500, but was repaired in time for the event on 31 May. There it was driven by George Hill, who retired after twenty-three laps with a broken water pump or drive. Barney put the Bugatti's performance in the 500 more succinctly when he colourfully recounted that: 'George Hill drove it at Indy but the boys went by him like he was tied to a brick house.'[21]

All trace of these two transatlantic cars have disappeared. By contrast, examples of Ettore's 855cc design, which emerged as the Bébé Peugeot in 1912, have survived in large numbers. Its correct designation has long been considered as Type 19 though I inspected a drawing of the model's distinctive transmission, held by the Bugatti Trust, which is marked Type 16 . . .

With this design Bugatti created the concept of the baby as opposed to the light car. Today it can be seen as the progenitor of the small four-cylinder model: the original large car in miniature.

Ironically Bugatti is best remembered for his exquisite racing cars of the 1920s, but their novel and idiosyncratic features rarely influenced his contemporaries. The Bébé Peugeot theme, by contrast, remains with us to this day.

The Bébé Grows Up

Not only did the Bébé provide Peugeot with a best selling model, but it can be seen as the pioneer of the small four in pre-1914 France. This approach was initially copied by Le Zèbre and Clément-Bayard and, after the war, was exploited to great effect by the new Citroën company with its 856cc 5CV of 1922. Peugeot had itself perpetuated the concept in 1921 by introducing its 668cc Quadrilette, with curious tandem seating – a model in the spirit of the Bébé but mechanically unrelated to it. This Peugeot inspired Austin in Britain to produce the smallest capacity four-cylinder home-built car of its day, the immortal 747cc Seven of 1922, which

---

[20] 'Letter from Simon Moore' in *Bugantics*, vol. 38, no. 3 (autumn 1975).

[21] William F Nolan, *Barney Oldfield* (quoted by Moore).

The prototype Bébé Peugeot of 1911. It is accordingly fitted with a cut-down Bugatti radiator and it would be two years before the reversed quarter elliptic springs were fitted to a Bugatti road car.

endured until 1939. Fiat further refined the theme in 1936 with its ingenious 500 model. This led, in turn, to the numerous small capacity fours of the post-World War II years of which the 848cc Mini is one of the best known examples.

It was in 1911, as Friderich has pointed out, that '... the Chief, always on the alert, had installed himself in a completely separate workshop with three workers and myself. There we assembled a new 7hp car' and W F Bradley states that it 'was the development of a machine he had produced in 1907, but had been brought up to date and somewhat modified for cheaper and bigger production'.

Bradley's statement does seem plausible because the car's tiny 855cc engine is a T head design, a concept that was starting to become outdated in Europe in 1911 and was being replaced by the more efficient and cheaper L head unit which only required one side-mounted camshaft, rather than the earlier design's two. The tiny engine had a 55 x 90mm bore and stroke and the crankshaft ran in two ball bearings, while lubrication was by Ettore's familiar drip feed system. Externally, the carburettor and magneto were on the right-hand side while the exhaust manifold had clear echoes of Ettore's familiar 'bunch of bananas'.

The engine was not mounted directly to the simple channel section chassis but was ingeniously attached to a robust undertray which added to the rigidity to the forward end of the car. There were

The production Bébé Peugeot, announced in 1912, was built in respectable numbers until 1916.

half-elliptic springs at the front, while the reversed quarter-elliptic springs made their first appearance on a Bugatti-designed road car. The gearing was particularly ingenious. The drive was conveyed, via a simple cone clutch, to two concentric propeller shafts, the final drive pinions of which were in continuous engagement with two sets of teeth on the rear axle mounted crown wheel. Each respective shaft was engaged by operating a conventional gear lever which actuated sliding cog clutches. There was also a reverse gear train at the forward end of the lower gear drive shaft.

The prototype car was fitted with a simple open two-seater body which was temporarily graced with a cut down oval radiator of the type that had already featured on the 1910 Prince Henry model and would appear on the eight-valve Bugattis from 1912. Ettore clearly had no intention of producing the model himself because the completed car (with the ever diligent Friderich at the wheel) was dispatched to other motor manufacturing companies to see whether they would express an interest in buying it. Once committed, the firm would then no doubt fit a radiator of its own design. Friderich recounts that he presented it 'to the Wander [er] works at Chemnitz in Saxony'. It seems likely, however, that this motor cycle manufacturer rejected the 855cc baby on the grounds that it was planning to enter the car market with its own 1,145cc model, which was popularly known as Puppchen (little doll), and appeared later in 1911.

After leaving Germany, Friderich infers that his next port of call was Beaulieu-Valentigney in France, only about 120km (75 miles) south-west of Molsheim, where the Lion-Peugeot car was being built. Only in the previous year of 1910 had the two arms of the Peugeot company been reunited as *Automobiles et Cycles Peugeot*. Robert Peugeot had originally broken with the family business in 1903, had begun building motor cycles at Beaulieu and cars there from 1906. Friderich says that, following his visit to this factory, he left to show the car to M. Gudorge, Peugeot's general manager, who was based further to the south at Aix-les-Bains. 'I left it with him and returned to Molsheim by train: it was essential to lose as little time as possible, as production could not be allowed to suffer as a result of this considerable travelling.'

The contract between Bugatti and Peugeot for the production of the car was signed in August 1911 and, after evaluation, the prototype was returned to Molsheim and can today be seen at the *Musée national* at Mulhouse. The car proper was announced in October 1912, and Peugeot chose a provincial motor show to launch the model. The venue was Bourges, about 200km (125 miles) south of Paris, though 'rumours . . . have been current in the trade for six months.'[22] The car, which revived the name of Peugeot's best selling model of the 1902/1905 era, was fitted with an open two-seater body as standard, while a scaled-down version of the radiator used on larger Peugeot models was fitted.

[22] *The Motor*, 8 October 1912.

The Bébé only weighed about 360kg (7.1cwt) and, on its announcement, was reported to be '. . . so low built and has such small wheels, that one feels tempted to kneel down in order to examine it'. This was highlighted at Bourges when 'as soon as an enquirer showed a purchasing interest . . . the starting handle would be pulled over . . . one of the salesmen would lift up the front – motor still running – swing the car round, set it down, and it would be driven through the hall into the street for a trial trip.' *The Motor*, probably represented by W F Bradley, reported that the 'motor had a most healthy bark, a true Bugatti-like roar'.

The Beaulieu-built little car had a top speed of about 56kph (35 mph), sold complete for 4,250fr. (£169) and proved to be a great sales success. In 1913 the Bébé helped Peugeot to move ahead of Renault, which that year produced 4,704 cars, to become France's largest car maker, with about 5,000 automobiles manufactured. Unfortunately, this impressive progress was cut short by the outbreak of World War I in 1914. The model continued in essentially unchanged form, apart from the introduction of an additional gear, until 1916 when it was discontinued. By this time 3,095 examples had been built, making it the best selling Bugatti-designed car.

In addition to the Bébé, Peugeot also commissioned a second design from Bugatti, set down in a contract of 16 November 1911. It was an enlarged version of the Bébé design and was accordingly a four-seater designated the Type 20: a medium-sized 1,500cc car with T head engine and cone clutch. It entered production and one example survives in France.

From Eight Valves to Sixteen

The contact between Peugeot and Bugatti was, in all probability, a two-way dialogue. While Molsheim directly affected the affairs of the Peugeot company in the design of the Bébé, Bugatti's new Type 27 featured four valves per cylinder in the manner of Peugeot's celebrated L76 racer, though it still retained a single overhead camshaft. It should be made clear that, at this time, Molsheim was continuing to manufacture its eight-valve cars, which would briefly continue after production restarted following World War I. Plans, however, were made to increase the engine's original dimensions of 65 x 100mm to 68 x 108mm, giving 1,570cc for 1914 and a few engines of this new capacity may have been built. Also, the Type 22 and 23 variants would have become the Type 25 and 26 respectively, though the outbreak of war put paid to these initiatives.

For reasons that will soon become apparent, no complete 16-valve cars were built prior to the 1914 conflict. But first the Type 27 designation. Although this has previously been considered as only referring to the new engine, it must, in fact, relate to the complete car. Indeed, a rear axle drawing of a notably different design to the contemporary eight-valve cars and clearly titled *Type 27*, is held by the Bugatti Trust.

As for the engine, its dimensions were 68 x 100mm, giving 1,368mm to conform with the rather strange capacity limit of 1,400cc for the revived *Coupe des Voiturettes* race of 1914 which had been in abeyance since 1911. While the unit retained Bugatti's distinctive single overhead camshaft and banana-shaped tappets, curiously the design was such that the inlet and exhaust manifold were on opposing sides to those of the eight-valve cars. The 16-valve engine was the only Molsheim model with the exhaust manifold on the right and carburettor on the left. Its bottom half was essentially unchanged and the three plain bearing crankshaft was retained. However, the modifications to the top end of the engine had the effect of almost doubling output, from around 15 to 30bhp.

The *voiturette* race was to be held on the demanding and hilly 13km (8.5-mile) circuit at Auvergne in the volcanic district of Puy de Dôme to the west of Clermont-Ferrand, which had been last used for the Gordon Bennett event of 1905. It was intended to run the *Coupe de l'Auto* race for cars with a 2.5-litre capacity limit over the same course. Both events were scheduled for 23 August 1914 but, by then, World War I was into its third week. Germany had declared war on Russia on 1 August and, three days later, on the 3rd, on France. Bugatti had decided that he and his family should return to their native Milan and, on 2 August, the day before mobilization, they left Germany for their native Italy.

It must have been a difficult decision for Ettore to say farewell to his beloved Molsheim, after a mere four and a half years, but he was clearly ready to re-start car manufacture there after the war, regardless of its fate. We know this because, on 30 October 1914, Ettore wrote a brief handwritten letter to his friend, Dr Espanet. He ended it with a memorable postscript which perhaps encapsulates, not only his confidence in his own abilities but also the vitality of his spirit and sense of optimism for the future.

After being asked to be remembered to Madame Espanet, thirty-two-year-old Ettore wrote:

Mon usine est en bonne etat mais meme elle serait *rase au sol* un tete qui est mon capitale en fera une novelle. [sic.]

The English translation of this reads: 'My factory is in good shape but even if it were *razed to the ground* my head, which is my capital would build a new one.'

This long wheelbase Type Garros, pictured at Molsheim, is believed to be chassis 473, delivered to M. Heille of Schonlinde, who had previously been a customer for some of Ettore's pre 1910 models.

4
A Flight of Folly

'I was in the test room when the first [King-Bugatti] engine blew up... When the engine let go, all that was left of it were the mounting brackets... It was a miracle that no one was injured, much less killed.'

Cornelius van Ranst[1]

The four years of World War I represented a time of mixed fortunes for Ettore Bugatti. Cut off from Molsheim, his emotional commitment to France saw him gravitate to Paris where he created two aero engines for use by the allied powers. It has to be said that these were not particularly successful concepts, and perhaps highlight the limitations of his essentially empirical approach to design. No Bugatti-powered aeroplane saw action during hostilities. On a more positive front, however, Ettore was able to see the concept of the straight-eight engine which he had nurtured in pre-war days, come to fruition. He thus bequeathed the configuration to the cars that he was to build in the 1920s, which were unquestionably the finest of his career.

As mentioned in the previous chapter, mobilization saw Bugatti and his family decide to leave for Italy. Ettore's daughter L'Ébé, points out that there was no need for him to leave Molsheim. Bugatti's difficulty was that, although he was a German resident, he was, at heart, a Frenchman. 'At Molsheim we spoke French between ourselves; German was the language employed in the workshops', she says. While Ettore 'respected and admired many aspects of the German character, [he] was not interested in its culture, being instinctively drawn by nature and upbringing to French civilization'.

The Molsheim workshop force was therefore dispersed and Ernest Friderich, who was an Alsatian and a French national, rejoined his regiment, the 8th Artillery, having reached their Luneville headquarters after travelling through neutral Switzerland. On the evening of Sunday 2 August, his wife bade farewell to the Bugatti family at Molsheim's railway station.

L'Ébé Bugatti recounts: 'We left everything behind. All that my mother was able to take with her in the way of funds to meet living expenses were her own savings, about DM50,000 [£2,500], and a few jewels, in a small canvas bag; for the bank accounts had been frozen.' To get to Italy they first had to travel east and managed to get the last train to Stuttgart:

> Our departure was so hurried that we young children were still wearing nightdresses under our top coats. We travelled with the Kreisdirector (head official) of Molsheim, who had been called up; while continuing to advise my father to stay at Molsheim, he took the sleeping Jean on his knees and looked after him until we reached Stuttgart.

Once there they found the frontiers had been closed. Fortunately Ettore learned that Count Ferdinand von Zeppelin, – whom says L'Ébé, Bugatti 'knew quite well' – was in the German city of Constance, 150km (90 miles) to the south. The Count,

---

[1] Griffith Borgeson, *The Golden Age of the American Racing Car* (W.W. Norton, 1986).

who created and gave his name to the rigid airship, was responsible for giving Ettore and his family safe conduct to travel to his headquarters at Friderichshaven on the shores of Lake Constance. From there, they crossed the lake to the safety of Switzerland. Then it was a relatively simple matter for the five Bugattis to travel to Italy and so to Milan, where they took up residence at their home at 14, Via Trieste. By October Esister were also there.

In the following month of September, Bugatti briefly returned to Molsheim. By then it was apparent that, for the time being at least, Italy was to remain neutral – indeed Italy would not declare war on Germany until August 1916. His visit is recorded by Friderich, who says that: 'Three . . . cars were left at Molsheim where the camshafts were buried'.[2] These were the 16-valve blocks, or parts, of the cars which had been prepared for the aborted *voiturette* race of only two months previously and would remain safely interred until after the war. Ettore must have returned to Molsheim with an assistant because he made the journey back to Italy with 'two racing cars by road'. We now know that one of these was an unused Type Garros chassis (715). They were stored in a Milan cellar, though one was probably forwarded to Ettore when he later moved to Paris.

Once back in Italy, Ettore was able to reflect on their flight from Germany. 'Our journey was not too bad'[3] though all his horses were gone. 'You can imagine how sorry I am and how much I miss them', he told Dr Espanet in a letter of 18 November 1914. Dr Espanet was by then test pilot for the Nieuport Aircraft Company in Paris. But luck was of little concern to him. 'Italy is quiet, but even if she enters this terrible war I should not be called up, and as a volunteer I should not be in much danger. [!]' Ettore also gave Gabriel Espanet news of the whereabouts of other members of his family informing him that his brother, Rembrandt, 'was in Milan'.

The Other Bugattis

The youngest of Thérèse and Carlo Bugatti's three children had just celebrated his twenty-ninth birthday and had been widely acclaimed for his magnificent animal sculpture. Rembrandt had gone with his parents to Paris in 1904 and, in July of that year, had signed an agreement with Adrien A Hébrard, who had opened a gallery in the rue Royale to display the work of contemporary artists. There are echoes here of Ettore's contract of two years previously with de Dietrich, because it had to be countersigned by Carlo. Like his elder brother, Rembrandt was under age, being nineteen at the time. But there the resemblance between the two agreements ends because a feature of Ettore's contracts was that he could legitimately pursue his own projects, while being retained by the various firms for whom he worked, whereas Rembrandt was paid a regular fee to undertake his sculpture in return for its sole use by Hébrard, plus a royalty on each piece sold.

In 1907 the younger Bugatti moved to Antwerp, where he became a regular visitor to the local zoo and used the animals there as the subjects of some of his finest work. When World War I broke out, the shy, sensitive Rembrandt decided to stay in the city, although as a neutral he could have left. Instead, he remained with his artist friend, Walter Vaes, joining the Zoo's branch of the Red Cross. There, 'he worked tirelessly as a stretcher bearer and helping to care for the wounded laid out in rows in the [Zoo's] famous Marble Hall'[4] but, by November, he was back in Milan. Not surprisingly, Ettore says that his brother 'naturally saw terrible events at Antwerp'.

The only member of the Bugatti family not to be in Italy at this time was Carlo, who remained at Pierrefonds, about 75km (50 miles) north-east of Paris where he and his wife had moved in about 1910. Ettore reminds Espanet that: 'It is far away but we have good news. If by chance you go in that neighbourhood, do go and see him, he is quite

---

[2] Friderich, 'How the firm of Bugatti was born'.

[3] Conway, 'Ettore Bugatti: Correspondence 1913–16'.

[4] Mary Harvey, *The Bronzes of Rembrandt Bugatti* (Palanquin Publishing, 1979).

close to the property of Mr Clément.' [Adolphe Clément, of Clément-Bayard fame.]

Since his arrival in Paris, the talented Carlo had entered into the third phase of his remarkable artistic career. He took up the design of silverware and some of his exquisite products appeared at a 1907 exhibition at the prestigious Hébrard gallery, which also featured sculpture by his son, Rembrandt. There Carlo displayed a tea service incorporating, apparently, tusked elephant heads in their design though L'Ébé has subsequently told us that they were not but '*Ses Bêtes*, his creatures'. Carlo Bugatti was 'not satisfied merely to stylize existing animals, but preferring to take parts of them that he deemed beautiful and to combine them to create a ménagerie both picturesque and novel'.[5]

By this time Carlo was not making artefacts himself. The work, of formidable quality, was undertaken in Hébrard's workshops, to his designs. A silver tea pot of 1910, decorated with gothic themes, has also survived on which 'the hinged cover, on opening, clears the handle by a hair's breath.'[6] This creation by Carlo perfectly displays his talent for expressing proportion and precision of line, a gift which was clearly inherited by his eldest son and is to be found in so many facets of the Bugatti motor car.

With Carlo's departure to Pierrefonds – the move was made because of the fragile health of his wife who found the location more agreeable than the metropolis – the third phase of his career effectively came to an end. During World War I he became Mayor of Pierrefonds, in what was his adopted country. L'Ébé Bugatti says her grandfather was long remembered for his work there, tending the refugees and the wounded during hostilities. When the German advance threatened, Carlo defiantly exhibited a clay sculpture in the window of his studio, showing a Gallic cock, with wings outstretched, triumphantly perched on a German spiked helmet. After the war, Carlo continued to paint and, over the next two decades, occasionally produced pieces of furniture. These were similar in style to his pre-1902 work and have presented connoisseurs with formidable dating problems ever since!

Back in October 1914, Ettore Bugatti was resident in Milan and it was not long before he was at work on the design of what we believe was his first aero engine. Apart from the obvious demand for such power units during wartime, Bugatti was well acquainted with the world of aviation, as there was a considerable intermingling of automotive and aircraft circles in France and Germany in pre-World War I days. Ettore was already well known to such notables as Blériot, Voisin and Zeppelin. His friend Dr Espanet was closely involved in the aviation industry and, of course, there was Roland Garros. W F Bradley records that on Garros's first aerial visit to Molsheim in 1913, he had 'confided to Bugatti that he was interested in attempting a flight across the Atlantic. The immediate reply was "Why not?" They at once tackled the problem, Garros occupying himself with the aeroplane and Bugatti concentrating on two engines, which were to be based on the experience gained with the racing car model.'[7]

This would refer, appropriately enough, to the Type Garros engine, and the resulting unit would have therefore been a straight-eight. It will be recalled that Bugatti had created a crude eight-cylinder single-seater in 1912 by joining two of his eight-valve engines together. He clearly had plans for producing a big eight-cylinder car (*see* Chapter 9), because on 11 April 1913, prior to Garros's visit, he informed Dr Espanet: 'As for the eight-cylinder, it is on the drawing board, but not yet in production.'[8] This project was probably set aside with the outbreak of World War I, but the eight-cylinder aero engine was another matter.

We do not know how far its design had advanced following Bugatti's dialogue with Roland Garros. It seems likely, however, that he continued work on an eight-cylinder aero engine following his arrival in Italy. Ettore was assisted in this work by a hand-

[5] L'Ébé Bugatti, *The Bugatti Story*.

[6] Garner, 'Carlo Bugatti.'

[7] Bradley, *Ettore Bugatti*.

[8] Conway, 'Ettore Bugatti: Correspondence 1913–16'.

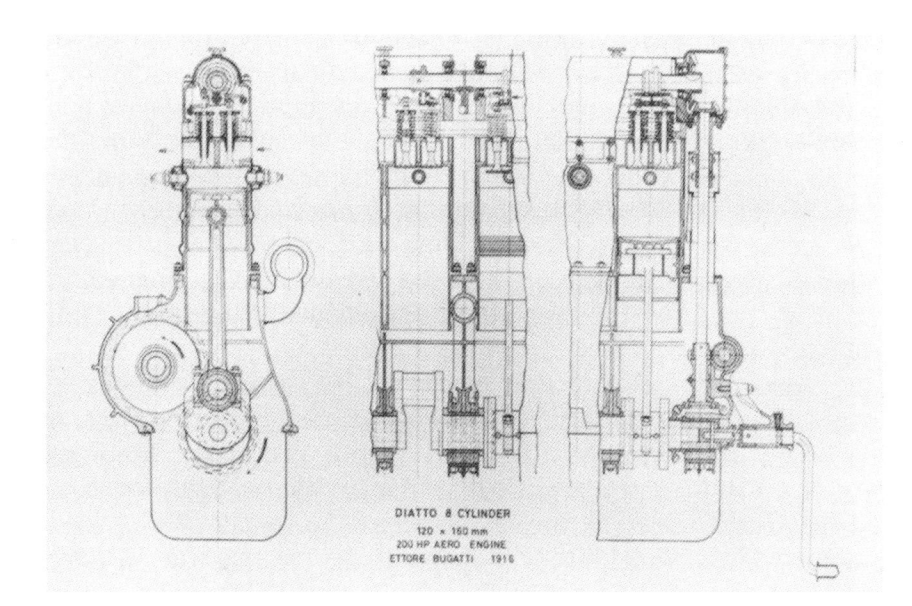

The first aero engine that Bugatti designed during the 1914–1918 war was this 200hp unit for which Diatto acquired a licence. It featured four valves per cylinder though does not appear to have entered production.

ful of draughtsmen, whom W F Bradley informs us had followed him from Alsace to Milan. Maybe Bugatti was thinking of spending the war in Italy because he soon began negotiations with the Turin-based Diatto car company to produce the completed engine.

A Move to Paris

Despite the outbreak of hostilities in August 1914, Italy was at this time still neutral and would remain so until 23 May 1915. It was only then that she declared war on Austro–Hungary though was clearly apprehensive of Germany and did not extend a commitment against that country until well over a year later, on 27 August 1916. When in the autumn of 1914, Bugatti submitted two of his inventions to the Italian authorities of a 'special rapid-firing gun which ejected shells acting first as shrapnel and then as a high explosive bomb',[9] this may have been the reason why he did not receive a positive response. It seems likely that following this rebuff, in November 1914 he left Milan for Paris, where he was known, and perhaps hoped for a more receptive atmosphere to his idea. Not only was he familiar with the French capital since his childhood but he felt a spiritual commitment to that country and was no doubt anxious to offer his services to the war effort.

Once in Paris and never one to do anything by halves, Ettore and his wife took rooms in one of the city's finest hotels, the Grand Hôtel in the rue Scribe, adjoining the Opera House. Soon after his arrival, on 3 January 1915, he informed Dr Espanet that 'I am still in Paris, I do all possible to be used to the defence of our *belle* France.'[10] Ettore told his friend that he had offered his services to the Panhard and Levassor company, which was based on the south side of the Seine in the Avenue d'Ivry. He says that he was collaborating with the firm 'in the construction of a machine gun' and also makes reference to 'a pretty interesting bomb'. This may be the same invention that had been so arbitrarily dismissed by the Italian authorities.

The outcome of his association with Panhard is not known but it seems likely that it came to nothing. Meanwhile, Ettore was pressing ahead with the design of his aero engine. This was undertaken in his rooms at the Grand Hôtel, a practice that has some echoes of the occasion in 1904 when he designed the Hermes car for Émile Mathis at

[9] Conway, *op. cit.*

[10] *Ibid.*

Hôtel de Paris in Strasburg. Ettore did not operate in the Grand in complete isolation because a small nucleus of essential draughtsmen followed him from Italy, or had reached Paris from Molsheim in other ways. These were 'united in the hotel drawing office, and a few mechanics were installed in a workshop Bugatti had been able to secure in the rue Jean Jaurès at Puteaux',[11] which is on the western outskirts of Paris.

On 16 January 1915, in a letter to his doctor friend, Ettore once again reverts to the subject of his aero engine. He says that it is an eight-cylinder unit, which he hoped would develop 200hp and weigh between 300 and 350kg (5.9 and 6.8cwt) and adds that 'the cylinders are one behind the other in line [as opposed to the V configuration] with a speed reducer which is not mounted on the camshaft, it is a special hollow shaft which can permit you to fire through it.'[12] By this time Ettore was clearly apprehensive about re-establishing his business in Alsace. 'We will succeed certainly but very probably we will never return to Molsheim. I am beginning even to dislike my house.'

Bugatti was expecting a car from Italy for, on 1 March, he says that it 'has not yet arrived'. Up until 23 May, when Italy joined the Allies, Bugatti was operating as a free agent but from that date, he was 'mobilized in France, at the order of the *Section technique.*'

The *Section technique de l'Aéronautique militaire* (Army Aeronautical Technical Section, or STA) was based to the south-west of Paris at Chalais-Meudon. The Count de Guiche, who was based there and would prove to be of great assistance to Ettore during these war years, says that 'Major Doran, who was in charge . . . gave him a warm welcome.'[13] Bugatti no doubt provided details of his proposed aero engine but more practical assistance came from the Count, who was later to become Duc de Gramont. L'Ébé Bugatti says that Rembrandt Bugatti 'knew the Count . . . through having sculpted a crucifix for him', though we do not know whether he was instrumental in arranging a meeting between the French aristocrat and his brother. As a qualified scientist, the Count de Guiche had been recalled from the Front to undertake work on fighter planes and later recounted that his first meeting with Ettore was 'at Chalais-Meudon , . . From the very outset, I was impressed by the strength of his creative imagination. [He] often came to see me there.'

Fortunately for Ettore, the Count was keenly interested in the science of aerodynamics. He was also the first to recognize that, apart from his Puteaux workshop, Bugatti did not really possess suitable premises where he could develop and build his straight-eight. In response to an offer from de Guiche, who says that Ettore 'began to assemble his first aero engine in my laboratory at Levallois-Perret', Bugatti was to remain at these premises, at 86 rue Chaptal, in the north-west of Paris, for the remainder of the war. In the meantime he had left the Grand and moved into a flat at 20, rue Boissière, not far from the Arc de Triomphe.

The Eight is Tested

Bugatti must have completed work on his engine by the autumn of 1915, because on 16 November the Delaunay-Belleville company with premises at St Denis, just north of Paris, purchased the French rights for it. Ettore received 150,000fr. (£5,454) for his invention. But first it had to complete its all-important 50-hour test and the eight appears to have undergone such an evaluation on 6 January 1916, 'which seems to have been for the duration of ten hours'.[14]

Bugatti clearly had some misgivings about his eight because, on 14 March 1916, he informed Dr Espanet that although he was 'content from the business point of view . . I do not see the engine coming out for a long time.'[15] He was apparently

---

[11] Bradley, *op. cit.*

[12] Conway, *op. cit.*

[13] L'Ébé Bugatti, *op. cit.*

[14] *Ibid.*

[15] *Ibid.*

During the war Ettore was based in the Levallois district of Paris in premises owned by his friend, the Duc de Gramont. Shown here is an aero engine under test.

awaiting a trial engine from Delaunay-Belleville but it 'has not arrived'. He then says that he was going 'to make a new engine . . . of 150 to 170hp' and here hints on the shortcoming of the 200hp unit because he adds: 'I hope to achieve such a difference in weight and consumption . . . that it will be difficult for it to escape attention.'

On 9 May, Delaunay-Belleville wrote to Bugatti, stating that 'the first engine, now re-assembled, is again ready to make the 50-hour test'[16] which suggests that it had yet to run for the allotted time. Once again the outcome is unclear and it seems unlikely that the eight performed as it should. In fact, the only documented 50-hour Paris test was not successfully completed until over a year later, in July 1917.

What is more certain is that this unit apparently did run for the necessary hours in Turin in September 1916 as, it will be recalled, Ettore had sold the Italian rights to his engine to the Diatto company, following his arrival in Italy in the autumn of 1914.

This is an appropriate moment to consider the design of the completed engine. As mentioned earlier, it can be described as two blocks of the enlarged Type Garros variety mounted on an aluminium crankcase. However, instead of three valves per cylinder, it was fitted with four in the manner of his Type 27 car engine. It had a 120 x 160mm bore and stroke, giving a capacity of 14.5 litres with the camshaft driven, as usual, from the front of the forward-mounted block, where it also drove twin magnetos to serve the two sparking plugs, one on either side, of the cylinder. The water pump was traditionally mounted on the left of the shaft. The crankshaft ran in nine main bearings and we can only assume that the lubrication was of Ettore's traditional non-pressurized variety.

Diatto appears to have had more success with the engine than Delaunay-Belleville but we do not know whether it was a Turin-built unit, or whether the Italian firm relied on one from France; be it Ettore himself or his French licensee. But whatever the source of the engine, little more was subsequently heard of it.

A Variety of Aero Engines

Before moving on to consider Bugatti's best known aero engine, the 500hp U16 unit, let us look at the

[16] Conway, *op. cit.*

types of engine in production by the opposing factions during World War I. Writing soon after the ending of the hostilities, Harry (later Sir Harry) Ricardo, summed up the respective engineering philosophies of the opposing powers in the following way. As far as Germany was concerned; 'From the start, [it] decided to restrict development almost entirely to the six-cylinder straight-line water cooled engine, on the grounds that this type of engine, though heavy, would give the maximum of reliability and fuel economy and permit . . . the largest production with limited manufacturing resources.'[17] These engines were initially of between 160 and 300hp and this clear-sighted approach dramatically contrasted with that of France.

Ricardo states that she 'possessed comparatively, a very large number of aeroplanes propelled by every conceivable type of engine, including air-cooled, water-cooled, fixed radial and rotating radial, four-, six-, eight-, and twelve-cylinder stationary types, in fact a heterogeneous collection . . . but apparently without policy as to which types to perpetuate for immediate military purposes.'

Most of the French-designed engines built during hostilities were, inevitably, of pre-war design. This applied to the radials of Clèrget, Gnome and Rhone, while Renault's popular but heavy air-cooled V8 dated back to 1908. Fortunately for the Allies in 1914 the Spanish/Franco, Hispano-Suiza company introduced its first aero engine. This was appropriately designated the Type A: a liquid-cooled V8 of 150hp, with a single overhead camshaft per bank, progressively employing enclosed valve gear, and was designed by graduate engineer, Marc Birkigt. This 11.7-litre short and compact aluminium unit, stove-enamelled to combat porosity, was to remain in production throughout hostilities. The engine progressed to develop 220hp by 1916, while the larger 18.4-litre model H of the same year was a 308hp unit and ultimately attained 340.

Probably the most outstanding allied aero engine of the war, it nevertheless suffered from poor cylinder cooling and distortion of the exhaust valves. On the other hand, the high quality of its specifications made great demands on its licensees, which were legion – the V8 being built in Britain, America, Italy, Japan and even Russia. There were no less than thirteen firms in France alone engaged in its manufacture, one being Delaunay-Belleville which also had a licence for the Bugatti eight. Yet despite these impressive credentials, the Hispano-Suiza was only capable of about twenty hours reliable running, provided that it was carefully maintained between overhauls.

And so to Sixteen

Ettore was attempting to produce a far larger and more powerful engine than the Hispano-Suiza V8 and, while it has to be said that Bugatti's unit was greatly inferior to it, his sixteen was probably no worse than many of the second generation of more powerful aero engines that appeared towards the end of the war. Having said that, it was probably just as well for Ettore's future reputation that no Bugatti-powered aircraft saw action during the war.

In the last surviving letter of their war-time correspondence, on 19 June 1916 Ettore briefly informed his friend Dr Espanet, that 'I am making a new engine, very good.'[18] Only three months previously, in March, Bugatti had made reference to a new engine of 150 to 170hp, though we do not know how far this had advanced. Perhaps the latter reference was to the U16 and we know that the concept had been in Ettore's mind since at least 1915, for a double-banked parallel engine appeared in his British patent specification (101534) of 15 September of that year.

With this unit, Ettore applied the modular concept that reached back to his tricycling days at the turn of the century and perhaps reflected his self-taught approach to design and lack of theoretical education. Then, as in 1916, he strove for more power, not designing a brand new engine, but

---

[17] Harry Ricardo, *The Internal Combustion Engine Volume II*, (Blackie, 1923).

[18] Conway, 'Ettore Bugatti: Correspondence 1913–16'.

adopting the cautious approach of duplicating an existing, proven unit. In 1915 his two Type Garros fours became an eight and in 1916 he doubled the configuration once again to sixteen cylinders.

What Bugatti did, in effect, was to mount two of his existing eight-cylinder aero engines, complete with their respective nine bearing crankshafts driving a common propellor shaft, on a single crankcase. Each eight was, in reality, two interconnected fours and their respective crankshafts were therefore in two halves, connected at the centre by a taper and key, shrunk and drawn up by a nut. The two sections were assembled at right angles to the other and Bugatti was to perpetuate this arrangement on his eight-cylinder car engines. The U16 differed from the eight in that it was far closer to his Type Garros concept and accordingly employed the three-valve layout of two inlet and one exhaust, which Ettore had introduced on that engine in 1911. Each four-cylinder 'engine' was fitted with its own updraught carburettor, making a total of four.

Unlike the eight-cylinder aero engine, on the U16 each cylinder bank had its single overhead camshaft driven by a shaft positioned between the two cast iron blocks. Centrally located twin magnetos on either side of the cylinder blocks were driven by bevel gears at right angles to the shaft and fired one sparking plug per cylinder. Bugatti also later replaced the front camshaft drive of his eight-cylinder engine to this central location. One water pump was fitted, driven at engine speed off the left-hand crankshaft at the rear of the engine. On the U16, Ettore reduced the engine's bore 120 to 110mm though the stroke remained at 160mm. This resulted in a capacity of 24.3 litres. The engine weighed a not unreasonable 483kg (1,607lb) and was a reflection of Bugatti's proven impeccable visual judgement in this regard. Less desirably, the lubrication system was unpressurized and relied on Bugatti's usual crude 'gravity and splash' arrangement.

The new Bugatti aero engine ran for the first time at the rue Chaptal laboratory on 23 October 1917. Runing time lasted about fifteen minutes and preparations were at once put in hand for the crucial 50-hour test. This was to be supervised by the French STA and also officials of what had become known as the Bolling Commission, for the Americans, who had entered the war six months previously, planned to mass-produce the Bugatti U16 in the United States.

When the USA had declared war on Germany on 6 April 1917, it was sadly deficient in its supply of aircraft. Fourteen years after the Wright brothers had given the world powered flight, America had only built around 700 aeroplanes and only rated fourteenth in the league of world production. Its most sophisticated machine was the Curtiss Jenny, which could only manage about 120kph (75mph) on a good day, and that had been created to apprehend Pancho Villa in his ill-fated Mexican expedition of 1916.

The American Commitment

In the month after the declaration, on 16 May, Secretary of War, Newton D Baker, hastily created the Air Production Board (APB). As the intention was to overcome the country's lack of aircraft and aero engines as quickly as possible, the government considered that the solution could only be found by what was then known as 'The American System of Manufacture', in other words mass production, and that meant recruiting experts from the automobile industry. It was for this reason that Baker appointed Howard Coffin, president of the Hudson Motor Car Company, to head the board and this, in turn, created an Aeronautical Commission, which was to visit Europe and choose designs that could be brought back to America, built in quantity there and then exported to help win the European air war.

The Commission was headed by a lawyer, Major (later Colonel) Raynal C Bolling, legal counsel to the US Steel Corporation and, significantly, nephew of the American president Woodrow Wilson. Another member was Howard C Marmon, who had gained his engineering training in the family flour-milling business and had gone on to produce his own advanced and highly individual

Indianapolis-built car from 1902. Also included was a talented West Point officer, Captain E S Gorrell. The Commission sailed from New York for Europe on 17 June 1917.

Its brief was to 'sample airplanes, engines and data necessary for their production'.[19] The American party first visited Britain and there arranged for the newly introduced, single-engined two-seater light bomber and reconnaisance aircraft, the de Havilland 4, to also be built in America. It was to become the sole US-built combat plane of the conflict. The Commission's week in Britain was followed by ten days in Italy, while a similar projected spell in Paris (where Bolling and his team arrived early in August) was extended to the duration of the war when they decided to make their headquarters there.

On arrival in the French capital, Major Bolling wasted little time in seeking out an English-speaking expert on the automobile and aeronautical industries and found him in the shape of W F Bradley who, for the previous two years, had been advising Howard Coffin, in his capacity as a member of the National Defense Committee, on how cars performed under wartime conditions. Bradley was immediately retained by the Commission and it was he who effected the introduction with Bugatti.

At this time, Ettore's earlier straight-eight aero engine had completed its all-important 50-hour test (on 16 July) though it would be a further three months before the U16 would be ready for its first trial run. But the Americans were clearly excited by its potential. The eight was rated at 210hp, so they calculated that double that was 420. As a result of his contacts with Ettore, on 15 August Bolling informed his Washington-based superior: 'The Bugatti engine appears perhaps to offer the most interesting future development for light weight per horsepower and ease of quantity production.'[20]

W F Bradley, who based much of his biography of Ettore's on the latter's personal reminiscences, says that it was 'Howard Marmon, who was at once impressed with the possibilities of the Bugatti double eight.' In the later Congressional inquiry into the affair, it was Captain Gorrell who accepted responsibility for the decision, maintaining that 'the Bugatti was far and away the lightest engine and was a trifle smaller in size, gave less wind resistance, and at the same time produced 100 more horsepower than the Eagle Rolls-Royce engine.'

Although the Commission had originally been required to make payment to the French government for aircraft and engines, there were undoubtedly complications because of Bugatti's Italian citizenship and he is reputed to have been personally paid the handsome fee of $100,000 (£21,008) by the Americans for his invention. Bradley confirms that Ettore received 'a substantial cheque signed "R C Bolling." Somewhat innocently he took it to the bank and asked if it was valid. "As many as you like, and as much as you like, with that signature on it", was the teller's reply.'[21]

In the Name of Liberty

On 13 August, just two days prior to Bolling telling the APB about the Bugatti U16, across the Atlantic in America the prototype V12 of that country's own purpose-built Liberty aero engine had run for the first time. This unit had its origins in a V12 designed by Packard's Colonel Jesse Vincent in 1915 and, following the US declaration of war, had been updated in May 1917 by its designer and Elbert John Hall of the Hall-Scott Motor Company.

The original intention was to produce it as a 45deg 18,025cc (1,100cu in) V8, with a single overhead camshaft per cylinder bank. Draughtsmen had begun the drawings for the design on 4 June and had completed them a mere five days later, on the 9th. The first engine was completed at Packard's Detroit factory on 3 July and it was accordingly dubbed the Liberty in recognition of its arrival on Independence Day. It ran under its own power for the first time on 25 July and was found to develop 214hp though Vincent had hoped for 275hp.

[19] Borgeson, *Bugatti by Borgeson*.

[20] *Ibid*.

[21] Bradley, *Ettore Bugatti*.

The Bolling Commission's recommendation that the DH4 aircraft be built in America, plus news from the Front which suggested that more power would be needed, meant that the engine had to be enlarged to an anticipated V12 configuration of 28,677cc (1,705cu in). This development was speedily effected by perpetuating the V8's 127mm (5in) and 177mm (7in) stroke. After its successful trial of 13 August (it actually ran for 55 hours), the 346hp the V12 recorded fell somewhat short of an anticipated 400bhp. However, further testing saw the engine achieve an average of 404hp by February 1918. This was after the first operational engine had been delivered in November 1917.

It was during the same November that, in France, Bugatti's U16 was undergoing its first sustained test, with the Americans no doubt anxious to see it achieve its all important 50 hours of continuous running. The initial start-up on 23 October has already been noted and this formal evaluation started at the rue Chaptal on 9 November. It began well enough but, after seven hours, the test was halted by the failure of the water pump's drive shaft. This was replaced but the same component broke again after a further eight hours. Once again the shaft was changed but again failed, this time after nine hours. The replacement operation was repeated and the next run proved to be the longest of the series but, at 10hr 20min, the test was aborted after a piston and gudgeon pin broke.

A week after the first series of tests, the programme was resumed but on 16 November, after approximately five hours, it was marred by a horrific

The U16, as refined by C B King and manufactured in America by Duesenberg.

accident. According to Bradley, 'an American sergeant passed in front of the engine, was sucked in by the propeller and before any of the assistants could intervene, was horribly mutilated. As he was the first member of the American Air Force to be killed in France, it was decided that his body should be sent home at the same time as the engine.'[22] The accident had splintered the U16's propeller and it was deemed advisable to strip the engine itself down to see whether it had been damaged by the incident though it was found to be 'good as new'.

This tragic accident was perhaps responsible for masking the U16's shortcomings though they would be exposed by lengthy running in America. The engine also suffered minor maladies of the sort that might be expected from a new power unit, certainly one of the complexity of the U16. 'Carburettor fires occurred twice, a fuel line broke once, and "oil cooling" became clogged once.'[23] The constant interruptions caused by the breakages of the water pump shaft were said to have been caused by the spontaneous unscrewing of the packing gland nut but a more likely cause was that it was over-stressed through running at engine speed. The pump had proven to be relatively trouble-free on the straight-eight but had then been geared down from the camshaft drive.

When all the engine's interrupted running bouts had been totted up, they amounted to 37 hours, 'most of that time at 1,880rpm and 375bhp but occasionally [the engine] was run for a very few minutes at as much as 1,980rpm, at which it yielded 432bhp'.[24] With a weight of 480kg (1,067lb) this was equivalent of 2.5lb per bhp, which was considered good for its day.

The Mission to America

Little wonder that much was expected of the Bugatti engine in America. Assistant US Secretary of War of the day, Benedict Crowell, has recounted that the Aviation Section of the Signal Corps, which had responsibility for the engine's production in America, 'prepared to proceed with the Bugatti on a scale that promised to make its development as spectacular as that of the Liberty'.[25]

Clearly time was of the essence and it was decided to remove the engine from France and ship it to America, where the testing programme could be resumed and the U16 prepared for quantity production. Therefore on 14 December, the Bugatti Mission left Bordeaux for New York, complete with the engine and its straight-eight predecessor. Bugatti himself did not join the party but it included none other than his old friend and authority on aviation, Dr Gabriel Espanet. The latter was retained as a civilian employee of the US government at the generous salary of $750 (£157) a month as a 'technical expert'.

Then there was Ernest Friderich, who received a not inconsiderable $15 (£3.15) a day, which was the equivalent of $105 (£22.05) a week. In 1915 Ettore had requested that Friderich be withdrawn from the Front, after distinguished military service, to assist him with his aero engine work. Friderich had taken part in the fighting around Nancy, later in the Arras sector and, in May 1915, had been awarded the Croix de Guerre.

The party was completed by the presence of aircraft designer, Captain George Le Père and Lionel de Marmier. There was also a pair of Alsatian mechanics, Shopfer and Lucke and, of course, the corpse of the unfortunate American Air Force sergeant who had died during the rue Chaptal testing. On the same boat was the French government's mission to America though it differed from the Bugatti one, which was funded by the US government.

We know from W F Bradley that the boat reached 'New York just before Christmas, the Bugatti Mission went first to Dayton'. This was the headquarters of the Aviation Section of the Air Corps. One of the main contractors for building the DH4 aircraft in America, the recently established

---

[22] Bradley, *Ettore Bugatti*.

[23] Borgeson, *Bugatti by Borgeson*.

[24] *Ibid*.

[25] *American Munitions, 1917–18*.

Production, as such, did not get underway until the First World War was nearly over. However, this is a batch of completed engines at the Duesenberg factory.

Dayton Wright Airplane Company, was also located there. (The other two recipients of the contract were Fisher Body in Detroit and the Standard Corporation.) After its brief visit to Ohio, the party divided into two sections. Gabriel Espanet and George Le Père departed for Detroit and the Packard factory at East Grand Boulevard, where the V12 Liberty engine was under construction. It was also the place where a projected, all-important Le Père-designed aircraft would be built. One variant being tailored for the Liberty engine and the other for the Bugatti U16 unit.

The rest of the party – Friderich received his directive on 5 January 1918 – was to proceed to Elizabeth, New Jersey, where the Duesenberg Motors Corporation had opened a brand new factory to build the Bugatti engine.

Originally, Duesenberg was to have been involved with the Liberty project 'and was even tooling up' for it but, such were the hopes for the U16, Assistant War Secretary Crowell says that it was, 'directed . . . to assume leadership in the production of Bugattis'. While the Liberty project was centred around Detroit, 'we now prepared to establish a new aviation district in the East, associating it in such concerns at the Fiat plant at Schenectady, NY, the Herschell-Spillman Co, of North Tonawanda, NY and several others.'[26]

The U16 engine arrived at Duesenberg's North and Newark Avenues plant in January and was first dismantled so that it could be inspected by Charles Brady King, aeronautical mechanical engineer of the US Signal Corps. His brief was to evaluate the design in relation to its mass production potential and also to gauge its likely airworthiness. After inspection, the engine was reassembled and the U16 ran for the first time in America on 19 February 1918. The testing proper began three days later, on 22 February, and although a number of runs were made, misfiring was so bad that no figures could be taken. The Champion plugs, of American manufacture, were therefore removed and replaced by Bosch ones though it was later found that their insulation was badly cracked.

Trouble with Testing

On the fifth run, the U16 began to vibrate badly and the engine was shut down when a noise was

[26] *American Munitions, 1917–18.*

A Duesenberg-built Bugatti U16 aero engine as displayed at the Smithsonian Institution in Washington. The cambox, complete with its shaft, has been removed from the forward bank to reveal the three valves per cylinder.

heard. On dismantling, it was found that the number three connecting rod on the left cylinder bank had broken near the big end, number two rod had suffered from overheating through bearing failure and number three main bearing had burnt out. Further inspection revealed that pistons number two and three were broken beyond repair and the upper crankcase webbing had fractured, while this internal catastrophe had also damaged two of the cylinder blocks. In addition, the propeller shaft bearing had overheated and was unusable for future running.

This was clearly a serious set-back though Friderich considered that the trouble had been due to the 'Motor running without propeller was not cooled in any way, thus causing the crankcase to attain a high temperature in a very short time'.[27] It later emerged that the problem had been caused by a 'simple [sic] failure of lubrication due to a blockage of an orifice in the crankshaft', which highlighted the limitations of the engine's non-pressurized lubrication system. In view of problems he had experienced, Charles King reported, with commendable understatement, that the Bugatti 16 'was not a commercial engine when delivered to this government'.[28]

There was nothing for it but for King to introduce some reliability into the U16 by remedying some of its more obvious shortcomings. The most serious deficiency was the primitive lubrication system, which Ettore was also to employ on his car engines up until 1928. One of its principal limitations was that 'the pressure [is] determined by the diameter of the open orifices . . . heavy or cold oil will take the path of least resistance and will not travel to the remote ends of oil leads. The small openings in bearings become clogged with sediment, waste and coagulated oil. Owing to the pressure not being sufficient to clear these passages, trouble can be expected.'[29] It seems likely that this 'trouble' caused the aborted test of 22 February. A proper pressurized system was designed, which not only served the crankshafts, but also the camshafts.

King's Quest for Reliability

King had also detected overheating problems with the cylinder head though Bugatti was to perpetuate this shortcoming on his later single overhead camshaft car engines. The heads of the French engine were sectioned and cracking was found

---

[27] Conway, *Bugatti, 'Le pur-sang des automobiles'*.

[28] Borgeson, *Bugatti by Borgeson*.

[29] Conway, *op. cit.*

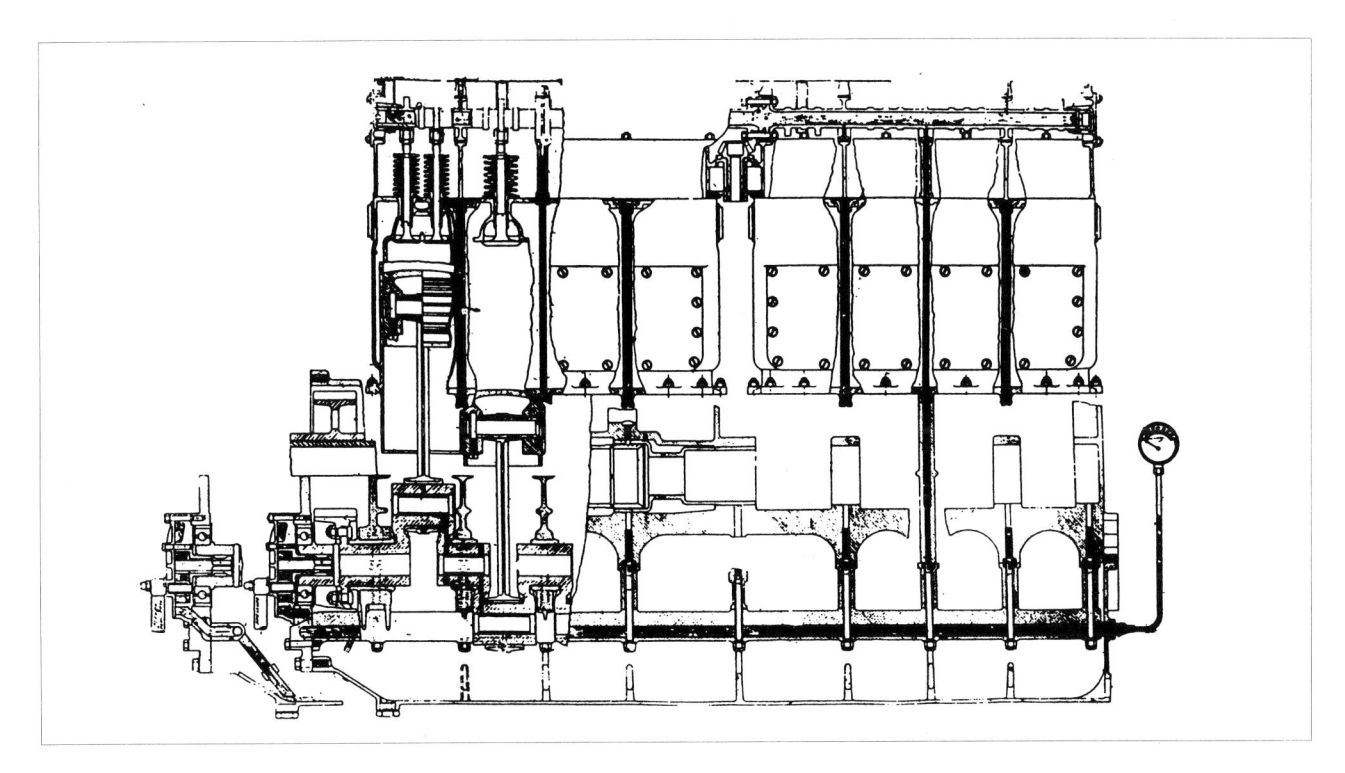

One bank of the U16 engine showing the pressurized lubrication system introduced by King.

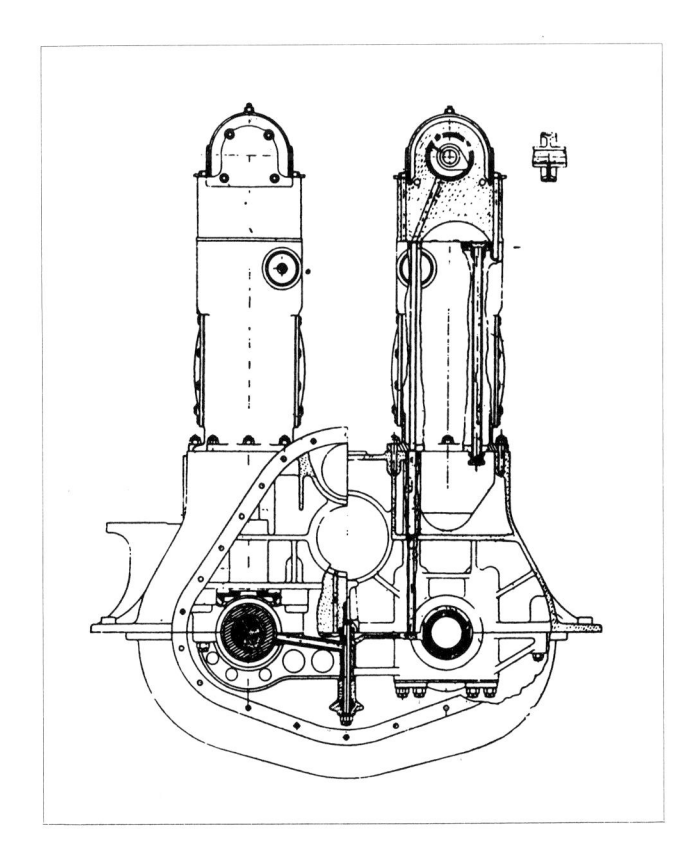

A cross-section of the U16 showing its lubrication system.

between the inlets and exhaust valves. King accordingly modified the design to improve the water circulation, while the shapes on the inlet and exhaust ducting was changed from Bugatti's original angular lines to more rounded contours. Changes were made to the jacketing of the blocks and King also introduced water cooling to the inlet manifolds. The water circulation system was simplified to eliminate the various pipes and connections of the original.

King also took the chance to re-design the water pump for, in addition to the problems already experienced, it possessed a further deficiency in that, if it leaked, the coolant tended to enter the engine's sump. The pump was moved back on the American engine, its support improved, so that if the pump did begin to weep, it would do so on the outside of the engine rather than into it. The front of the U16 was also modified so that the 'very complicated double bearings used on the original French model'[30] were replaced by a series of Hess-Bright compound annular and radial thrust bearings.

[30] Conway, *op. cit.*

The aero engine's porting, for two inlet and one exhaust valves, as revised by King.

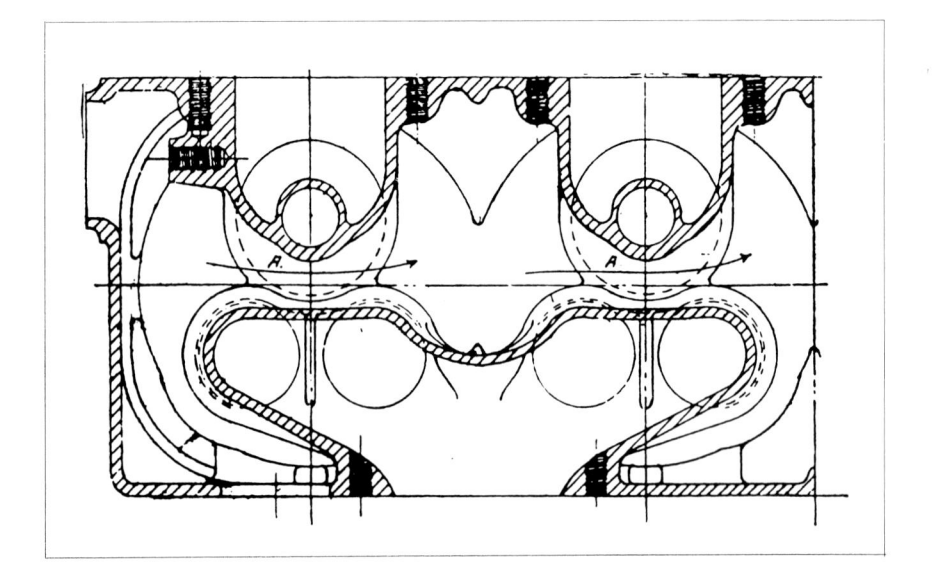

It also made sense to fit the U16 with American proprietary components and Miller carburettors and pumps were accordingly specified for the unit. The Los Angeles-based Harry Miller had been producing his Master carburettor in that city from 1907 to 1908 though in 1918 he 'went East to set up a plant in connection with the war effort, principally to supply carburettors to the Duesenberg aero engine facility in Elizabeth, New Jersey'.[31]

This plant, or depot, under the banner of 'Miller Products Company' was based at 109 West 64th Street, New York, though the facility was closed down with the coming of peace and Miller and his associate Fred Offenhauser returned to Los Angeles in February 1919. Miller was on the verge of producing his exquisite twin overhead camshaft straight-eight engines and equally formidable competition cars that would make him 'the greatest single figure in American motor-racing history'.[32] Miller's and Bugatti's paths were destined once again to cross within a short space of years.

Friderich no doubt kept Bugatti informed of the modifications made to his aero engine because, in a letter of 11 April to the Duesenberg 'Director', he regrets 'that these [changes] have been made without me warned beforehand.'[33] He did not want to question the competence of the engineers concerned but he considered his drip and splash lubrication system 'more certain' than a pressurized one. He had contemplated changing the water jacketing three years previously but had rejected the idea. He had used the cylinder head design on his cars for ten years – he disliked the water-cooled inlet manifold on the grounds that water might leak into the engine but was later to incorporate the feature in his own engines . . .

On 24 June a French engine was successfully tested. It is not known whether it incorporated King's modifications though it seems likely. It must have run for a respectable time because a figure of 460hp at 2,190rpm was attained. Only two days later, on 26 June, American unit number four, no doubt including King's changes, ran for 30 hours, a record for the U16. The 50-hour test was once again attempted in August but the engine was totally destroyed when a connecting rod broke after 20 hours' running. It might have been on this, or another occasion, that a crankshaft snapped, this malady having taken place on 28 August.

But King and his team persisted. In September

---

[31] Mark L. Dees, *The Miller Dynasty* (Barnes Publishing, 1981).

[32] Griffith Borgeson, 'Henry Miller' in *Automobile Design: Great Designers and their Work* (David and Charles, 1970).

[33] Conway, *op. cit.*

another U16 recorded 503hp at 2,400rpm before a crankshaft once again fractured. It was this component which Duesenberg's Cornelius van Ranst considered to be a 'very weak part of the original design.'[34]

Not until 4 October, less than six weeks before the signing of the Armistice and close on a year since it had first run in France, did Bugatti's U16 complete its vital 50-hour test 'at an average rated power of 410bhp'. Conversely, King's essential ministrations had put the U16's weight up by 60kg (133lb) to 544kg (1,200lb) which began to cancel out one of the unit's principal advantages over its rivals.

U16 Manufacture Begins

Production could now begin and Arthur A Schupp, who was 'in charge of the experimental work' at the Elizabeth factory, 'took the first three motors to France' which he delivered to Bugatti. This American's contact with Ettore must have made an indelible impression on him because, nearly forty years later, in 1957, he recounted that he went 'to see him and was sent into his engineering department and drafting room. There must have been twenty to twenty-five men hard at work on drafting tables . . . Up front, Bugatti had a nude female model and was busily engaged chiselling her statute from a piece of rock.'[35]

The Armistice was signed on 11 November and back in America according to Crowell, 'a total of only eleven' engines had been delivered. This contrasted with the original production programme recorded by Friderich which specified fifteen engines completed by October; fifty in November and 2,000 by April 1919. But he recognized that, if output did not begin until November (which it did), it would not have been possible to reach the 2,000 figure by June 1919.

Despite the ending of the war, a pilot run of about forty-five engines was completed. A surviving identification plate from what must have been this batch, indicates engine number J60 having been built on 1.15.19, which suggests that it was completed on 15 January 1919. Incidentally, it simply refers to the U16 as 'Bugatti U.S. Model'. There is no reference to King's contribution though a wooden mock-up clearly proclaims the U16 as the 'King-Bugatti', and contemporary literature refers to it as such.

he arrival of peace also put an end to the Le Père two-seater fighter, then under construction in Detroit by Packard. Two of the Liberty-engined LUSAC-11 (Le Père United States Army Combat) planes 'were in France undergoing trials – and Packard was busily completing twenty-seven more when the war ended'.[36] By contrast, only three examples of the Bugatti-powered version, designated the LUSAC-21, were built. Bradley optimistically informs us that 'Dr Espanet made the first flights, attaining the very satisfactory speed, for the period, of 150mph [241kph].' Whatever the truth, we do know that in November 1918, George Le Père himself piloted a LUSAC-21 but the flight was short-lived. The U16 engine overheated badly, burning the aircraft's wooden structure, and at this point the test was curtailed and there was not another one. It was perhaps the symbolic end of the U16 misadventure which had cost the American taxpayer no less than $4,934,798 (£1,036,722).

A French Revival

The French public, of course, knew nothing of these tribulations. As far as it was concerned, Bugatti's U16 had been bought by the Americans but its production had been curtailed by the ending of the war. Even if some whispers of problems reached France Ettore, with some justification, could have claimed that US authorities had taken the U16 before the design had been completed. The truth was only to be found on the other side of the Atlantic and it was not until 1958 that Hugh

---

[34] Borgeson, *Bugatti by Borgeson*.

[35] *Bugantics*, vol. 20, no. 3 (autumn 1957).

[36] Beverley Rae Kimes, ed., *Packard, A History of the Motor Car and the Company* (Automobile Quarterly, 1978).

After the war the Bréguet company took over the U16. The blocks have been removed from this example so that an overhaul may be undertaken.

Conway began to uncover the real truth of the U16 fiasco.

In view of this sorry story, it comes as some surprise to find that the Paris-based Bréguet aircraft company took up the design of the U16, which suggests that it did not know the truth of the matter either. The Bréguet brothers of Louis and Jacques were pioneer aviators whose eighteenth-century forbear was one of the world's finest watch makers and Ettore was to become a keen collector of Bréguet time pieces. In September 1907, the brothers, in conjunction with Professor Charles Richet, built a machine which resembled a collection of apparently unrelated, horizontally-positioned step-ladders, powered by a 40hp Antoinette engine. The Bréguet-Richet No 1 was the first to raise itself by vertical take-off but, as four helpers had to hold it at arms' height without letting go, this did not qualify as a free flight.

The brothers let others develop what was to become the helicopter and, in 1911, Louis Bréguet founded the company to bear his name. While he had clearly established a reputation for the quirky and offbeat, the Bréguet 14 nevertheless proved a significant World War I bomber. At the 1919 Paris Air Show, no less than two examples of the renamed Bréguet-Bugatti U16s were displayed, coupled together in 32-cylinder tandem form.

If outward appearances are anything to go by, they were of the pre-1918 French type, which suggests that Ettore may have obtained some rights to the design, and we can only speculate as to whether they incorporated any of King's all-important modifications. Internal and external changes were later made to the engine by Bréguet, one of which was to reduce the bore from 110 to 108mm to compensate for a weakness in the engine's cylinder wall. This massive combined unit was to have powered Bréguet's projected and substantial Leviathan transporter. The unit was positioned in its own engine-room, rather like a ship's, to drive this biplane's single propeller. Perhaps fortuitously, the idea proved to be a non-starter.

Although Bugatti was originally retained in a consultancy role, he was later replaced by an engineer named Vullierme. The U16s were used in another smaller biplane and this flew in June 1922. Afterwards 'it was noted with pride that the engines had logged 70 hours of operation, on the bench and in the air.'[37] But the U16 jinx persisted when, in the *Grand Prix des Avions de Transport* in November, a bank of the aircraft's cylinders failed and later a further eight cut out. This forced the machine to the ground though, fortunately, without damage to the craft or pilot.

Two years later, at the 1924 Paris Air Show, Bréguet revealed the ultimate manifestation of the breed, the extraordinary *quadrimoteur*. This was an H32, effectively consisting of two U16 units, mounted crankshaft to crankshaft. By this time steel had replaced cast iron and there were four valves per cylinder. But the 950hp Bréguet – the Bugatti name had been dropped by this time – proved to be too heavy at 1,090kg (2,403lb), or just over a ton, for its own good. At this point Bréguet decided to curtail his aero engine activities, though aircraft manufacture continued, and no more was heard of the U16.

The Tragedy of Rembrandt

Although the war had enabled Ettore to create the straight-eight engine and brought him some financial security, as far as his personal affairs were concerned, on 8 January 1916 he had been devastated by the suicide of his younger brother.

In November 1914, Rembrandt had joined most members of the Bugatti family in Milan. But by the end of the year, he was 'worn out and infinitely depressed . . . His mood was made worse by personal worries. He was growing deaf, due to a series of neglected inflammation of his ears, a condition which made him even more cut off from the world.'[38] As a sculptor he had also suffered financially from the collapse of the international art market brought about by the war.

As if this was not enough, while in Antwerp, 'he had been made very unhappy when an artist friend's daughter, with whom he thought he had an understanding, married someone else.'[39] Later he followed Ettore to Paris, where he had an atelier at 3, rue Joseph-Bara, Montparnasse. But there 'a brief love affair with a childhood friend ended unhappily, and he was rejected for the second time in two years.' After attending Mass at the Madeleine early on the morning of Saturday 8 January 1916, he bought a bunch of violets from a flower seller on the church steps. On returning to his studio, Rembrandt wrote letters to Ettore and the Superintendent of Police and, next to the Legion of Honour medal he had been awarded in 1911, he laid the bunch of violets 'as a last discreet homage to the women he loved'.[40] Dressed in his best suit, he stopped up any gaps in the bedroom and turned on the gas. He was discovered during the afternoon still alive but died *en route* to the nearby Laennec Hospital. He was aged thirty.

In view of the testimony of the flower seller, who had also regularly sold flowers to Ettore and his daughters, and had been the last person to see Rembrandt alive, the Archbishop of Paris authorized a religious funeral, which was held in the church of Notre Dame des Champs. Rembrandt was later buried at the Bugatti family vault at Dorlisheim, near Molsheim.

---

[37] Borgeson, *Bugatti by Borgeson*.

[38] Harvey, *The Bronzes of Rembrandt Bugatti*.

[39] *Ibid.*

[40] *Ibid.*

Ettore's father Carlo, who was to become mayor of Pierrefonds during the First World War. He would eventually join his son at Molsheim.

A death mask of Rembrandt was ordered by Ettore and he also signed his brothers' last work, a bronze of a Lioness and Serpent which had been at the Valsuani foundry when he died. Over two months later, on 14 March 1916, Ettore confided in a letter to Dr Espanet, then with the French Squadron in Venice, that he was 'really knocked about [*bouscoulé*] after all this unhappiness... My brother, this brother who was for me different from all brothers it is impossible to express how his absence upsets me. Even now I cannot find a reason. Fact it must be and [I must] support this grief which will never leave me and will grow the more with time. I did not know that such unhappiness could exist.'[41] Once back at Molsheim, Ettore created a small museum containing examples of Rembrandt's work, in memory of his brother.

The Death of Garros

Just over a month before the Armistice of November 1918, Ettore received a second blow with the death of his great friend, Roland Garros, who was killed in action on 5 October 1918. Many of Ettore's surviving war-time letters to their mutual friend, Dr Espanet, are peppered with enquiries regarding Garros's safety. Ironically France's most famous aviator had been a German prisoner of war for twenty-two months but had escaped and made his way back to Paris via neutral Holland in February 1918. Ettore had been to see him some weeks after his return and L'Ébé Bugatti records that: 'There he noticed some recent photographs of him on a table. Bugatti asked to have one, and Garros wrote on it: "To Ettore Bugatti, the incomparable artist who alone knows how to give life to steel." It was inscribed "in admiration and friendship, Roland Garros".'

When Barbara and Ettore's fourth child, a boy, was born on 28 August 1922, he was christened Roland Cesare Maria Carlo, his first name in memory of France's illustrious pioneer aviator and one of his father's best friends.

---

[41] Conway, 'Ettore Bugatti: Correspondence 1913-16'.

5
Breakthrough at Brescia

'Thus did Bugatti score that famous grand slam which had been denied him at Le Mans in 1920 . . . The performance caught the imagination of the motoring world.'

Kent Karslake

It should be remembered that Bugatti cars had only been built for a mere four years prior to the outbreak of war. After hostilities ceased, production was resumed and, almost from the outset, Bugatti adopted a selective and initially successful *voiturette* racing programme to publicize his products, which set the pace for even greater Grand Prix triumphs later in the decade. Although the original pre-war four-cylinder models were perpetuated, in 1922 Molsheim introduced its first straight-eight car and the Bugatti marque would be forever identified with that engine configuration.

Ettore was determined to resume car production with the coming of peace. During hostilities, when the outcome of the war was still in doubt, he had 'purchased a plot of land opposite the historic château of La Malmaison, near Paris'.[1] If for any reason Bugatti and his family had been unable to return to Alsace, he could have built a new factory on this site, which was not far from the Hurtu car works at Rueil, then a small village on the north-western outskirts of the French capital and only about 10km (6 miles) from the laboratory at Levallois-Perret that he had occupied during the war years. Such a location made geographical sense because the bulk of the French motor industry was located in and around Paris, as were component suppliers, though somehow a Rueil-built Bugatti does not quite have the same cachet as a Molsheim manufactured one . . .

Fortunately the war was won by France and her allies though at great cost in terms of lives; she had lost about one and a half million men in the conflict. But for Bugatti the really significant outcome of peace was that Alsace–Lorraine, which had been annexed by Germany following the Franco–Prussian war, was returned to France. This had been one of the American president Woodrow Wilson's famous fourteen points of January 1918, number eight reading: 'All French territory should be freed and the invaded portions restored, and the wrong done to France by Prussia in 1871 in the matter of Alsace–Lorraine . . . should be righted.' These were adopted by the Allies to form the basis of the peace negotiations and were confirmed by the Treaty of Versailles, signed on 28 June 1919. Bugatti could have only been delighted that his beloved Molsheim was now once more in French territory and, from thereafter, his racing cars would be no longer painted in the white of Germany but in France's racing blue.

In December 1918 Ettore was joined by Ernest Friderich, who had been in America at Duesenberg's Elizabeth, New York factory, working on the U16 aero engine. On 6 December, he had received instructions to return to Paris for demobilization, had left America by boat on 10 December and was back in the city on 19 December. There he

---

[1] Bradley, *Ettore Bugatti*.

'rejoined the Chief . . . and we went to Molsheim where the works was at a standstill, and the workers were not very reliable. Accordingly, the Chief decided to dismiss the whole staff and to start again with reliable workers.'[2] This should not perhaps be taken too literally but Bradley says that the plant itself was, thankfully, 'not seriously damaged. The factory had been used for valve production, the machinery was the worse for wear, but there had been no wanton destruction.' So in January 1919 Ettore and Barbara Bugatti returned to Molsheim after an absence of over four years.

The First Post-War Cars

Ettore's first priority was to get production restarted and this began later in 1919. The cars were effectively those that he had been building in 1914; the eight-valve Types 13, 22 and 23, though he did not pursue the Type 25 and 26 replacements he had contemplated in 1914. Bugatti even considered resuming production of his 5-litre model but no examples were built.

The rejuvenation process was a slow one and it was not until July 1919 that the first eight-valve chassis (number 764) was delivered and output continued intermittently into the following year. In July 1920, Ettore was informing a British friend, Llewellyn Scholte, 'I've got a lot of trouble getting production going again as a result of the lack of raw materials, but I hope to be able to get a regular number of chassis out every month.'[3] The final eight-valve car (chassis number 843), was completed two months later, in September 1920.

While the eight-valve cars were still being produced, Bugatti introduced his new 16-valve model early in 1920. This was powered by a 68 x 100mm, 1,453cc engine, which was fitted in the Type 22 and Type 23 chassis, with the racing version being allocated to the shorter Type 13 wheelbase. The first batch of 16-valve cars, built in February 1920, amounted to nine chassis (numbers 905 to 913) and it should be noted that Bugatti chassis numbers, although they ran consecutively were not completed in a chronological order. Chassis 901, the first number of the series, did not leave Molsheim until the following month of March.

As ever, Ettore was looking for racing to provide the engine of publicity to sell his cars. One of his first acts on his return to Molsheim was to retrieve the 16-valve blocks that he had so carefully buried in September 1914. As these were intended for the 1914 *Coupe de l'Auto* race, for cars of up to the slightly curious capacity of 1,400cc, they had cylinder dimensions of 66 x 100mm, giving 1,368cc to fall within the required limit. They were disinterred and mounted in Type 13 chassis and no less than four of them (maybe two of these 1,368cc cars had been built in 1920) were entered for the *Coupe des Voiturettes* race. This was to be held at the Le Mans circuit on 29 August, just over six years to the month since the 1914 event had been cancelled because of the outbreak of World War I.

Because the event was organized by the Sarthe-based Automobile Club de l'Ouest, it was held near the town of Le Mans on a short triangular 17km (10.5-mile) course. The cars would cover twenty-four circuits, making a total of 408km (255 miles).

Unfortunately France's roads had suffered badly during the war and Sarthe's highways were no exception. In addition, 'the whole region around Le Mans had been an important American army centre, and the grandstand occupied the place of one of the biggest American camps.' This was why the organizers had chosen a relatively short course which had been treated 'with calcium chloride brought from England at the cost of £40 per ton, the sides of the road were loose and dust was thrown up whenever two cars passed one another'.[4]

Opportunity at Le Mans

The cars that Bugatti entered for the event were 16-valve Type 13s and accordingly had a 2,000mm

---

[2] Friderich, 'How the firm of Bugatti was born'.

[3] Scholte/Bugatti Correspondence (held by Farnham Museum, Farnham, Surrey).

[4] *The Autocar*, 4 September 1920.

(6ft 6in) wheelbase. Very rudimentary bodywork was fitted, no doubt in the interests of weight saving; there was a combined bonnet and scuttle, side panels alongside the driver and his mechanic, and no doors. The rear-mounted bolster petrol tank had 'Bugatti' badges on either side, so that spectators could be in no doubt as to the cars' origins.

Although four Type 13s appeared on the entry list for the race, reports of the event only refer to three cars, so perhaps the Bugatti race number 18, which was due to be driven by Ferdinand de Vizcaya, was not completed in time. The other cars were driven by Ernest Friderich, with Étien as his mechanic, Pierre de Vizcaya, son of the Baron, with Emil Mischall, while Michele Baccoli, who worked at Molsheim, was attended by Lutz. The cars were mechanically similar, though the usually accurate French magazine *Omnia*, pointed out that Friderich's car had two plugs per cylinder. A similar car was served by a specially modified magneto, positioned in the customary position at the front of the engine, so it seems likely this had a similar layout. But in the race, *Omnia* noted that Vizcaya's car, which was not so equipped, proved to be slightly faster.

Ettore Bugatti himself was in control of his team at Le Mans, while he set the pattern for Bugatti facilities at race meetings during the decade by providing a champagne lunch which was laid out in the pits.

The Bugattis were clear favourites and the only other overhead camshaft cars present were four Majolas. However, a clear threat was presented by three Bignan-Sports, which boasted four wheel brakes. There was a 1,400cc Sizaire Naudin, which had probably been prepared for the 1914 event, while a single-cylinder Corre-la Licorne was a survivor from a past era. A curious entry was the two-stroke Major, with two pistons operating in a single open-ended cylinder and a central combustion chamber . . . The British sent a team of three Eric Campbells and there was a lone GN. A further Bugatti-designed car was also present in the shape of a Bébé Peugeot, which was probably too heavy for the cycle car race of the previous day!

The event itself began at 9.30 a.m. and the weather was 'dull and cold . . . with every indication that rain might fall . . . however this fear was not justified'.[5] The cars set off in pairs, at half-minute intervals, and the first to leave was de Vizcaya and the GN. Only ten minutes after the start the Bugatti was in the lead, Friderich was only 31 seconds behind, the Major next, while Baccoli, in the third Type 13, was fourth.

It did not take long for everyone to see that the Bugattis were the fastest cars present, with the only real opposition coming from the Major, Majolas and Bignans. Later the Major withdrew with overheating and before long, at the half-way point, the Molsheim cars were running first, second and third. By the time that 20 of the 24 laps had been covered, Friderich was in front, followed by de Vizcaya and Baccoli. With the end clearly not far off, the spectators were no doubt preparing for a grand slam victory for Bugatti, when Vizcaya stopped at the pits, by all accounts, for some oil. He apparently received some and was just moving off when, Ettore 'shouted [for him] to stop, ran after the car and started to unscrew the radiator cap'.[6] As a result, the car was disqualified as the race rules specified that only the driver and his mechanic could touch the car.

This was assumed to be a serious mistake on Ettore's part and so it appeared until 1958, when de Vizcaya's mechanic, Emil Mischall, revealed the truth of the matter. Many years after the event he told how, during practice, de Vizcaya's car had suffered from an overheated rear main bearing and it was decided to remove the engine and inspect the crankshaft and white metal. Mischall was 'about to replace the connecting rod bearing and sump' when Ernest Friderich asked him 'to do another job' and he would complete the work. Although de Vizcaya had been ordered 'to drive in a particular way but . . . he had decided to go out to win'.

During the race, according to Mischall, the car once again suffered, not only from bearing trouble, but 'a connecting rod had broken! We stopped beyond the pits and opened the bonnet for Bugatti

---

[5] *The Autocar*, 4 September 1920.

[6] Karslake, *Racing Voiturettes*.

The Coupe des Voiturettes race at Le Mans on 29 August 1920. Here, two of the three sixteen-valve Type 13s – Baccoli is on the left next to de Vizcaya's car – receive attention in the pits.

who had arrived on the scene. One glance was enough – and at a quick sign from him we closed the bonnet, while he walked round the front of the car and started to unscrew the radiator cap. No wonder he was reported to have accepted disqualification without comment!'[7]

Victory for Friderich

Bugatti had acted deliberately in having the car disqualified rather than reveal a mechanical malady. The only possible anomaly in this story is that, according to Kent Karslake's authoritative *Racing Voiturettes*, after the radiator cap incident, 'de Vizcaya went off at last with his same mechanic, and was greeted when he came round next time by

[7] Emil Mischall, 'Le Mans 1920', in *Bugantics*, vol. 21, no. 4 (winter 1958).

an official with a red flag. The crowd very properly hissed; but de Vizcaya was out of the race.' This risky further lap is also confirmed by W F Bradley in his biography of Bugatti. In any event, Bradley probably reported the race for his new employers, *The Autocar*, who had recently offered him double the rates being paid by their rival weekly *The Motor* for whom he had previously worked.

Despite this incident, Friderich went on to win the event at 92.7kph (57.6mph) but Baccoli, in the surviving Bugatti, had difficulty in restarting his engine after a plug change and came fifth. Bignan-Sports were second and third. Nevertheless the Bugatti victory was a great success for the marque, showing that it could combine speed with impressive roadholding, combined with the all-important ingredient of reliability.

This last named facility was underlined after the Le Mans event when Mischall was able to satisfy himself that de Vizcaya's car had not suffered from

William Fletcher Bradley 1879-1971

W F Bradley, who represented English language motoring magazines in Paris from the early years of the century for the next fifty or so years, knew Bugatti from his earliest days as a car manufacturer, and was a regular visitor to Molsheim and destined to become Ettore's biographer. An outstanding reporter of the contemporary motoring scene, his words still retain their readability and vitality though he was, alas, less of a historian. As will become apparent, his *Ettore Bugatti, Portrait of a Man of Genius*, is strongly partisan in his subject's favour, good on some personalities and atmosphere though it should not be relied on too heavily for factual content.

Bradley was born, in far from comfortable circumstances, in Scarborough, Yorkshire on 8 March 1876. His father had tried his hand at plumbing and at baking and William only received elementary schooling. Later the family moved to Leeds, and to make ends meet, his mother took in lodgers. One was a reporter-cum-typesetter on the local newspaper and the business fired young Bradley's imagination. Very much left to his own devices, at the age of thirteen, Bradley became indentured with a printer for a seven-year apprenticeship. Recognizing the limitations of his schooling, he attended evening classes in a range of subjects including typesetting, shorthand, algebra and, fortuitously, French. He completed this all-important training in 1896 at the age of twenty, moved to London and, in November 1901, successfully answered an advertisement from a firm which published a small English language newspaper in Paris.

WF Bradley, third left, witnessing Bugatti's first victory in the 1925 Targa Florio. Meo Costantini can be seen enjoying his customary cigarette.

Once in the French capital, he married a Scottish girl whom he had known during his early years in Leeds. Paris was then the automotive hub of the world and Bradley had an English friend named Vingoe who had conceived the idea of translating items in the French motoring press into English and selling them to American newspapers. Vingoe opted for *Motor Age*, while Bradley chose *The Automobile* which was published in New York. That journal changed its name to *Automotive Industries* in 1917 and took over *Motor Age* in 1918. He was to remain its European correspondent for the next fifty-three years.

The famous Paris to Madrid race of 1903 marked the start of Bradley's career as a motor-racing reporter. However, his financial circumstances were greatly improved in the following year when the *Daily Mail* introduced a continental edition and W F was taken on as assistant proof reader. Soon afterwards, he was asked by George F Sharp, of the newly established, London-based weekly magazine, *The Motor*, to become its Paris correspondent.

With the outbreak of the First World War in 1914, Bradley served at different periods in the British, Belgian and Italian armies and as a second lieutenant in the latter force, he drove an ambulance in the Italian campaign and saw action at Gorizia.

Since 1915, two years prior to America's entry to the war, Bradley had been retained by the Hudson Motor Company's Howard Coffin to report to the National Defense Committee in Washington on how automobiles behaved in the field of battle. At the end of the war, Bradley was commissioned by Fiat to establish an English language news service in Paris. He was given impressive offices in the Champs Elysées, commuted to Turin on a monthly basis and became acquainted with many key members of that great organization, from Giovanni Agnelli downwards.

Bradley continued his association with the British motoring press but, in 1919, he was taken on by Iliffe's *The Autocar*, after that publication had agreed to double what *The Motor*, published by the rival Temple Press, was paying him. He was to write for the magazine for the rest of his working life. In those days of editorial anonymity, he was initially 'W F' though articles soon began appearing under his own name.

With the arrival of the Second World War and the occupation of France in 1940, Bradley and his wife were unable to escape to Britain and decided to sit the war out at their home. However, W F was imprisoned and remained in German custody until late in 1943 when he was released. The Liberation came in June 1944 and Bradley and his wife returned to Britain soon afterwards, first living in the country and then the sea side resort of Bournemouth.

It was in 1947 that W F conceived the idea of writing a biography of Ettore Bugatti although, 'on the day we should have met, Le Patron was reported to be ill'. Unhappily, his subject died in August though the slim volume was published in March of the following year.

Bradley and his wife remained in Britain until 1950 when they moved back to France and the town of Vaucresson not far from their original home. Then, unhappily, Mrs Bradley died in 1953 and W F later 'officially retired' at the age of eighty in 1956.

He continued to write and his last book, *Motoring Racing Memories*, was published in 1960. That year he travelled to London to receive the Guild of Motoring Writers' Harold Pemberton Trophy to the member 'considered to have made the most outstanding contribution to the cause of motoring during the year'.

What had become *Autocar* continued to publish Bradley articles. His last, on the 1921 French Grand Prix which he so vividly recalled, appeared in the magazine of 22 April 1971. He died, the undisputed doyen of motoring journalists, just over five months later, on 6 October, at the age of ninety-five.

The race was a triumph for Bugatti as Friderich won. Ettore in the foreground gives a characteristic gesture with his right hand and next to him is Friderich's young daughter.

a component fault when its connecting rod broke. After the engine had been removed, he inspected the damage and 'found that one big end had become unbolted. I could find no trace of split pins. Did my helpful colleague of the night before forget them?'[8]

Racing in Italy

Bugatti's success at Le Mans was highlighted by the fact that it was the only important motor race to be held in France that year and it was not until nine months later, on 22 May 1921, that the team once again participated in an international competition. This was held in Ettore's native land at Salo, near the town of Brescia in north-eastern Italy and was the *voiturette* section of the Lake Garda *Formule Libre* event.

It may have been the first occasion that Bartholomeo 'Meo' Constantini drove for Bugatti. He was to regularly do so until the end of 1926, when he then took over as team manager from Ettore. The other cars were driven by Baccoli and a new recruit, Eugenio Silvani. The opposition effectively consisted of two Chiribiris and the event proved a relatively easy Bugatti victory, with Silvani placed first and Constantini second. But the Type 13s were less lucky at the Grand Prix de Boulogne in July, when neither of the two Molsheim cars finished.

The next race, held at Brescia later in the year, on 8 September, was the first significant *voiturette* event to be held in Italy and was part of a speed week which also featured a *Formule Libre* race. Although the intention was to adopt Grand Prix rules for the *voiturette* race and only accept entries from factory teams, two of the cars, the Restelli and SB, were driven by their creators. Other privateers were relegated to a second stream *voiturette* event, scheduled for four days after the first one.

Sensibly, the capacity limit for these events was increased from 1,400 to 1,500cc and Molsheim accordingly responded by enlarging the size of the 16-valve engine to 1,453cc, achieved by upping the bore size from 66 to 68mm, so bringing the

[8] Mischall, *op. cit.*

An old friend of the Bugatti family, Giuseppe Ricordi at the wheel of one of the Le Mans cars with a young Jean Bugatti in the passenger's seat at Molsheim. On these Type 13s the front tie rod was located between the front chassis members. Whatever happened to E.B.1?

capacity into line with that of the production road cars. In addition, there were changes made to the crankshaft, in that the plain bearings were replaced by ball races for the rear and centre journals. This represented a tacit acknowledgement by Bugatti of the limitations of his splash lubrication, certainly as far as plain bearings in competition were concerned, as the arrangement worked better with the freer running but more complex and expensive rollers.

The location of the central roller meant that the crankshaft was made in two sections, secured in the middle by a split pinned nut. Because of the proximity of the camshaft drive, the original front plain bearing was retained while the big ends employed roller rather than bronze ones. There were two hollows in the crankshaft webs to receive oil squirted from a two-holed pipe on the right-hand side of the crankcase. The oil thereafter found its way, via drilled passages, to the big-end journals.

As in the case of some of the Le Mans cars of the previous year, twin sparking plugs were fitted though, instead of them being fired by a single double spark magneto, there were two of them driven off the rear of the camshaft. Each Bosch unit fired a separate set of plugs, one on each side of the block, and they were mounted, in the manner of the 1908 FE Isotta Fraschini, under the cars' scuttles and projected into the driving compartment. There they were removed from the heat of the engine, which would have been exacerbated by the Italian summer. If they did require attention, this could be rendered without lifting the bonnet.

All these ministrations made no difference to the power developed by the engine but perhaps it was thought that changes to the crankshaft would prevent the bearing trouble that had affected at least one of the cars at Le Mans in the previous year. The Type 13s were otherwise similar to the previous year's racing Bugattis.

Ettore obviously took the event very seriously and prepared a team of five cars for the race, though only four of them were entered. These were driven by the established team of Friderich, de Vizcaya and Baccoli, while there was a new recruit in the shape of Pierre Marco, who was to remain with Bugatti for many years and would be responsible for the day-to-day running of the firm in the post-World War II years.

The opposition was almost wholly Italian. The exception was Eugenio Silvani, who had driven a works Type 13 in the Lake Garda event of two months previously. At Brescia he drove his own Bugatti, fitted with a streamlined body of his own design, and was accordingly called the SB. The rest of the cars were side valvers; a quartet of 1,496cc OMs – on home ground, being manufactured in Brescia itself – while the three Chiribiris were Turin-built. There was also a lone Restelli, driven by its creator, of which little is known, and it made no impact on the race.

The Brescia Circuit

The short 17.2km (10.7-mile) course, consisted of a fast, straight, flat triangle of roads and, as a result, was 'probably the fastest road circuit that had ever been used in Europe'.[9] Competitors had to complete twenty circuits of the course, making a total of 345km (215 miles), which would place great demands on the reliability of the entrants' respective engines.

As was the fashion of the day the cars started in pairs at one-minute intervals, and it soon became apparent that the Bugattis and the SB, with their overhead camshaft engines, were the fastest cars present. The heavier but slower OMs were nevertheless reliable, while the Chiribiris were said to show their worth on corners, of which there were few.

By the time that five laps had been completed, Friderich was in the lead, with de Vizcaya 21 seconds behind, followed by Silvani. The remaining

[9] Karslake, *Racing Voiturettes*.

Bugattis were fourth and fifth. Then Friderich burst a tyre and Silvani moved briefly into the lead. The SB, it was noted, was faster on the straights than the works Bugattis on account of its streamlined body, a facility that might not have escaped Ettore's attention. By the half-way point, it was de Vizcaya that had moved ahead though he was being hard pressed by Silvani, who was only 5 seconds behind and by the three Bugattis, followed by a Chiribiri. Silvani, who had put up an excellent challenge, dropped out shortly afterwards with 'burned out bearings'. This was perhaps an endorsement of Bugatti's decision to switch from plain to roller bearing on the works Type 13s. By this time, the other entrants had fallen by the wayside, apart from the OMs.

Victory in the Sun

This left the Bugattis in the first four places and, with three-quarters of the race completed, a little less than 8 minutes separated de Vizcaya, the leader, from Marco in fourth position. During the final few laps, Friderich (as ever playing a waiting game) moved into the lead and came in to win, in 2hr 59min 18sec, having averaged 115.47kph (71.75mph). He clearly considered the changes to the cars' crankshaft bearings a key factor in his victory, later recounting that Bugatti 'had decided to mount the crankshaft in ball bearings. It was after a few successful innovations of the same kind that I won the Grand Prix at Brescia . . .' Next was de Vizcaya, followed by Baccoli and Marco and the four OMs, the final car coming in half an hour behind the winner. But there was not, of course, any French opposition at Brescia though this would become all too apparent at the all-important *Coupe des Voiturettes* event later in the month.

Despite the fact that this victory had been achieved in the face of a handful of little-known Italian cars, in an inspired piece of marketing, Ettore decided to offer a supplementary 16-valve model, fitted with the modifications which had proved to be so successful in the Italian event. Available for the 1922 season, the car was appropriately

When the Type 13 became the Brescia. Ernest Friderich pictured at Brescia in the car in which he won the Voiturette Grand Prix on 8 September 1921 at 115.47kph (71.75mph). Bugattis were also placed second, third and fourth. These cars outwardly differed from the previous year in having larger shock absorbers with a straight tie rod located between the two.

called the Brescia and not only was there a Type 13 version but the longer 22 and 23 frames were also fitted with the roller bearing engine though with plain big ends and associated twin magnetos but with a 69mm bore, rather than a 68mm one, giving 1,496cc. Customers were accordingly able to purchase a pukka *voiturette* racing car. This innovative sales approach proved to be devastatingly effective when Bugatti later came to introduce his superlative Type 35 Grand Prix car.

The first production Brescia was completed in March 1922. About thirty-nine were built during the year, the last car (1599), being delivered in January 1923. Of these, the Type 23 Brescia was the most popular. There were sixteen of these, the cramped facilities of the Type 13 being not, perhaps, to everyone's taste. There were eleven of these and ten Type 22s. One of the first true Brescia to reach Britain went to twenty-one-year-old Raymond Mays, who had just come down from Cambridge and was destined for an impressive career as a racing driver and constructor. Mays later christened his Brescia *Cordon Rouge* and its history, along with that of its *Cordon Bleu* sister car, is charted separately. The model's successor, the *Brescia Modifiée*, will be considered later in this chapter.

Back in 1921, after his success at Brescia, Bugatti had entered four cars for that year's prestigious *Coupe des Voiturettes* race, which was held ten days later, on 18 September. Once again the venue was the Le Mans circuit and three weeks prior to the event, it was reported that 'the Bugattis were already on the course.'[10] But Ettore later caused

[10] Karslake, *op. cit.*

Raymond Mays at the wheel of his Brescia, 1318, which he christened Cordon Rouge *and purchased from B S Marshall Ltd, in 1922. He is pictured here, giving an aircraft a run for its money, at Porthcawl Sands in 1922.*

Raymond Mays, right, with Amherst Villiers, who was responsible for tuning Cordon Rouge. *They are pictured at Mays' home, Eastgate House, Bourne, in Lincolnshire in 1923. Villiers is holding an experimental Brescia four-bolt connecting rod while on the bench is the substitute radiator in the form of a modified Castrol tin which Mays was to use in his unsuccessful attempt to break the Test Hill record at Brooklands later in the year. It involved removing the bonnet and even the radiator to make the car as light as possible! On the right are a pair of Brescia wheels while, on the wall, is the spacer which, when positioned between the block and crankcase, increased the engine's compression ratio.*

Mays pictured in his other Brescia, Cordon Bleu, *chassis number 2059 which he bought in 1923, in the Paddock at his beloved Shelsley Walsh in 1924. The car is distinguished by the flutes that Mays has added to the bonnet. They were also made for* Cordon Rouge *but perhaps not fitted.*

In the spring of 1925 Mays sold both his Bugattis. Cordon Bleu *went to a young Oxford undergraduate named Francis Giveen and* Cordon Rouge *to Frank Taylor who later sold Bugattis in Birmingham. He is seen here at the paddock at Shelsley running with single rather than twin magnetos, in 1926.*

uproar when he withdrew from the race just prior to the event. 'It was held', thundered *The Autocar*, that, 'having officially entered, a competitor should be obliged to start in a race.' It went on to perceptively point out that 'perhaps it was a case of proverbial discretion. The opposition this year was very formidable to the Alsatian *voiturettes*.'[11]

[11] Karslake, *op. cit.*

Ettore's excuse was that there had not been time to prepare his cars after the Italian race but he was often to employ such a tactic when, on weighing up the opposition, a Bugatti victory was in question. He was no doubt concerned that a successful challenge to his cars on the circuits would damage all-important sales and the adverse publicity from a late withdrawal would have less impact than the report of a victory by a rival make . . .

A 1922 Brescia, chassis number 1321, being driven by Eddie Hall in a rather wet hill climb.

B H Austin, chairman of the Disabled Drivers' Association, pictured at the wheel of his 1923 Type 22, chassis number 1612, with home-made engine-turned open two-seater bodywork. Its engine, no. 1, was the first of the standardized ball-bearing crankshaft units. The front axle was bound with tape, in the interests of aerodynamic efficiency. Curiously the badge on the radiator reads La Corrida. *This car survives in Britain.*

14 EVENTS.
15 PLACES.

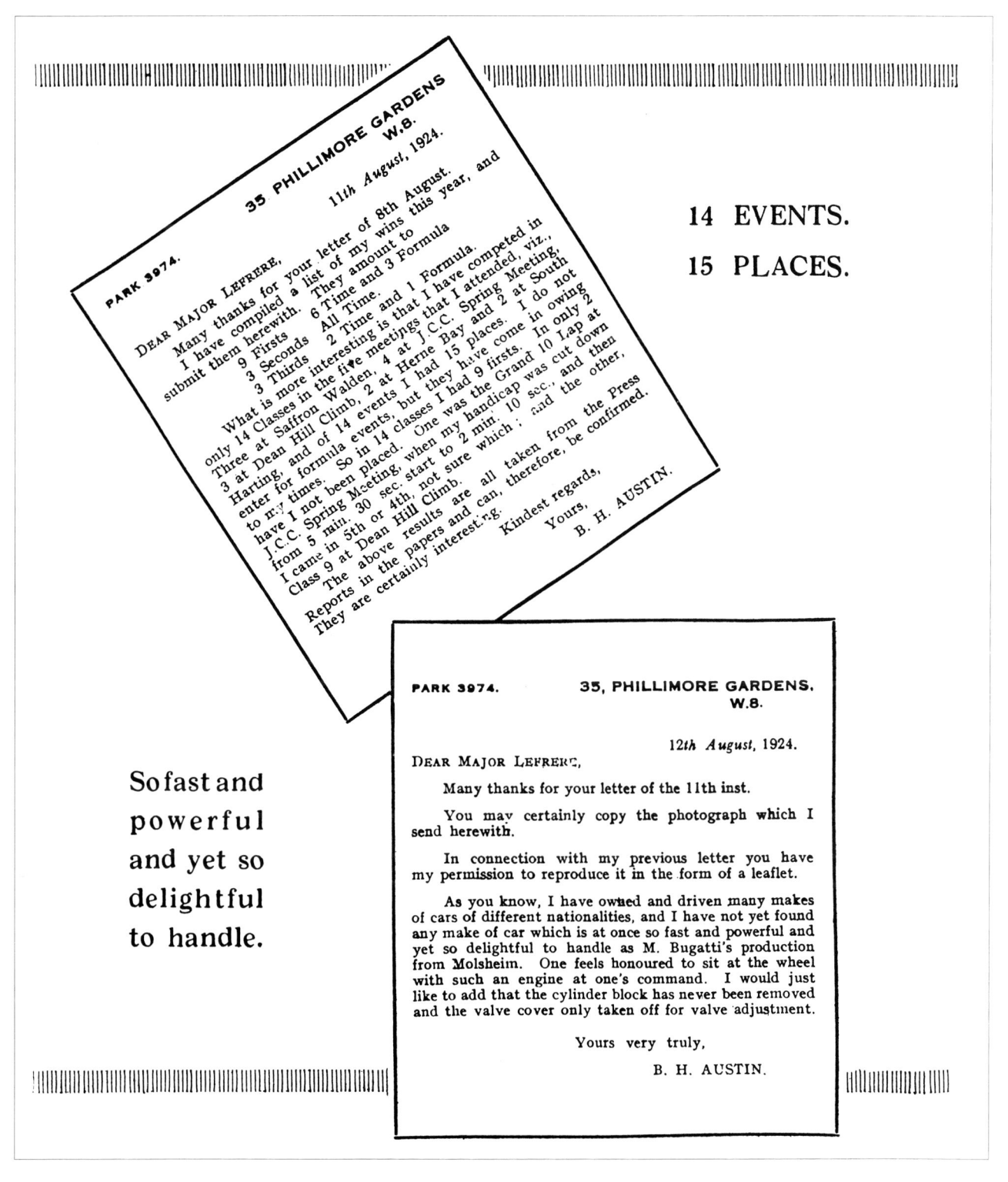

PARK 3974.

35, PHILLIMORE GARDENS,
W.8.

11th August, 1924.

DEAR MAJOR LEFRERE,

Many thanks for your letter of 8th August. I have compiled a list of my wins this year, and submit them herewith. They amount to

9 Firsts   6 Time and 3 Formula
3 Seconds   2 Time and 1 Formula All Time.
3 Thirds

What is more interesting is that I have competed, viz., only 14 Classes in the five meetings that I attended, viz., Three at Saffron Walden, 4 at J.C.C. Spring Meeting, 3 at Dean Hill Climb, 2 at Herne Bay and 2 at South Harting, and of 14 events I had 15 places. I do not enter for formula events, but they have come in owing to my times. So in 14 classes I had 9 firsts. In only 2 have I not been placed. One was the Grand 10 Lap at J.C.C. Spring Meeting, when my handicap was cut down from 5 min. 30 sec. start to 2 min. 10 sec., and then I came in 5th or 4th, not sure which; and the other, Class 9 at Dean Hill Climb.

The above results are all taken from the Press Reports in the papers and can, therefore, be confirmed. They are certainly interesting.

Kindest regards,

Yours,

B. H. AUSTIN.

So fast and powerful and yet so delightful to handle.

PARK 3974.   **35, PHILLIMORE GARDENS,**
W.8.

12th August, 1924.

DEAR MAJOR LEFRERE,

Many thanks for your letter of the 11th inst.

You may certainly copy the photograph which I send herewith.

In connection with my previous letter you have my permission to reproduce it in the form of a leaflet.

As you know, I have owned and driven many makes of cars of different nationalities, and I have not yet found any make of car which is at once so fast and powerful and yet so delightful to handle as M. Bugatti's production from Molsheim. One feels honoured to sit at the wheel with such an engine at one's command. I would just like to add that the cylinder block has never been removed and the valve cover only taken off for valve adjustment.

Yours very truly,

B. H. AUSTIN.

Austin successfully campaigned his car in Britain in 1924 and, as these letters to Major Lefrère of Chas Jarrott and Letts show, the firm was pleased to publicize his achievements.

The Invincible Talbots

At Le Mans the challenge was a very real one and came in the shape of a trio of new Talbot–Darracqs,[12] which were making their competition debut. The design was extraordinarily modern, the cars being powered by a twin overhead camshaft, dry-sump lubricated, four-cylinder engine, with the four-speed gearbox being mounted in unit with the aluminium, steel-linered block. Four wheel brakes were employed. The power unit was effectively half that of the 3-litre straight-eight which had appeared at the 500 Miles Indianapolis race earlier in the year under the Sunbeam name. This was because the Darracq and Talbot companies, which had become allied in 1919, had merged with Sunbeam to create the unwieldy Anglo–French conglomerate, STD Motors, in 1920. The lovely little *voiturette*'s 1,486cc four valves per cylinder unit produced over 50bhp at 4,000rpm. By contrast, Bugatti could only muster a little more than 30bhp from his single cam four, and when the two makes did meet on the track, the Type 13, along with its contemporaries, always played second fiddle to the cars from the Surèsnes district of Paris.

Ettore's forebodings at Le Mans were vindicated when the Talbot–Darracqs swept all before them in the 1921 *voiturette* race and achieved the hat trick that had cruelly eluded him in the previous year. René Thomas, who drove the victorious car at 115kph (72mph), albeit on an improved road surface, was a good 24kph (15mph) faster than Friderich's winning average. So complete was the Talbot's victory that the cars remained unbeaten in every event in which they competed for the next two racing seasons, so earning the memorable 'Invincible' appellation.

A month after Le Mans, on 16 October, Bugatti consoled himself with a win in a second-stream Spanish event: the *voiturette* race on the Villafranca circuit near Barcelona. Molsheim entered a team of three cars though the only two recorded drivers both had appropriately Iberian connections in the shape of Pierre de Vizcaya and Jacques Monès–Maury. The latter was a Spanish aristocrat who was making his debut as a Bugatti works driver. De Vizcaya won, with his team-mate second, ahead of a French Le Père.

The inevitable contest between the Bugattis and Talbot–Darracqs came not in France, but only six days later, on 22 October, in Britain. The occasion was the Junior Car Club's 200-Mile Race at Brooklands. The cars were entered by Major E G A Lefrère of Jarrott and Letts, which was Bugatti's London distributor. Although the Type 13s were inferior in relation to the Talbot–Darracqs as far as the power of their engines were concerned, they were also at a secondary disadvantage because of their crude, bolster-tanked, pre-war inspired bodywork. 'According to ear-witnesses, the Bugattis sounded as if they had as much power [as the Talbots] but lost 3 or 4mph [5 or 6kph] through their lack of streamlining.'[13] These handicaps conspired to relegate de Vizcaya and Monès-Maury to fourth and sixth positions, behind the Talbot–Darracqs which took the first three places. As a result of this drubbing, Ettore ensured that his factory cars never again met the Talbots in competition, apart from on one subsequent occasion. That was at the following year's Tourist Trophy race, and inevitably, the result was much the same.

The British Bugatti

Jarrott and Letts was to play a role in an unlikely licensing agreement that Bugatti made in November 1921 with the Manchester-based Crossley Motors. William Letts (knighted in 1922), the firm's managing director, had established with racing driver Charles Jarrott the London-based motor sales business that bore their names in 1903. Wanting a prestigious British car to sell, the enterprising duo had approached Sir William Crossley to make one for them. This respected company had, from 1866, produced the Otto gas engine in

---

[12] The name of a Milanese engineer, Edmund Moglia, has been linked with these cars. (See W F Bradley, *Autocar*, 4 May 1967.) He subsequently joined Ballot and was later to be responsible for supercharging Bugatti's Type 35 engine.

[13] Karslake, *Racing Voiturettes*.

Britain, obtaining the licence from the same Deutz-based firm that was to employ Bugatti in the 1907–1909 era. Car production began at Manchester in 1904 and six years later, in 1910, Crossley Motors was established, incorporating Jarrott and Letts. At this point, however, Charles Jarrott withdrew from the firm.

These corporate changes made little outward difference to its prestigious London showrooms just off Regent Street at 45 Great Marlborough Street, where Crossley cars were, of course, the mainstream marque. But these sturdy products of industrial Manchester were soon being sold alongside a newly introduced small car from Germany, Jarrott and Letts having obtained the Bugatti concession for Britain and Ireland just prior to the outbreak of World War I.

Alas by 1921 Crossley, like the rest of British industry, was suffering from a terrible year of depression. The company's difficulties were exacerbated by it being solely reliant on the sales of large capacity models; a 25/30hp which dated back, in essence, to 1908 and a newly introduced 3.8-litre 19.6hp car. Motor manufacturers were consequently turning to smaller, more economical cars to sell and Austin's famous Seven, for instance, was created at this time. The 1,496cc Bugatti was a light car and the firm's connections with Molsheim, via Jarrott and Letts, were well established. In this gloomy economic climate, it no doubt made sense for Crossley to contemplate manufacturing a proven design. This could then be sold at a cheaper price, since it would not be penalized by the 33⅓ per cent McKenna Duty paid by the genuine, imported product. William Letts revealed Crossley's intentions 'of manufacturing a British edition of the Bugatti' at the 1921 Motor Show, with the Type 13s Brescia success still only two months old.

As the economy began to pick up in 1922, in the words of *The Automotor Journal*'s Edgar Duffield, 'the poor little Bugatti was left standing on one leg.'[14] So instead of manufacturing the complete car at its Gorton works, Crossley imported unmachined components from Molsheim and 'set aside a bay or two . . . for machining and assembly'. Consequently it was not until two years after the announcement, in the autumn of 1923, that the Crossley Bugatti made its appearance. Duffield, a Bugatti owner himself, was clearly impressed by the model, the Type 22, and although it 'seems to lack some of the sting of the Molsheim product . . . I really regard it as a better car . . . capable of much smoother slow running.' Ironically there was no difference in the chassis price of the Crossley Bugatti and the genuine article. Both sold for £350 in 1924, compared with the latter's £650 in 1921, a result of Bugatti reducing his prices and the falling value of the franc.

Hugh Conway has pointed out[15] that the reason for the model's quiet running was that Crossley increased the width of the gears in the gearbox. Also, the outer end of the rear axle half shaft was enlarged, a modification that prevented the loss of a wheel, which had occasionally occurred on the French cars. Externally, the Crossley Bugatti could be identified by its tie bar, located between the spring eyes, rather than the shock absorbers on the Molsheim cars. The metric threads were replaced by BSF ones and the cars were allocated their own series of numbers, from 1600 to 1625. These twenty-five examples were prefixed CM for Crossley Motors. Clearly the idea was not a success for either party. The price differential had evaporated, with British customers no doubt preferring to buy a Bugatti built at Molsheim rather than Manchester.

An Italian Licence

The Crossley liaison had been the third of Bugatti's licensing agreements for his 16-valve cars, and although none of these was particularly successful, they probably provided Ettore with funds at a time when he needed money most. The first arrangement was with the Italian Diatto company, and the car first appeared at the 1919 Paris Motor Show. This was the same firm that had obtained a licence

[14] *The Autocar*, 20 August 1921.

[15] Conway, *Bugatti, 'Le pur-sang des automobiles'*.

to manufacture his straight-eight aero engine during the war and Bugatti himself also mentions that it bought a licence for his U16 aero engine, though it appears that none were made. Ettore also says that he sold 'three types of cars to Diatto' which suggests the Types 13, 22 and 23.

Diatto was originally a railway engineering company which began building cars in Turin in 1905. Like most Italian firms of the day, these were French cars made under licence, in this instance the Clément. This continued until 1916 when the firm switched to war work and the firms of John Newton of Turin, Scacchi of Chivasso and a plant owned by the French Gnome Rhône aero engine company. The latter works was intended for air frame construction and was where the Bugatti aero engine had been tested.

With the coming of peace, the firm was reorganized as Automobili Diatto. The Bugatti was designated the Model 30 though the Paris Show car, which was fitted with a commodious saloon body, appeared before Molsheim had begun production of its own 16-valvers. It retained a Diatto radiator shell though a Bugatti one was later fitted. The French company's lasting influence on the marque was that Diatto took the opportunity of dispensing with the white circular badge that it had hitherto employed. This was replaced by a rather Bugatti-type red ellipse, which was fitted to all Diatto cars from thereon, regardless of their origins. Otherwise, it was the aero engine all over again and only a few cars were made. Molsheim dispatched a 16-valve chassis (number 930) to Turin in May 1920 and a further two (902 and 903), nearly a year later, in March 1921.

Of the handful of cars built, one put up a good account of itself when it won Brescia's second *voiturette* race of 12 August, before Bugattis had swept the board in the competition at the beginning of the city's 1921 speed week. Run over the same course, the entry included future Bugatti driver Tazio Nuvolari, who was to become one of Europe's most celebrated racing drivers, in an Ansaldo tourer. He soon dropped out and Franz Conelli, in the Diatto-built Bugatti, won at an average speed of 109.14kph (67.82mph), which was only 6kph (4mph) slower than Friderich's winning speed in a works Type 13. Conelli also put up the fastest lap of the day at 116.41kph (72.34mph).

In the following year of 1922, Diatto introduced its own overhead camshaft car, the 2-litre Tipo 20, which originally developed 40hp, though the subsequent ministrations of the firm's development engineer, Alfieri Maserati, saw this boosted to 70bhp. Diatto ceased car production in 1927 though Maserati and his brothers went on to greater things . . .

The Rabag Cars

The second of Bugatti's trio of licensees appear to have been the most successful, certainly in regard to the number of cars built.

Rheinische Automobilbau AG of Düsseldorf, better known as 'Rabag', announced in August 1921 that it had acquired 'the manufacturing and sales rights of the Bugatti cars for nearly all countries'.[16]

This firm had been founded by the Funke brothers in December 1920 though later became part of the group headed by Hugo Stinnes, which also included the AGA, Dinos and Stolle marques. Rabag did not manufacture the Bugatti itself but probably assembled it from components produced by sub-contractors. The chassis and engine were the responsibility of Union Kleinauto-Werke of Mannheim, its title indicating that in 1921 it had built the small Bravo car.

Some changes were made to the Bugatti engine, which was produced in 1,453 and 1,496cc forms. In one or both versions, the four valves per cylinder were reduced to two and the handsome cam cover replaced by a rather anonymous semi-circular one. Conversely, the inlet manifold was an improvement on the original Molsheim product. Bodies were the work of Bendikt Rock of Nordstrasse, Düsseldorf and the Rabag made its debut at the 1923 Berlin Show.

The car was available in Type 22 and 23 forms, with a rather Bugatti-like radiator, though with a

---

[16] *The Autocar*, 20 August 1921.

Sisters under the skin: the Rabag Bugatti was built in Germany between 1923 and about 1925.

pronounced top lip, and appears to have been built in open and closed forms. There were also examples fitted with sporting bodies, which were to be seen participating in German speed events and races during the 1920s. However, the firm's affairs were thrown into confusion in 1924 by the untimely death of Hugo Stinnes and, by the following year, Rabag had merged with AGA of Berlin. Unfortunately bankruptcy proceedings for this company began in November 1925 and that was effectively the end of the Rabag.

A Tourist Trophy Team

It is not known how many of these German-made Bugattis were built, probably about forty, which was rather more than the twenty-five Crossley-Bugattis which appeared in 1923 and 1924. In 1922 Molsheim had fielded a team of 16-valve cars under this name for that year's Tourist Trophy race, the first since 1914, to be held on 22 June and staged, as ever, in the Isle of Man. Despite the Crossley liaison, the three cars (chassis numbers 1397/8/9) were Molsheim-built and the tactic may have been adopted as a face-saving expedient in the event of the expected Talbot–Darracq victory. It was to be the second and final occasion that the Bugattis were to meet the unbeatable cars from Suresnes on the race track.

The extensively modified Rabag engine was fitted with eight rather than sixteen valves.

The Marquis de Casa Maury, who raced works Bugattis as Jacques Monès-Maury. He is pictured at the International 1500 Trophy race, run concurrently with the Tourist Trophy in the Isle of Man on 22 June 1922. The yellow-painted cars were entered as Crossley Bugattis and his was the first Molsheim car home in third place behind two Talbot-Darracqs.

The TT was split into two classes that year. The 3-litre Grand Prix formula had finished at the end of 1921 and was replaced by one for 2-litre cars, yet the RAC, in an astonishing bout of insularity, perpetuated the 3-litre limit into 1922 for this event. Not surprisingly, the entry list was only British and somewhat limited, so a '1500' Trophy, for cars of up to 1.5 litres, was instigated to attract the *voiturettes* and this elicited a response from the front-line European makes.

The two races were to be run concurrently, with the larger cars completing eight laps of the 60.75km (37.75-mile) circuit, while the smaller ones ran for only six, making a 428km (266-mile) contest.

Three Bugattis were entered. They were resplendent in lemon-yellow bodywork, rather than their customary French blue, perhaps to underline their 'Crossley' associations. Less apparent were internal changes to the engine, which was now fitted with slightly shorter connecting rods of the sort that had been developed for Molsheim's stillborn Type 28 straight-eight, which we will encounter in the next chapter, Enter the Eight. This meant that the piston height had to be increased from 61 to about 71mm, giving them a rather curious top heavy appearance. The Bugattis were driven by Pierre de Vizcaya and Jacques Monès-Maury while British laurels were upheld by the presence of 'Bunny' Marshall, driving a works Bugatti for the first time.

In addition to the STD Motors-entered Talbots, Aston Martin also fielded a trio of cars. On race day it rained and few spectators bothered to watch the

Another Brescia at the TT, this time driven by Bertram (Bunny) Marshall, who was placed sixth. Marshall was a Bugatti agent for the London area and based in Hanover Square. The mechanic is Harry Mundon. Note that these cars were fitted with stone guards.

start. 'I think 1922 was the wettest [TT] of them all', remembered W O Bentley, who was driving one of his own 3-litre cars, and recalled the presence of 'the leaping, fleet little Bugattis.'[17] Over three years later, Bentley's path and that of one of the Molsheim drivers would cross. Late in 1925, W O would have to share the managing directorship of Bentley Motors with Monès-Maury – who also happened to be the Marquis of Casa Maury – after the latter's friend, Woolf Barnato, had taken the firm over.

At the Isle of Man, Talbots took the first two places in the 1500 Trophy. De Vizcaya in particular drove an excellent race but he hit a wall on the last lap, resulting in a puncture, and he dropped back to finish fourth behind Monès-Maury. Marshall was sixth in the third of the Molsheim cars, so Bugatti had the satisfaction of winning the team prize, the third Talbot having crashed on the first lap. But it was to be six years before the next TT which, in 1928, moved to Ards, in Northern Ireland.

[17] W O Bentley, *W. O.* (Hutchinson, 1958).

Segrave's Bugatti

Following the 1922 race, the three TT Brescias were returned to the mainland but were subsequently sold off by the factory. Such former works Bugattis were rare in Britain at the time. The best publicized example had been the car driven by Henry Segrave and purported to be the Type 13 in which Friderich had won the 1920 *Coupe des Voiturettes* race. It is now known not to be the car, but the one which was probably the 16-valver that Baccoli had driven into fifth place in the same event.

After the TT, Monès-Maury's car was driven by Marshall in the 200-Mile Race at Brooklands on 19 August, with Leon Cushman at the wheel of another Brescia. He came in fifth and Marshall was sixth behind the winning Talbot–Darracq and another in third place. Second was an Aston Martin and an Enfield-Allday came fourth. Cushman did better in the following year's 200-Mile Race and came second in an alcohol-fuelled car behind an Alvis at the extraordinary speed of 146.60kph (91.1mph).

The third of the 1922 Isle of Man cars, chassis number 1317, registered XK 9542. By this time the front tie rod is curved which must have made it difficult to start on the handle! Leon Cushman is at the wheel. He was an employee of Jarrott and Letts, which held the Bugatti concession for the British market between 1913 and 1925. He is pictured at Brooklands in 1922. The car was subsequently . . .

. . . fitted with a streamlined radiator cowl and enlarged fuel tank. It is pictured here at the Southsea Speed Trials.

It was early in 1923 that Bugatti discontinued the production of the plain bearing 16-valve engine. Cars that had been hitherto so equipped were now called by the Brescia Modifiée name and were accordingly cheaper than the racing and so-called 'Full' Brescia, with twin magnetos on the Type 13 chassis, which was continued. All cars were thereafter fitted with the same two-piece, two bearing crankshaft though with circular rather than flat webs. The 1496cc capacity and 69 × 100mm dimensions were thus perpetuated. The Modifiée was outwardly similar to the earlier version and the first example (chassis number 1612), was dispatched in February 1923.

The Model Evolves

Molsheim also took the opportunity to make changes to the gearbox of the 16-valve cars. This had a stronger casting, with wider gears of the sort used on the Bugatti's newly introduced Type 30 straight-eight. A further underbonnet change came when the original worm and nut steering-box was replaced by a worm and wheel unit. It had previously been attached to the chassis, behind the offside engine bearer, and was moved forward and incorporated in the bearer itself. This saved space inside the car and improved passenger accommodation.

Perhaps the most obvious external modification to the 16-valve range came in 1923 when the radiator was changed and became closer in appearance to its Type 30 contemporary. It will be recalled that the original egg-shaped radiator, which dated back to the eight-valve cars of 1912, had effectively been perpetuated after the war. However, the post-1919 one was slightly narrower; or to continue the culinary simile, rather more 'pear' than 'egg'. This was the shape which endured for four years, the new radiator following much the same external contours of its predecessor, though the honeycomb now went to the edges of the shell, noticeably at the top, where an area had been retained on its predecessors for mounting the badge. On the new

Bertram Marshall incorporated parts of the TT car in a longer Type 23 frame and rectified the absence of front wheel brakes by fitting these Perrot units.

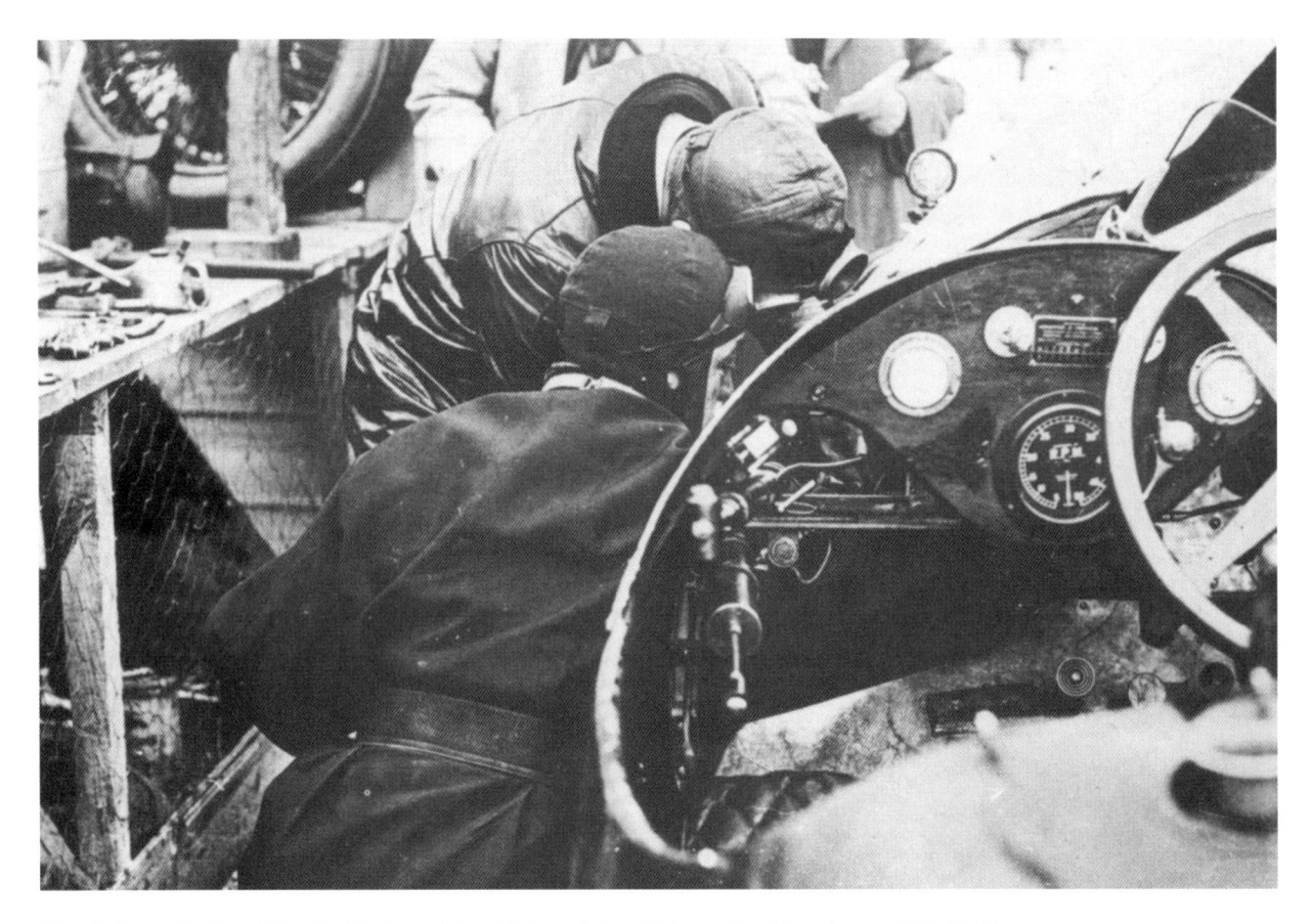

Marshall ran the Type 23 at the Boulogne Grand Prix, held on 30 August 1924, and won at 87.66kph (54.47mph). Here the engine is receiving attention in the pits.

radiator, it was attached directly to the block, was flatter along its base and easier to make because there was no cut out for the starting handle which was relegated to beneath the unit. New front and rear axles, courtesy of the Type 30, were fitted at the same time.

The arrival of the new radiator saw the model uprated by the fitment of a cast aluminium bulkhead, which was also a Type 30 inheritance. Similarly it included a housing for the dynamo, which was now driven by a belt from a pulley attached to an extension at the rear of the camshaft.

Then in 1925, which was the penultimate year of 16-valve production, the cars belatedly received front wheel brakes. As *The Autocar* diplomatically put it, 'it has been known for a long time that the

In 1925 Marshall reverted to a Bugatti axle, though now with front brakes, and once again won at Boulogne, this time at 103.49kph (64.31mph).

Early in 1923 the Type 22 and 23s' oval-shaped radiators were replaced by one similar in design to that used on the Type 30. The two were outwardly identical though that of the 16-valve cars had less depth.

This Type 23 fiacre style saloon dates from the early 1920s and is fitted with this unusual rear door.

A pre-1923 Type 23 four-seater tourer, pictured at a motor show. Is that Ettore's bowler hat hanging up behind?

car is extremely fast for its size, but that the available power could not be exploited fully because the brakes were scarcely sufficiently powerful for the capabilities of the engine.'[18] Prior to this date the braking system had been basic to say the least, the footbrake operating a Ferodo-lined transmission brake, just behind the gearbox, while the handbrake activated cast iron shoes which acted directly on the rear drums.

The new brakes were cable operated and a simplified version of the system Bugatti had already employed on his racing cars. The pedal actuated the front and rear drums via pairs of miniature chain wheels through which was threaded a short length of bicycle chain let into the cable. On the front drums, the actuating cam was not traditionally mounted on the top of the back plate but ahead of the front axle. When the front wheel brakes appeared early in 1925, the transmission brake was eliminated.

The Modifiée sold in Britain for a chassis price of £330 in 1925, while the Brescia itself sold for £55 more at £385. With a top speed approaching 112kph (70mph). This put it on a par with a car of the likes of the 12/50 Alvis and the Bugatti represented excellent value when compared, for example, with a 1.5-litre Aston Martin with a chassis price of £695, which was over £300 more.

The 16-valve models finally ceased production in June 1926, the final chassis number being 2906.

[18] *The Autocar*, 27 February 1925.

This late Brescia with front wheel brakes dates from 1925. They belatedly arrived that year and were initially optional at £25 extra but were soon standardized.

How the Brescia's front brakes worked, with the operating cam ahead of the front axle rather than at the top of the drum, as was the norm.

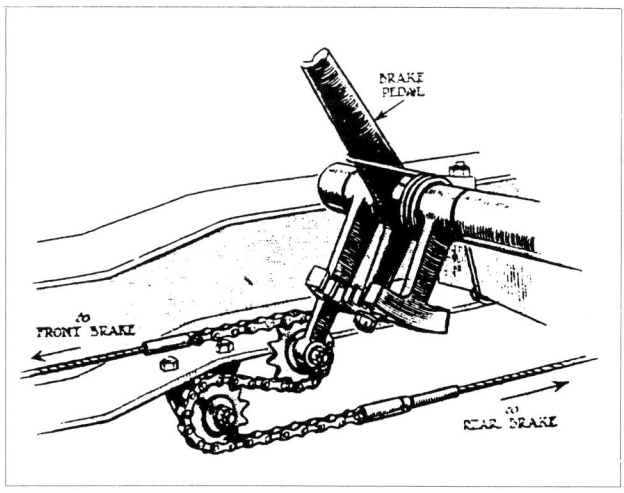

The brake compensating mechanism. There was one each side of the car, incorporating a short length of bicycle chain.

Bugattis would dominate the Targa Florio between 1925 and 1929 though this private entry dates from the 1923 event. Renzo Lenti's Brescia had the distinction of being the first Bugatti to finish in the race, Tornaco entered a car in 1922 but did not start. Lenti's Bugatti has been fitted with Perrot front-wheel brakes as the model did not offer them until 1925. He did well to finish in seventh place.

A 16-valve saloon of 1925 with front wheel brakes. Somehow an English coachbuilder has managed to squeeze a four-door saloon body on to a Type 23 frame.

The Bugatti factory hack, a Brescia fitted with a pick-up body and pictured in the late 1920s carrying a pair of Type 52s on its rear platform. Non-standard front-wheel brakes are fitted and the Bugatti *sign at the front of the radiator would have made little difference to the car's performance. The model usually ran over-cooled!*

These little cars were the most numerically successful of Bugatti's products. About 2,000 examples were built, which was a remarkable achievement in view of the fact that the design dated back, in essence, to 1910.

Inevitably, by this time the small Bugatti was no longer a front-line racer as the all-conquering Talbot–Darracqs had continued to dominate the *voiturette* field, with the original 1921 cars having been replaced by an equally successful design in 1923. Yet by 1926, the cars from Molsheim had achieved supremacy in the field of front line Grand Prix competition, which was reflected by Bugatti having become that year's World Champion through the success of his superlative Type 35 racing car.

However, before contemplating what was, without question, Ettore's masterpiece, it is necessary to see how he became wedded to the concept of the eight-cylinder engine, and his pioneering role in adopting the configuration is considered in the next chapter.

6
Enter the Eight

'The six-cylinder must be abandoned for racing purposes on account of the inherent difficulties in balancing it . . . Rolls-Royce . . . could get better results . . . by the use of eight cylinders in line.'

Ettore Bugatti[1]

In the years immediately before the outbreak of World War I, practically all Grand Prix cars were powered by four-cylinder engines. There were occasional sixes but that configuration was far more popular on road cars. The eight was looked upon as an eccentricity and it was not until 1919 that Ballot in Europe and the American Duesenberg company introduced their epochal straight-eight racing cars. The latter were to render previous configurations obsolete so that, by 1924, practically all front line racing cars were so powered, a state of affairs that would endure for the next thirty or so years.

The concept was also extended to road cars with Isotta Fraschini in Italy, Duesenberg and Bugatti all pioneering the concept and becoming wedded to it – a commitment which, in the longer term, would play a role in the demise of all three firms.

It will be recalled that in about 1912 Ettore had experimented with a car powered by a crude straight-eight engine. This consisted of two fours in tandem as an expression of his modular approach to design. This had been followed by an eight-cylinder wartime aero engine and his subsequent U16 unit consisted of two in-line eights mounted in parallel. So Bugatti was in the forefront of the straight-eight revolution and its post-war evolution is therefore of great relevance to his story.

In pre-1914 days there had been such unorthodox mechanical expressions as the 1903 CGV and the Weigel of 1907. The first serious contenders, however, appeared at the 1919 Indianapolis 500 Miles Race, held on 30 May, and subtitled 'The Victory Stakes'. There, straight-eight cars from the old and new worlds met for the first time in the shape of the French Ballot and the American Duesenberg.

The Paris-based Ballot company had been established in 1906. Its founder Ernest Ballot had been an engineer's first mate in the French merchant navy and therefore adopted an anchor flanked by his initials as his trademark. He began, appropriately, by producing marine engines and then proprietary units for the motor industry, supplying such firms as Delage, La Licorne and Mass. The firm was recapitalized, to the tune of 1.8 million francs, in 1911 and renamed *Établissements Ballot*. This saw a group of financial high flyers buy into Ballot, headed by motoring and aeronautical pioneer Adolphe Clément, his son-in-law Fernand Charron, Pierre Forgeot, a lawyer and future Minister of Public works, with Marc Birkigt and Jean Lacoste from Hispano–Suiza.

Ballot prospered during World War I, and produced thousands of proprietary engines and associated compressors, generating sets and Hispano–Suiza aero engines. When the war came to an end, the firm decided to enter the automobile

[1] *The Autocar*, 7 May 1921.

market with a car in the Hispano–Suiza mould. It would not be until 1921 that its first roadgoing model, the 2LS, made its appearance.

The firm no doubt decided that the perfect publicity for a new make would be success on the race track but the only major event scheduled for 1919 was across the Atlantic in America in the shape of the 500 Miles Indianapolis Race. Therefore on Christmas Eve 1918, the firm signed a contract with Ernest Henry, who had been closely concerned with the creation of the pioneering twin cam racing Peugeot of 1912, to build the first Grand Prix Ballot. As time was of the essence and the 4,917cc (300cu in) formula was current in America, Henry effectively took a pair of pre-war 2.5-litre four-cylinder Peugeot engines of the type that had been created for the aborted *Coupe de l'Auto* race of 1914, and mounted them on a common crankcase.

The resulting power unit was Europe's first significant straight-eight racing engine. Of 4,894cc capacity, it had four valves per cylinder, while the four-piece crankshaft ran in five roller main bearings though Henry adopted a complex plain bearing construction for the big ends. There was a separate four-speed gearbox. The chassis was essentially Grand Prix Peugeot and although the car's innovative engine was noteworthy, the body with rear bolster tank, like the frame, clearly harked back to pre-war days. The four cars, which left France for America on 26 April (a staggering 120 days after they had been conceived) are reputed to have cost Ballot no less than £30,000.

The Duesenberg Eight

Like Ballot, the German-born American-domiciled Duesenberg brothers also began by building proprietary engines, in their case in 1912. Their power units were distinguished by horizontally mounted valves, actuated by long, side-mounted rockers, which were given the curious 'walking-beam' name, which is steam engine nomenclature. Their use in racing cars reflected the brothers' interests in that direction, and although their power units were used in other manufacturers' chassis, prior to World War I a few Duesenberg racing cars were also built. From 1913, the brothers were based in a factory at St Paul, Minnesota, and that year Commander James A Pugh commissioned them to build two twelve-cylinder motor boat engines for his hydroplane 'Disturber IV' in which he intended to participate in the 1914 Harmsworth Trophy race. The event was cancelled because of the outbreak of the war.

As a result, in 1915, the Duesenbergs announced six and, above all, straight-eight marine engines of the same walking-beam design and soon after the end of World War I, late in 1918, they decided to perpetuate the eight-cylinder theme by building a radically different, single overhead camshaft racing engine. Much has been made in the past of the brothers' exposure to the Bugatti U16 aero engine, when the company that bore their name played host to it during the war. However, Duesenberg historian, Fred Roe, is of the opinion that its presence was an almost incidental factor. For chief engineer Fred Duesenberg's 'strong concern with simplicity led him to favour the single overhead camshaft design instead of the double with many more parts.'[2] He would have encountered this feature in a number of outstanding European designs, namely on Daimler's formidable Mercedes racing car and aero units, and Fred adopted their distinctive welded-on water jacketing for his own monster V16 aero engine, built in 1919.

The brothers would also have been exposed to the excellent Hispano–Suiza V8 aero engine, with a single overhead camshaft per bank, being built under licence at nearby Trenton, New Jersey and 'there was the freest possible exchange of technical information between the two plants, which were just 50 miles [80km] apart.'[3] In such company, the presence of Bugatti's eccentric and fragile single overhead camshaft U16 aero engine would have only represented an endorsement of an established design feature.

[2] Fred Roe, *Duesenberg, the Pursuit of Perfection* (Dalton Watson, 1982).

[3] Borgeson, *The Golden Age of the American Racing Car.*

The straight-eight that the Duesenbergs produced for the 1919 500 Miles Race was of 4,850cc (296cu in), with dimensions of 76 x 133mm (3.0 x 5.25in), while the shaft-driven single overhead camshaft operated three valves per cylinder. These, it should be said, were the reverse of Bugatti practice, for there were one inlet and two exhausts – their number and allocation may have derived from the aforementioned V16 walking-beam aero engine the Duesenbergs had designed in 1918. Unlike European practice, the racing eight had a detachable cylinder head, with a three bearing crankshaft, the front and centre being plain, along with the big ends, with a ball race at the rear. Only 92bhp from 3,800rpm was claimed for this pioneering unit, compared with the Ballot's 140bhp at 3,000rpm. A separate three-speed gearbox was employed.

The Duesenbergs were well aware of the innovative nature of their work though only one car was ready in time for the 1919 race at Indianapolis. In 1925, Augie Duesenberg reflected that 'they had expected their new engines to be a complete surprise there',[4] while the presence of the Ballot team also caught them unawares.

The magnificent, low-built French racers did not, alas, come up to expectations. Two crashed, one came fourth and another tenth, which was particularly disappointing after one of the Ballots had put up a record qualifying lap of 168.49kph (104.70mph). The lone Duesenberg was equally unsuccessful, lasting for fifty laps before the car was withdrawn with bearing failure, highlighting the limitations of its splash lubrication system, derived from the earlier walking-beam fours. Ironically, the race itself was won by an old 1914 four-cylinder twin overhead camshaft Grand Prix Peugeot...

Despite the new eights' relatively poor showing, Duesenberg for one, and the other front-line racing car contructors on both sides of the Atlantic, recognized the viability of the concept. This meant designing completely new engines for 1920 because the 1919 Indianapolis race was the last to be held under the 300cu in formula. The obvious cost disadvantage, not always a factor in such technology, of the doubling up of components was more than offset by the overriding advantages of 'perfect balance, better gas distribution, a valuable increase in piston area, excellent torque, a higher permissible rate of revolutions owing to smaller pistons and lighter reciprocating parts, plus the opportunity to dispense with the heavy flywheel'.[5] These attributes were to exercise a great influence on racing car designers with the soon to be supercharged straight-eight engine holding sway until the demise of the blower in the 1950s.

Straight-Eight Road Cars

The straight-eight passenger car was not far behind its racing first cousin. It was in late 1920 that Isotta Fraschini, which had discontinued all its other models, began deliveries of its Tipo 8, the world's first series production straight-eight. The Milan company had developed the concept with its 19-litre V5 eight-cylinder aero engine of 1915, though its nine bearing 5.8-litre pushrod engine 'was more noted for long life and reliability than for power and refinement, in the marine rather than the aviation idiom'. It also boasted four wheel brakes, a feature of the make since 1910, and its formidable 3,700mm (12ft 2in) wheelbase made it ideal for the carriage trade for which it had essentially been created. The Tipo 8 and its derivatives, of which there were about 1,650, lasted until 1935. In that year, Isotta Fraschini ceased car production; a celebrated victim of the world depression and its own straight-eight, one model policy.

The Milan company was not alone; most of those firms, and there were eventually scores of them, which opted for the configuration ultimately suffered the same fate as Isotta Fraschini. As Lord Montagu has pointed out: 'There is to my mind a ... relation between the eight-in-line and the bankruptcy courts' and a great many firms that

---

[4] Roe, *op. cit.*

[5] 'Cyril Posthumus' in *The Racing Car Development and Design* by Clutton, Posthumus and Jenkinson (Batsford, 1956).

chose to tread the eight-cylinder path 'breathed their last to the silken purr of eight cylinders . . .'[6]

In Britain, Leyland introduced its exclusive 7.5-litre, single overhead camshaft Eight at the 1920 Motor Show though only eighteen examples had been built by the time that production ceased in 1923. Prospective customers were no doubt put off by an all-in price of £3,050 and the firm reverted to more profitable commercial vehicle production.

In America it was once again the Duesenberg brothers who were in the vanguard of developing the straight-eight road car. Its Model A was first displayed at the New York Show of November 1920, though was not powered by the single overhead camshaft engine of the racing cars but by a new variety of a fixed head 4,260cc (260cu in) walking-beam unit. But by the time deliveries began from a new Indianapolis factory in late 1921, this engine had been sidelined and replaced by an overhead camshaft unit related to the racing eight, though with two, rather than three, valves per cylinder. Yet another feature of the 1920 Show car was its innovative fitment of four-wheel hydraulic brakes, which will also have their part to play in the Bugatti story.

By 1924 Duesenberg was in financial difficulties, and the fiscal stalemate was only broken by the firm being bought by Errett Lobban Cord in 1926. He brought flair and marketing expertise to the firm, though Duesenberg failed, along with the rest of the Cord Corporation, in 1937.

As will have already been apparent, Bugatti had been committed to the straight-eight engine since at least 1912 when he produced his twin four-cylinder engined single-seater. His wartime aero engine of the same configuration was a perpetuation of the concept and, as will emerge, although his eight-cylinder Type 28 of 1921 was not built, it did point the way to the Type 30 of the following year, which was the first production eight-cylinder Bugatti. Four-cylinder cars were also built until 1933 and, as Ettore could not emotionally contemplate designing a six, from thereon only eights were marketed. The reality is that had Bugatti been solely dependent on car sales, he would have followed his celebrated contemporaries into oblivion during the decade. As it happened, Molsheim was busy producing railcars for the French railways, their engines being derived from his greatest eight-cylinder folly: the Royale.

Ettore had contemplated the production of a straight-eight luxury car in pre-war days, a concept which ultimately manifested itself in the shape of the 12.8-litre leviathan.

Ettore's First Luxury Car

On his return to Molsheim at the beginning of 1919, Bugatti immediately instructed his drawing office to begin work on what he expressed as 'the problem of eight-cylinder in line engines'.[7] The resulting unit was to power a high quality touring car, designated the Type 28, but all the difficulties associated with post-war production meant that it did not appear until nearly three years later. It was unveiled in a partially completed though acceptable state at the Paris and London Motor Shows of October and November 1921. The fact that it was of 3 litres capacity, so coinciding with the racing formula that came into force at the beginning of 1920 and only lasted until the end of 1921, suggests that the engine could have also formed the basis of a Grand Prix car, had it appeared earlier.

The power unit can be considered as a synthesis of Bugatti's wartime aero engine experience. It was a straight-eight consisting of two four-cylinder blocks, their 69 x 100mm dimensions echoing that of the current 1,496cc 16-valvers. These two fours were mounted on a common aluminium crankcase, with the camshaft drive positioned between them in the manner of the U16. But as this was intended as an expensive, maybe closed model, Ettore went to some lengths to keep this otherwise noisy drive quiet. Consequently the upper camshaft bevels were duplicated, differing by one tooth, which accordingly ran at slightly differing speeds, and were held together by a friction clutch. The controlled

---

[6] Lord Montagu of Beaulieu, *Lost Causes of Motoring* (Cassell, 1960).

[7] L'Ébé Bugatti, *The Bugatti Story*.

Ettore's first eight-cylinder car, the 3-litre Type 28, when the distinctive angular lines of a Bugatti engine first appeared. Despite being displayed at the Paris and London Motor Shows of 1921, it never entered production.

slip that resulted reduced any chatter produced by the meshing bevels. Ettore was to use a similar device when he came to design the Royale engine.

On the right of the power unit, the drive shaft drove the magneto via a cross shaft, and there were two plugs per cylinder, all on this same side. This was balanced by a centrally located water pump on the left. A nine bearing crankshaft was employed and, according to at least one contemporary authority, 'lubrication is under pressure throughout.'[8] If this is correct, then the Type 28 would be the first Bugatti engine to be so equipped.

Unlike the contemporary four-cylinder 16-valve cars, the straight-eight employed three valves per cylinder of the type that Ettore had pioneered on

[8] *The Autocar*, 15 October 1921.

his 5-litre car of 1912 and had been perpetuated on the U16. But instead of the banana-shaped tappets of his fours, on this engine Ettore introduced fingers between the camshaft lobes and the tops of the valves, a feature which he would thereafter adopt on all his single cam engines and would be perpetuated on the twin cam Type 57 of 1934. This 2,991cc unit developed 90bhp at 3,400rpm.

The most significant visual aspect of this, Bugatti's first eight-cylinder car engine, was that the curvilinear elements that had been apparent on all his fours prior to this date were replaced by clean, linear outlines. These were accentuated, it should be said, by the Type 28's partially completed state, and was to be such a distinctive feature of Bugatti's engine from thereon. The eight's proportions were highlighted by the apparent absence of its camshaft drive though the reality was that it was

The driving compartment of the Type 28, the gearbox of which was located in the rear axle. Note the twin steering and track rods. This chassis was later fitted with a fiacre-styled body.

enclosed and even the magneto was concealed beneath an angular, boxed cover. The geometrical themes were perpetuated by the presence of the circular covers of two Bugatti-designed carburettors, each supplying its respective cylinder block, with the four horizontally located rods connecting the two, carefully positioned in the same plane as the top of the cambox.

A Cubist Connection?

We do not have to look very far for the inspiration for such distinctive under-bonnet architecture: it represented an automotive interpretation of the themes and proportions so effectively expressed by Ettore's father, Carlo, on his extraordinary and idiosyncratic furniture of nearly twenty years previously. It was an approach maybe triggered by the growing influence of the formalized shapes of the German Bauhaus, a school which, incidentally, placed great emphasis on the industrial designer of the future having a workshop grounding, just like Carlo and Ettore Bugatti's. What today we know as the art deco style also had its cubist elements and Ettore's son, Roland, has said that the Cubists also thought that his father had been influenced by their work. 'They once invited [him] to be a guest at one of their gatherings in Paris. He enjoyed their food, drink and company, and he made no bones about telling them what lousy artists he thought they were.'[9]

The drive from the engine passed through Ettore's familiar multiplate clutch, with the propeller shaft running at engine speed between two wood-reinforced torque arms. For the first time on a Bugatti, the gearbox was mounted in unit with the differential on the rear axle, and the two speeds were actuated by a centrally positioned gear lever.

The chassis, with a 3,150mm (10ft 4in) wheelbase, was conventional (for Bugatti) and suspension was by the customary reversed quarter-elliptic springs at the rear. Double front half-elliptics, in the manner of the 5-litre cars, were fitted. The two-spoked steering-wheel was unusual, being adjustable for reach, and the four instruments were contained in an aircraft-like binnacle visible within the arc. The steering-box was insulated from its column by a leather coupling, which would become a familar Bugatti feature from thereon. Unexpectedly, the drag link consisted of two parallel rods, universally jointed at either end. The track rods

[9] Borgeson, *Bugatti by Borgeson*.

were also twin-tubed – another instance of Ettore 'doubling up' – and attached to the hubs via a leather block in a bid by its inventor to create a maintenance-free joint.

Cable-operated brakes featured at the rear – on its announcement, the Type 28 car was said to be 'bristling with original features'[10] and one of these was that it was to be fitted with hydraulic front brakes, although they did not appear on the show car. The unseen hydraulics were probably similar to those fitted to the subsequent Type 30 model, so a description of them will be reserved for that model.

It seems likely that Duesenberg was the inspiration for their inclusion on the Type 28 because their presence had been a decisive factor in the make's celebrated victory in the 1921 French Grand Prix. The car from Indianapolis was fitted with four-wheel hydraulic brakes and helped Duesenberg to become the only American car to have won the event. It is highly likely that Ettore was present at Le Mans for the race and, even if he had not been, he could have gleaned details of their fitment in the contemporary press.

Although the Type 28 only appeared in chassis form, its price in Britain was £1,400, which was about £300 more than the contemporary 3-litre Bentley one. Surviving factory drawings show it fitted with a Sedanca de Ville body in Ettore's familiar but archaic *fiacre* style, but the car never entered production. This may have been a reflection of the fact that 1921 was an economically bleak year and 1922 could have represented a hostile climate for such an expensive car. Another factor was that the 3-litre racing formula ended at the end of 1921, so there could be no cross-pollination of types. Also, the Type 28 had a plain bearing crankshaft and, although Bugatti's racing cars had initially been similarly equipped, these were discontinued with the introduction of the roller bearing crank. The latter had appeared on the Type 13s campaigned in the celebrated Brescia race of September 1921, held only three months prior to the Type 28's debut.

The Grand Prix Car

Despite the fact that the Type 28 had been cancelled, the time and work expended on the project had not been in vain. Its blocks, pistons and connecting rods formed the basis of the later Type 37 racing and Type 40 road cars, and the entire engine was to form the starting point of the Type 44 touring car, introduced in 1927.

Bugatti had participated in competition almost from the marque's birth and most of his successes had been in *voiturette* events. The closest he had come to producing a front-line racing car had been his celebrated 5-litre model of 1912. But in 1922 he planned to introduce his first Grand Prix car and it was naturally powered by a straight-eight engine which was also employed to power a road car.

This was the year in which the 3-litre formula was replaced by a 2-litre one of greater duration and was to last for a total of four years, until the end of 1925. It would herald what was later termed as a 'Golden Age' for European motor racing. Not only did it 'stimulate the keenest competition among rival members of the recovering motor industry',[11] but the new formula triggered the development of the small capacity engine which was coupled with a high degree of technical development.

The rules were uncomplicated. Cars were to have a minimum capacity of 2,000cc, they should be two-seaters and not weigh more than 650kg (1,433lb). Ettore's new Type 30 was constructed in accordance with these specifications.

Four examples were entered for the 1922 French Grand Prix held near Strasburg, only 6km (4 miles) from the Bugatti factory at Molsheim. The cars represented a new departure for Ettore, not only because they were powered by his first production straight-eight engine but their unconventional bodywork represented another radical change from previous practice. But first the mechanicals.

The 1,991cc engine had a completely new set of cylinder dimensions for a Bugatti, with a bore and stroke of 60 x 88mm. Its clean, architectural lines

---

[10] *The Autocar*, 15 October 1921.

[11] 'Cyril Posthumus' in *The Racing Car Development and Design*.

The team of four Type 30 Bugattis entered for the 1922 Grand Prix held at Strasburg on 16 July. From left to right: Marco (22), Monès-Maury (18), de Vizcaya (12) and Friderich (5).

The car which Friderich drove at Strasburg. The front brakes are hydraulic. The unusual circular bodywork, no doubt adopted for its apparent wind-cheating qualities, was said to be the work of a Strasburg coachbuilder. The exhaust pipe emerged from the tail.

were similar to the earlier 3-litre unit though it differed from it in that the camshaft drive was now located in the front of the engine, as in the case of the eight- and 16-valve cars. However, like the 28, it was similarly enclosed, resulting in a very neat-looking engine. It did, however, differ from the earlier eight in that the oil pump was driven on the right with the water pump on the left. The presence of the former did not, alas, herald the arrival of a pressurized lubrication system. It was, in all probability, Ettore's familiar arrangement in that oil was squirted from jets into grooves in the crankshaft, whereby it found its way, via drilled holes, by centrifugal action to the big ends. The two eight-cylinder magnetos were driven from the rear of the camshaft and located under the scuttle in the manner of the Brescias. There were two sparking plugs per cylinder, located on the right-hand side of the engine. Twin Zenith carburettors were fitted.

Bugatti perpetuated the three-valve layout, with fingers between the cam lobes and valve stems, as introduced on the Type 28. At the other end of the engine, radical changes were made to the circular webbed crankshaft. It was not carried in the nine plain bearings of the Type 28 but by a mere three ball races and resembled a scaled-up version of the Brescia arrangement. There were three double row Skefco self-aligning bearings, the shaft being split centrally to fit the middle one. Plain big-end bearings were employed. Unlike the 16-valve cars, which had a split crankcase, the Type 30's was a single casting with detachable sump. This new engine developed 88bhp at 4,000rpm. The four-speed gearbox, of the type used on the 16-valve cars, was separately mounted behind the engine.

The chassis retained the 2,400mm (7ft 10in) wheelbase of the Type 22 and Bugatti was to perpetuate these dimensions on most of his later racing cars, notably the celebrated Type 35. The front brakes represented an inheritance from the stillborn Type 28 in that they were hydraulically operated. The Duesenberg that had run at the 1921 French Grand Prix was fitted with brakes designed by Malcolm Loughead, though the spelling was subsequently altered to the Four Wheel Hydraulic Brake Company, which manufactured them, to the more phonetically acceptable Lockheed name.

On the Type 30 Bugatti designed his own brakes, though the master cylinder differed from the Duesenberg's in that he anticipated that pressure would be maintained by the close fit of the piston in the plunger and there was a single leather seal at the end of the plunger. This did not prove particularly satisfactory and required plenty of pumping of the pedal by the driver. The fluid was probably a water/glycerine mix and reached, via flexible tubing, the backplate-mounted single cylinders. These in turn actuated simple, strap-like brake linings inside the front finned drums. The latter were, incidentally, noticeably smaller than the rears though their internal construction ensured that they would not operate in reverse!

New Body for the Race

This quartet of Type 30s, chassis numbers 4001, 4002, 4003 and 4004, were driven by the familiar team of Friderich, de Vizcaya, Monès-Maury and Marco. The all-important bodywork was to be of the traditional variety though was changed just prior to the race, it is said, at the suggestion of Pierre de Vizcaya. Pierre convinced Ettore of the wisdom of adopting what were then viewed to be more aerodynamically acceptable lines. As there was not yet a bodyshop at Molsheim, Bugatti got a Strasburg coachbuilder to produce essentially circular bodies resembling those of the three Ballots entered. Cowls completely enclosed the radiators of the Bugattis though their caps projected for water replenishment while the exhaust pipes emerged through the open tails.

Bugatti had become closely involved with Ballot prior to the race when it seems that he helped him out when his cars experienced problems with their twin overhead camshafts. Apparently Ettore 'put some of his men to work late at his factory on making . . . camshafts, fairly large in diameter which were more suitable'.[12]

[12] L'Ébé Bugatti, *The Bugatti Story*.

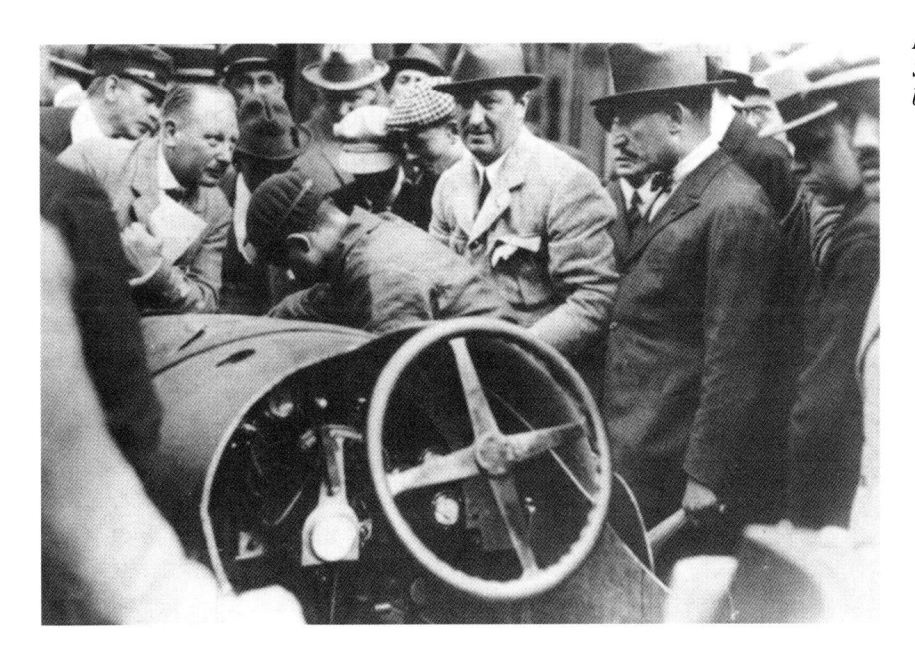

Ettore Bugatti (centre) with one of the Strasburg cars. The twin magnetos can be seen projecting through the dashboard.

For this race Ballot had forsaken its pioneering straight-eights, the previous year having introduced a four-cylinder twin overhead camshaft Henry-designed engine. Aston Martin and Sunbeam also fielded fours and Bugatti's choice of a straight-eight engine was only shared with Rolland-Pilain, which was also running with Duesenberg-inspired hydraulic front brakes. These all employed twin overhead engines with only Bugatti opting for a single cam unit. One of the Sunbeams, which were also the work of Ernest Henry, was to be driven by Henry Segrave, who clearly did not take Bugatti's cars, with their single overhead camshafts, seriously when he noted that they were: 'really only modified touring cars'.[13] But Ettore was vindicated when all but one of his cars lasted for the duration of the race and the trio of faster, more complex Sunbeams broke their inlet valves and all retired, as did the two twin cam Aston Martins.

The field at Strasburg was completed by a team of three formidable Type 804 Fiats. The premier Italian make had effectively taken over the technological baton from the Mercedes team, since German and Austrian cars were barred from the prestigious French Grand Prix, along with other major French and Belgian events – a state of affairs that remained in force from 1921 until 1924. The Fiat's design was of great relevance, not only because it represented the most technically outstanding car at Strasburg, but also because it was to have a considerable influence on contemporary racing design, including those cars created by Ettore Bugatti.

The Formidable Fiats

The Fiat team was headed by graduate engineer Giulio Cesare Cappa who, it will be recalled, had been responsible for the advanced Aquila-Italiana of 1906 and had joined the Turin firm in 1914. After the war, the company assembled a formidable battery of engineering talent with exclusive responsibility for racing car design. These new racing Fiats, which made their debut at Strasburg, were designated the 804–404 and powered by 1,991cc twin overhead camshaft engines though with six rather than eight cylinders. It would appear that this configuration had been adopted for reasons of convenience because the 65mm bore, with

[13] Henry Segrave, ghosted by James Wentworth Day, *The Lure of Speed* (Hutchinson, 1928).

reduced 100mm stroke, was shared with that of the previous year's 3-litre straight-eight 801–402.

This new engine, which was constructed in two blocks of three had, like its predecessor, welded-on water jacketing in the Mercedes manner. There were twin shaft-driven overhead camshafts, and significantly, two valves per cylinder instead of the more customary four. Fiat had pioneered this innovation on the 801–402 of 1921, which employed them in conjunction with a hemispherical combustion chamber rather than the four valve/penthouse roof layout then almost universally employed. This new Fiat initiative would, in due course, consign the earlier design to oblivion. At the other end of the engine, the crankshaft ran in seven roller bearing and big ends. This six developed 112bhp at 5,000rpm, or an unprecedented 56bhp per litre, compared with Bugatti's 45, Sunbeam's 44 and the Aston Martin's 22.5.

The four-speed gearbox was mounted in unit with the engine, while drive was by torque tube. Suspension was the usual half-elliptic springs though, at the front, they passed through the tubular front axle in the manner of the 1914 Grand Prix Vauxhalls. Servo-assisted four wheel brakes were employed.

The 804's bodywork was equally revolutionary and was to represent a formative influence on racing car design for the next twelve or so years. The tear-drop wind-tunnel tested shape, with input from Fiat's aerodynamicist Rosatelli, was a refinement of the previous year's lines. It therefore had a cowled radiator, though the side-mounted, externally located exhaust pipe was carefully faired, while the distinctive flat-sided wedge tail – a design which Fiat had pioneered in 1919 – was a notable feature and the fitment of an undertray completed this low drag design. The 804 was compact and, at 662kg (12.2cwt), was the lightest front-line Grand Prix car of the inter-war years.

The 1922 French Grand Prix, held on 16 July, was staged in Alsace to underline the province's return to France. This idea had emanated from W F Bradley who had originally suggested it as the venue for the 1921 race. It was to be run over sixty laps, making a total of 803km (498 miles). Also for the first time there was a starting grid, decided by ballot, and a rolling start. Despite the fact that there had been several weeks of sunshine prior to the event, the race day was wet and the track did not begin to dry out before the first ten or so laps. It was the Fiats that soon dominated, though it was Friderich in one of the Bugattis who put up a brave showing and initially succeeded in keeping pace with them, leading the field on the third lap. This was perhaps a reflection of his familiarity with the circuit, the Bugatti team having practised in slower, four-cylinder Brescias. However, Friderich was unable to sustain this pace and withdrew on lap 14 with engine trouble.

Friderich's car (5) withdrew with engine trouble on the fourteenth lap. Alongside it is Guyot's Rolland-Pilain with Goux's Ballot behind.

Pierre de Vizcaya's Bugatti (12) viewed from the grandstand at the 1922 French Grand Prix.

Foresti in a Ballot leads de Vizcaya's Bugatti at Duttlenheim Corner during the 1922 French Grand Prix. The latter came in second.

As other competitors dropped out, the race was beginning to look like a Fiat procession but, on lap 51, young Biagio Nazzaro's car lost its rear wheel, and he hit a tree with the 804 turning end over end, killing its driver. Then, alas, two laps from the end, the same thing happened to Bordino though he was unhurt. The third Fiat kept going and the forty-one-year-old Felice Nazzaro attained his second French Grand Prix victory, having previously driven the winning Fiat in the 1907 event. This success, however, was clouded by the death of his nephew, who had previously been his riding mechanic.

Cracking was found on the older Nazzaro's rear axle casing and, if this had also fractured, Bugatti would have attained his first, albeit hollow, Grand Prix victory. For de Vizcaya came in second, nearly an hour after the winning Fiat, and he was half an hour ahead of Marco while Monès-Maury, the only other runner left on the course, was flagged off. Once again Bugatti had shown the impressive reliability that had been a key ingredient of the marque since its inception. As a relatively small manufacturer, he had been unable to match the resources of Fiat which, as one of Europe's largest car makers, was able to draw on unparalleled technical and financial resources.

The Wrong Ratios

The same team of four Strasburg Bugattis was entered for the second Italian Grand Prix, held at the newly completed Monza circuit, seven weeks later, on 3 September. The event had attracted an impressive entry from all over Europe but this was soon depleted, following the Fiats' obvious supremacy at the French Grand Prix. The Type 30s all arrived at Monza but 'three-quarters of an hour before the start, an official communiqué was issued to the effect that Bugatti, unable to face the competition, had withdrawn his cars.'[14]

The reason was, apparently, rather different in that the Molsheim cars had arrived at the circuit fitted with incorrect gear ratios. Then, just half an hour before the start, Fiat announced that it had lent Bugatti a set of four wheels and tyres and one car (maybe running with roller, rather than plain, bearing big ends and minus its French Grand Prix cowl) was driven by de Vizcaya. As expected, the Fiats were victorious on their home ground, with 804s coming in first and second, though Bordino and Nazzaro were followed by de Vizcaya's type 30 in a creditable third place.

[14] Hugh Conway, *Grand Prix Bugatti* (Foulis, 1983).

One of the Strasburg cars, chassis 4001, was afterwards sold to Prague banker, Vincent Junek and rebodied. This is his wife Elizabeth, at the wheel. One of the most successful lady racing drivers of her day, she began her career in her native country in 1924 and in 1927 she drove a Type 35B in the Targa Florio and was well placed until forced out with a steering fault. In the following year she was placed fifth, an impressive achievement, and would have done even better, had her car not succumbed to mechanical trouble. Alas, later in 1928 her husband was killed in the German Grand Prix on 15 July and she gave up racing.

The Type 30s ran with uncowled radiators for the Italian Grand Prix at Monza on 3 September 1922. Here the cars are seen, piled high with luggage, arriving in Italy.

Journey to Indianapolis

On the basis of these two relatively successful Grand Prix showings, Bugatti was approached by Martin de Alzaga Unzue, a wealthy South American enthusiast from Buenos Aires, known to his friends of whom de Vizcaya was one as 'Macoco'. De Alzaga Unzue wished to enter a team of three cars for the 1923 500 Miles Indianapolis Race, which was due to be held on 30 May. It was an event, I believe, that was to have far-reaching repercussions for the Bugatti marque, the legendary Type 35 racing car and even its celebrated horseshoe-shaped radiator.

According to de Alzaga:

> We had agreed verbally that our engines were to be 122cu in, with brand new inclined four-valve heads [is this a hint that Bugatti had contemplated a twin overhead camshaft head as early as 1922, or of memory being faulty, fifty-six years after the event?] and the crankshaft mounted on ball bearings, metal bearings only on the rods and of bigger diameter than the ones used in 1922 in Italy and, incidentally, only two speeds in the gearbox.[15]

The Argentinian had specified these modifications because he recognized that the Strasburg cars would have been 'totally inadequate for a track like Indianapolis where, at that time, one hardly took one's foot off the accelerator on each of the four corners of the track'. Pierre de Vizcaya was to be team captain, the other drivers being Paul Riganti, then champion of the Argentine, and de Alzaga himself.

Three Type 30s, chassis numbers 4014, 4015 and 4016, were allotted to the South American team and the cars were of particular interest because Bugatti himself was not responsible for their single-seater bodywork. This was the work of Louis Béchereau, of the Paris-based SPAD aircraft company, and reflected the importance then being attached to the relatively new science of aerodynamics which had been so effectively harnessed

---

[15] Martin de Alzaga Unzue, 'Indianapolis 1923' in *Bugantics*, vol. 41, no. 3 (autumn 1978).

The starting point of the Bugatti radiator can be seen in the shape of the scuttle on one of the cars being prepared for the 1923 Indianapolis Race.

This is the single seater, driven by Prince de Cystria at Indianapolis in 1923. Louis Béchereau of the SPAD aircraft concern was responsible for the lines of the single seater bodywork, complete with cowled radiator. But underneath . . .

by Fiat. Such *monoposto* bodies had been specified for the American race from 1923, nine years before the Europeans. However, as the width of the Type 30's chassis was unchanged, their right-hand sides were flush with the chassis and therefore offset to the left which was, in fact, the wrong way round for the anticlockwise circuit!

Whose Radiator?

The Type 30 cars were fitted with a new radiator and its shape is of particular significance because it subtly differed from the egg-shaped one used on the straight-eight racers of 1922. Although sometimes concealed by a cowl, these Indianapolis Type 30s were the first Molsheim cars to be fitted with the horseshoe-shaped radiator, so spectacularly refined by Bugatti on his Type 35 of the following year.

But did Ettore himself create the contours of these radiators, or was the shape demanded by Béchereau in what he believed to be the interests of aerodynamic efficiency? It seems that the latter was far more likely, because Hugh Conway had already told us that de Alzaga ordered these cars 'on 18 January 1923 (being invoiced on 23 June) all on the short chassis Type 22, with special high compression pistons, axle ratios of 14:54, and *without radiators*'[16] [my italics]. Contemporary photographs show that these echoed the horseshoe contours of the car's scuttles, so perhaps SPAD's chief technician can be credited with a key role in the creation of one of the world's most famous radiators?

Martin de Alzaga Unzue left France in December 1922 and the next occasion on which he saw the Bugattis that he had ordered was on his arrival at the Speedway on 20 May 1923. There, he was horrified to find that in addition to them being in an unfinished state, the cars were not what 'we had contracted for Indianapolis, but [were] simply modified single-seaters with the engine and chassis exactly the same as had been raced in Italy in 1922 . . . You can imagine my disappointment

[16] Conway, *Grand Prix Bugatti*.

. . . was this horseshoe-shaped radiator. Should Béchereau also be credited with the radiator which was superbly refined on the Type 35 in the following year?

when we unboxed the cars.'[17] The principal mechanical modification was the replacement of the twin Zenith carburettors with four units and the engine had been converted to run on benzol, with a 7.5:1 compression ratio. Power output was quoted at a respectable 104bhp.

In addition to these three specially prepared Bugattis, there were two further privately entered Béchereau-bodied Type 30s at Indianapolis in the shape of a pair of former Strasburg cars, chassis numbers 4002 and 4004. They were driven respectively by Prince Bernard de Cystria from France and British-domiciled Polish Count Louis Zborowski, who had first encountered the straight-eight Bugatti when he drove one of the Aston Martins he had financed at the 1922 French Grand Prix. Factory-trained mechanics had been allotted to Unzue's contingent but they also had to service the two privateers which stretched the Argentinian's available manpower. It was not a good start but 'after many arguments we decided to race the cars anyway, or stand to lose all the money invested; indeed I had no other choice.'

Once at the circuit it was found that the camber of the cars' front wheels was unsuitable for the track, so the local Allison Engineering Company modified the two Argentinians' cars. None of the other Bugattis were so altered and de Alzaga achieved an impressive speed of 156.79kph (97.43mph) in practice. Alas, a few unofficial laps at over 160kph (100mph) resulted in a broken connecting rod. The car was repaired by Allison, though the problem returned and the engine was once again restored to running order.

De Alzaga was determined to run in the race to receive the all-important $2,000 (£430) starting money. But a big-end bearing broke down after six laps and Riganti's car was to withdraw with a fractured petrol tank on lap 19. Zborowski's car did the same on lap 41 while de Vizcaya, who had held fifth place for much of the race, managed to last until the 166th of the 200-lap event before he withdrew with the same problem. Only Prince de Cystria finished, in tenth position, and he received a $1,500 (£322) prize. These failures underlined the vulnerability of Bugatti's primitive lubrication system, particularly when used in conjunction with plain big-end bearings. As it happened, back at Molsheim, Bugatti had already switched to full roller bearing crankshafts on his racing engines though this was little consolation to the Indianapolis drivers.

The 500 Miles event was won by Tommy Milton in an impeccably prepared, white-painted Miller 122, owned by Harry C Stutz, purchased by him only days before the race, and then run under the HCS Special name. Harry Miller had offered this particular car for de Alzaga to drive during the race and then pay for it after the event though 'my contract and a cable from Ettore Bugatti forbade me to accept it.' But one of the Bugatti drivers had already placed an order for a 122 Miller.

The 23rd May issue of the Los Angeles *Times* reported that Count Zborowski was buying one and, the contract having been signed, 'Miller phoned long distance to his plant in Los Angeles to start work immediately.' De Alzaga followed the Count's example and ordered no less than two Millers, all three cars being mechanically similar to the Indianapolis one, apart from the fitment of two-, rather than single-seater bodywork. This was in deference to European racing regulations, which still required the presence of a riding mechanic. It seems likely that these cars were not in fact new but re-engined two-seater 183 Millers. Six of these had been completed for the previous year's formula in the autumn of 1922 for Clifford Durant, son of the founder of General Motors, who ran them as Durant Specials.

The trio of Millers was in Europe by early August, in time to participate in the Italian Grand Prix at Monza on 9 September 1923. Monza was given added status by it also being the first European Grand Prix. There, continental Europe had its first opportunity to examine these extraordinary American racers. Although two blown Fiats won the event (the first occasion that a supercharged car won a Grand Prix), Jimmy Murphy did well to be placed third and although Zborowski threw a connecting rod, de Alzaga kept going and finished

---

[17] de Alzaga Unzue, 'Indianapolis 1923'.

Ernest Friderich in the prototype Type 32 at Molsheim with the Bugatti villa in the background. The wheel discs, employed for aerodynamic considerations, were not fitted for the race.

sixth. There were, incidentally, no Bugattis present at Monza, Molsheim no doubt wishing to keep its distance from the formidable Fiats running, as they were, on home ground.

A Tank Battle

It had been a different story two months previously at the French Grand Prix, held on a triangular 22.6km (14-mile) circuit at Tours on 2 July. There Bugatti fielded a team of four purpose-designed Grand Prix cars which fell within the chassis numbers 4057, 4058, 4059, 4060 and 4061. As they differed significantly from the production Type 30s, they carried their own Type 32 designation. Henry Segrave, who was driving a Sunbeam, noted that they 'looked more like a miniature tank than a motor car',[18] and they are universally known as such today.

However, it should be noted that other contemporary descriptions included tortoise, dish cover, beetle and even roller skate! This was because their bodies, which resembled the profile of an aircraft's wing, were as bizarre and memorable as the lines of the following year's Type 35s were supremely elegant. But while many racing car manufacturers of the day were looking into the science of aerodynamics, what made Bugatti take such a distinctive approach to the subject – which differed

[18] Segrave, *The Lure of Speed.*

so radically to those of the Fiat school – with blunt-nosed, long-tailed bodies echoing the lines of an airship?

Clearly Ettore's own idiosyncratic response to any problem was a factor. However, the path leads, I believe, to the Issy-les-Moulineaux laboratory of that extraordinary Parisian pioneer aviator and friend of Ettore's, Gabriel Voisin, who switched to car manufacture after World War I. Speaking of 'My friend Bugatti', Voisin would recall, 'I do not remember the occasion of our first meeting. I seem to have known this amazing engineer all my life.'[19]

Voisin began producing his own, highly distinctive automobiles in 1919, and in 1922 entered a team of three of his 18CV cars for the first *Grand Prix de Tourisme*, held after the Grand Prix proper at Strasburg. As fuel consumption was a key ingredient of the rules, Voisin's team of three cars were fitted with specially prepared touring bodies. The rules demanded that these be 1.30m (4ft 2in) wide and, while other competitors literally interpreted this, Voisin opted for coachwork of only 90cm (2ft 11in), with streamlined side pods accounting for a further 20cm (7.8in) each side.

For reasons of aerodynamic efficiency, wheel discs were also fitted and all these features, coupled with three months of careful preparation, resulted in the economical, sleeve-valve Voisins sweeping to victory, taking the first three places. The winning car averaged 107.kph (66.9mph) on a fuel allowance of 6km/l (16.6mpg) which left less than 9 litres (2 gallons) of petrol in its tank.

When the regulations for the 1923 race were published, Voisin was horrified to find that they demanded the use of conventional open bodywork to negate any advantages of streamlining. A somewhat miffed Gabriel responded by writing an open letter to the president of the ACF in which he stated that 'more practical results could be obtained by studying streamlining for six months than by working on engine details for six years.'[20] So rather than enter the Touring Grand Prix, he decided instead to run four cars in the French Grand Prix proper which offered no such bodywork constraints.

These racing Voisins were designed to make aerodynamics do the work of horsepower, for the reality was that Voisin's passenger cars were powered by double sleeve-valve engines which were hardly suitable for Grand Prix racing. Neither did the firm possess an appropriately sized unit for the 2-litre Grand Prix formula, so Voisin effectively added two extra cylinders to his C4 model's 1,243cc four-cylinder engine though the resulting 1,993cc six only developed 75bhp, or about 37hp per litre compared with 56 of the previous year's winning Fiat.

The Voisin Approach . . .

To offset this relatively low output, Voisin produced an aerofoil-shaped body of plywood and aluminium of pioneering monocoque construction which, less engine, weighed just 37kg (83.5lb). But the heavy sleeve-valve six and running gear pushed the cars' total weight up to a far from impressive 750kg (14.7cwt). Work on the project began in about January 1923. Aided by his 'spiritual son' André Lefèbvre – who was later to have overall responsibility for the revolutionary *Traction Avant* Citroën of 1934 – Voisin's quartet, titled *La Laboratoire* by their maker, were completed in time for the Tours race.

They were, in the words of *The Autocar*, 'unlike anything yet seen on wheels',[21] though the subsequent description could equally apply to Bugatti's Tank. The magazine wrote that: 'The Voisins may be described as having the profile of an aeroplane wing. The underside is absolutely flat and is brought as near to the road as possible, and the upper surface has a considerable camber.' The highpoint of the car's body, just behind the driver's head, was a mere 1,016mm (3ft 4in) from the ground.

Once again Voisin employed wheel discs and, while the front ones were exposed to the driver in

---

[19] L'Ébé Bugatti, *The Bugatti Story*.

[20] *The Autocar*, 22 June 1923.

[21] *The Autocar*, 22 June 1923.

Two of Gabriel Voisin's extraordinary cars that ran at Tours in 1923 and bore some similarities to Bugatti's Tanks.

the usual manner, the narrower rear ones were contained within the body. There was no differential. These Voisin cars, with only the outline of the engine interrupting the distinctive wedge shape, achieved a top speed of around 170kph (105mph), which made them the slowest cars present at Tours.

So much for the Voisins. But what of Bugatti's Type 32? I should stress that any pre-race association between Bugatti and Voisin is pure surmise on my part. There is, however, a striking similarity between the two cars, in overall concept, if not detailed execution. This, plus the fact that both firms did, to some extent, share some of the same mechanical limitations, does suggest such an interchange of ideas. By this time, Bugatti's 2-litre straight-eight was developing approximately 90bhp, which was a good 40bhp less than Fiat's 1923 cars, destined to attain a remarkable 130bhp at Tours. Ettore's use of the bulbous, allegedly more aerodynamic bodies on the Type 30s at the 1922 French Grand Prix, revealed, in the face of more powerful opposition, his attempts to make his single overhead camshaft cars more competitive by such experimentation.

. . . And Molsheim's Line

We can gain some insight into Ettore's thinking from a document he circulated prior to the race for the use of the press, in which the similarity to the Voisin is striking. Bugatti maintains that 'the thick aerofoil section of this little car has only been achieved by the chassis and all the rolling mechanism being designed to be totally enclosed by a small envelope . . . The closer a vehicle approaches the ground, the less will be its resistance to forward motion.'[22]

Ettore also hints at problems he may have experienced in testing his Tank when he wrote: 'There is never-the-less great difficulty in lowering the vehicle . . . One gains in stability when cornering but

---

[22] 'The Grand Prix of the Automobile Club of France 1923' in *Bugantics*, vol. 42, no. 3 (autumn 1979).

the roadholding at high speed and in a straight line becomes more difficult.' He also reveals: 'I made no test in a [wind] tunnel because I consider that it is impossible to obtain results without very special equipment to approach realistic conditions, and I have little faith in scale models.' In other words Ettore relied on visual judgement in producing the Type 32's stubby lines.

Unlike the Strasburg cars of the previous year, the Tanks' bodies were produced at Molsheim and this construction marks the beginnings of a bodyshop at the factory. This would expand throughout the decade as it became fully occupied producing coachwork for the Type 35 racing car, together with coachwork for the passenger models.

The wheelbase of the Tanks was only 2,000mm (6ft 6in), which echoed that of his 16-valve Brescias, and was therefore 500mm (1ft 7in) shorter than the Strasburg cars. The bodywork, unlike that of the Voisin, was all-enveloping and enclosed the front wheels. However, the highest point of the body was 800mm (2ft 7in) and ahead, rather than behind the driver, and was consequently lower than that of the *La Laboratoire*. As the Bugatti's shape must have produced a significant amount of lift at high speed, the engine cover and tail were well vented to prevent any excessive build-up of air, an undertray being fitted.

The suspension was also unique to the model, in that inclined quarter-elliptic springs were employed front and rear, both axles being underslung. The engine was essentially that of the Type 30 though the connecting rods, with detachable caps, were at long last fitted with roller bearings, 'allowing', in Bugatti's words, 'good performance at high revolutions'. A further modification, albeit a minor one, was made to the magneto drive. The step-up gears at the rear cambox were inverted, so bringing the top of the magneto in line with that of the cambox, which meant that the height of the bodywork could be reduced by about 25mm (10in).

As far as the gearbox was concerned, Ettore pointed out that this 'had been eliminated' but the reality was that there was not room for it on its usual mounting and it 'had been incorporated in the rear axle . . . the same arrangement as on the 3-litre car shown in 1921'. It was accordingly a three-speed unit and demanded a central gear lever, which was an unusual location for the day as this control was invariably mounted to the right of the driver. Alongside it was the handbrake, which operated the rear cable brakes while front hydraulics were employed. The steering gear, with its stubby column, was also peculiar to the model and there was no protective bulkhead between the engine and the cramped seating for the driver and his mechanic. The Type 32s weighed about the same as the Voisins, turning the scales at around 750kg (14.7cwt).

The prototype Type 32 (4057), with disc-covered wheels, first appeared six weeks prior to the Grand Prix, driven by de Vizcaya, at the *Bol d'Or* race on 20 and 21 May. He was also assigned to one of the Tanks at Tours, the other drivers being Friderich, de Cystria and Marco.

The Opposition at Tours

The Bugattis were one of three straight-eight designs present at Tours, though they were the only ones with single overhead camshaft engines. There were locally built Rolland-Pilains, similar to the previous year's cars, while the third of the eights was, potentially, the most formidable. This was the new 805–405 Fiat, a team of three arriving late at Tours, having been driven over the Alps by road from Turin.

Low, sleek and purposeful, they were slightly larger than the previous year's cars and the engines were of great significance as they were fitted with Wittig superchargers, which contributed to the eights developing 130bhp at 5,500rpm, or an unprecedented 65bhp per litre. Interestingly, their bodies represented the antithesis of the Bugatti/Voisin approach. Experiments at Turin had shown the inadequacy of having a body too close to the road surface, the undertray reducing in area as it tapered towards the tail to help dispel any air trapped there.

The only other significant entry came from a team of three six-cylinder Sunbeams, unkindly but perceptively dubbed 'Fiats in green paint' as they closely resembled the Italian 804s of the previous year and were designed by a duo of ex-Fiat engineers. Delage, meanwhile, celebrated its return to Grand Prix racing by entering a single V12-engined car which weighed in at a creditable 690kg (13.5cwt) making it noticeably lighter than the Bugattis and Voisins.

The four Tanks were late in arriving at Tours, Race day was hot and the cars had to complete thirty-five laps of the circuit, making a total of 798km (496 miles). Fiats initially set up a blistering pace though their engines soon succumbed to dirt and grit being sucked in through their unprotected supercharger inlets. The cars had been tested on the smooth concrete of the Monza circuit and the Fiat engineers were unprepared for the poor road surface around Tours. The first car dropped out after a mere 135km (84 miles), the second left the race at half distance with the third retiring while in the lead and only four laps away from the finish. This permitted the Sunbeams to take the first two places with Friderich in one of the Tanks in third position, breaking a Sunbeam Grand Slam. Guinness was accordingly fourth and Lefèbvre, in the surviving Voisin, fifth.

Friderich's Tank passing Enrico Giaccone's Fiat 805 which failed on the eighteenth lap after dust had entered its engine's unprotected supercharger intake.

How did the Bugattis perform? Far worse than in the previous year's French Grand Prix, though the cars were by no means slow, being capable of about 185kph (115mph). It was their short wheelbase, coupled with a tendency for them to lift at

Friderich in one of the Tanks at Tours. He was to do best in the race and was placed third. Behind him is Prince de Cystria's car (16) and Marco's (18) which crashed.

speed, that made them unpredictable mounts. The race was soon over for de Vizcaya, who crashed on the first corner of the first lap, only 6km (4 miles) from the start, apparently blinded by the other cars' dust. He hit a tree and unfortunately landed amongst a group of spectators. A boy suffered serious injury while a man was less badly hurt; de Vizcaya suffered a head wound and his mechanic was badly shaken. It could have been much worse.

Then Marco dropped out after four circuits and by this time, de Cystria was being lapped and pressed by Segrave's Sunbeam at the same *Le Membrolle* corner that had claimed de Vizcaya. De Cystria crashed there, though disentangled himself from the fence and carried on to the pits where he retired.

Only at Tours

A works Bugatti, to be driven by de Vizcaya, was entered for the first Spanish Grand Prix held four months later at the new Sitges autodrome, which was opened on 28 October by the son of the King of Spain. The race took place immediately after the opening ceremony and was won by Divo in a Sunbeam. Zborowski, in one of the Millers, was second – the other two '122s' having returned to America after their appearances at Monza in the previous month. The French Grand Prix turned out to be the Bugatti Tanks' only appearance in a front-line competition and the Bugatti did not participate at Sitges.

However, following the Tours race one of the Tanks, driven by a private owner, de l'Espée, took part in the Boillot Cup race at Boulogne on 30 August.[23] Pierre de Vizcaya also drove one in a speed trial on the Arpajon road in 1924 when he recorded a speed of 189kph (117mph) over a measured kilometre. He is listed in the Bugatti sales records as taking delivery of car number 4057, the first Tank to be completed, from Bugatti's Paris showrooms on 24 February of that year. In addition, soon after the Grand Prix, Friderich's car (4059), was offered for sale in the town of Tours, displaying a placard proclaiming that the sale was to the *Bénéfice des Laboratoires* (to finance the Laboratories) which may be another pointer to Voisin's involvement in the Tank project.

Not surprisingly there were no takers and three months later, in October, Ettore disposed of the car to the Czechoslovak banker Vincent Junek and his wife Elizabeth. They only retained it for around six months, returning it to Molsheim where it was traded in, with 85,000fr. (£1,120) credit, for a Type 35 in the following year. The fate of cars 4058 and 4060 is unknown, so they may have been the two Tanks which crashed at Tours. Bugatti retained 4061; 'I count on keeping it to carry out experiments' said Ettore, and today it can be seen at the *Musée national* at Mulhouse.

So ended Bugatti's brief and highly individual flirtation with aerodynamics. But this is not quite the end of the story because the concept was triumphantly reborn thirteen years later with the appearance, in 1936, of three Type 57 based cars, also unofficially known as Tanks and they subsequently went on to win not only that year's French Grand Prix but were also twice victorious in the Le Mans 24-hour race.

The Type 30 on the Road

As will have already been apparent, Bugatti differed from practically all his contemporaries in that his racers did double duty as road cars. At this time Ettore's front-line contemporaries were Fiat, Sunbeam, Rolland–Pilain and Delage and their passenger models were essentially unrelated to their racing cars. Not so Bugatti, who was still a relatively small manufacturer though it was a state of affairs made easier by him retaining a single overhead camshaft engine rather than opting for a more expensive twin cam layout.

This permitted Ettore to claim a close association between his racing cars and those offered for public sale. Writing of his 1923 Tanks, he proclaimed: 'The whole vehicle is built up from

---

[23] 'Letter from William Boddy' in *Bugantics*, vol. 41, no. 2 (summer 1978).

Elizabeth Junek pictured in 1969 at the National Motor Museum with a 1925 Type 30. This particular car, chassis number 4468, was only run for three years and laid up in 1928 after covering a mere 4500 miles (7241km). It was not used until the summer of 1959 and was displayed at the Museum for many years.

production parts'[24] and was to make similarly questionable statements as far as the mechanicals of the subsequent and more sophisticated Type 35 was concerned. Only across the Atlantic in America did Duesenberg offer a single overhead camshaft eight on its Model A road car, which entered production in 1921, powered by an engine which was basically similar to that of its racing cars. However, Duesenberg soon switched to twin cam heads from 1922.

The roadgoing version of the Type 30 appeared at the end of 1922, five months after the 2-litre eight-cylinder racers had made their debut at the French Grand Prix. As already mentioned, the Type 30 was far less mechanically sophisticated than the Type 28. In Britain, its chassis price was £950, still making it an expensive car though £450 cheaper than the stillborn 3-litre model.

Mechanically the roadgoing Type 30 was very similar to the racing version. It was therefore powered by much the same 2-litre straight-eight engine with a three-ball bearing crankshaft though with plain big ends fitted throughout its production life. Initially twin Zenith carburettors were employed, though these were later replaced by a single unit. Also, coil ignition took the place of the original magneto. The gearbox was similarly a strengthened Brescia unit. The model initially had a 2,550mm (8ft 4in) wheelbase chassis of similar dimensions to the eight-valvers but strengthened to cope with the extra weight. For 1926 it was also available with a longer 2,967mm (9ft 5in) frame.

The hydraulic front brakes were also carried over, the model therefore being the first roadgoing Bugatti to be so equipped with four wheel brakes. Their efficiency was perhaps questionable and at least one Type 30 was said to need 40m (132ft) to stop from just 65kph (40mph). If any Bugatti model justified Ettore's memorable aphorism that he 'made his cars to go, not to stop', it is this, his first production straight-eight car. The hydraulics were discontinued in late 1924 and replaced by mechanical brakes of the type employed on the Type 22/23 and many hydraulic-braked cars were subsequently so converted.

[24] 'The Grand Prix of the Automobile Club of France 1923' in *Bugantics*.

A Type 30 with French two-seater with dickey coachwork pictured in Britain in the late 1920s. It was owned by Captain S M Townsend, grandfather of Robin Townsend of Jarrot Engines, which today specializes in the restoration of Bugattis.

Similarly, the model began life with an egg-shaped radiator of the type that had been pioneered on the eight-valve cars. This was later changed to the larger and similar flat-bottomed radiator and effectively was a similar version to that employed on the long-running fours. Otherwise the car was conventional Molsheim, with half-elliptic springs at the front and reversed quarter rear elliptics. Although the model's arrival coincided with the establishment of a bodyshop at the factory this was initially used for the Grand Prix cars. The Type 30 was therefore bodied by a variety of specialist coachbuilders of which Lavocat and Marsaud of Paris and the Profilée company were amongst the most popular. The model remained in production until 1926, and its final chassis number seems to be 4818–9.

There were at least 585 examples built over a four-year period which was an impressive figure for a straight-eight car. The Type 30 was capable of a respectable 120kph (75mph) which put the model in the forefront of the *Grands Routiers* of the day, a term that could be applied to practically all its eight-cylinder successors. Such fast tourers were peculiar to France for 'Nothing quite like them existing in a class elsewhere . . . the car which would suit the temperament of the *voiturette* owner who had become richer, supposing that his sporting instincts were still strong . . . High gearing gave them their high cruising speed, usually about 60–65mph [95–105kph] in relation to the maximum speed of 70–75mph [110–120kph] . . . This kind of steady progress ate up the great open distances of France, Spain and Italy.'[25]

A very real limitation, as far as the Type 30's engine was concerned, was its three bearing crank-

[25] T R Nicholson, *The Vintage Car* (Batsford, 1966).

It was Ettore's daughter, Lidia, who is said to have been responsible for suggesting a horse symbol for the marque. This advertisement dates from 1924.

shaft, which sometimes tended to break, usually around the front, or back, main bearing and has cast a retrospective shadow across the model's reputation. Yet 120kph (75mph) or so from a 2-litre engine was remarkable for its day, as the sales figure reflect. Roadholding was good and this relatively softly sprung car was, with its long wheelbase, surprisingly comfortable. The single overhead camshaft straight-eight also represented the starting point for the engine of the most fabled of all Bugattis, the Type 35 racing car, which forms the subject of the next chapter.

7
Masterpiece

'[The Type 35s], which appeared for the first time in the 1924 French Grand Prix, and attracted enormous attention principally because they were the first racing cars to have the coat of paint and nickel plating usually associated with a production machine, and were the first racing cars to be produced in quantities.'

S C H Davis, sports editor of *The Autocar*[1]

Up until 1924, or for his first fourteen years at Molsheim, Ettore Bugatti could be considered to be a relatively small, if rather eccentric specialist car maker, who had enjoyed some successes in *voiturette* and Grand Prix racing and also produced sports and touring cars of note. But in that year came the Type 35 racing car, with its distinctive horseshoe-shaped radiator, finely proportioned body and memorable eight-spoked aluminium wheels, the result of Ettore lavishing his supreme artistic talents on a design that was destined to make him respected throughout the motoring world.

Indeed, much of the mystique associated with the marque to this day can be traced to this one model. Yet the 35 looked as it did because it was a completely different sort of racing car, as it was designed, from the very outset, to be produced in relatively large numbers and 340 examples were built in the seven years between 1924 and 1931. This was at a time when the production by a manufacturer of six Grand Prix cars in a particular year was tantamount to mass production!

If the car was going to appeal to a wide clientele, it would not only have to possess the already established Bugatti ingredients of superlative road-holding, handling and reliability, but would also have to be imbued with a strong visual appeal. The Type 35 was the first racing car to have its bodywork deliberately styled in a way that had previously been reserved for the coachwork of exclusive road cars of the day. Hitherto, racing cars of the 1920s appeared as they did for the essentially practical reasons of weight saving and aerodynamic efficiency within the confines set down by the regulating authorities. Appearance very much played a second fiddle to these parameters.

Mechanically, the Type 35 perpetuated the cautious, conservative approach that had characterized its Type 30 predecessor. Thus in an era of twin overhead camshaft engines, Bugatti persisted with his single cam layout. It was not until 1931, seven years after the Type 35 had appeared, that Ettore bowed to the inevitable and adopted twin overhead camshafts, nineteen years after the feature had first appeared on the celebrated Grand Prix Peugeot of 1912. Bugatti was similarly guarded in his attitude to supercharging, introduced by Fiat in 1923, and it did not appear on his Grand Prix cars until two years after the arrival of the Type 35. Indeed, by 1926 Bugatti could no longer competitively afford to ignore it. However,

[1] SCH Davis *Motor Racing*, Earl Howe, ed. (The Lonsdale Library, Seeley Service, 1939).

the design work necessary to so convert the single cam eight was not undertaken at Molsheim but was the work of a Paris-based consulting engineer.

By chance, the years of the Type 35's manufacturing life coincided with a dwindling of participating firms in front-line racing and a consequential slowing in the development of the Grand Prix car. However, the sheer number of Bugattis in circulation ensured that they would contribute to making the model one of the most successful racing cars ever built, even if it was not the most powerful.

The reality was that the Type 35 developed only 90bhp in its original unsupercharged form, and about 50bhp more in its ultimate, enlarged blown state. However, when faced with a superior design in the shape, for instance, of the 170bhp 1.5-litre Delage in 1927, Ettore followed past precedent and withdrew his works cars from some of that year's key Grands Prix. These incidents excepted, the success of the Type 35 gave Bugatti a measure of financial stability and security. The 1920s were a golden era for Molsheim, Bugatti and his exquisite racing cars.

The failure of his Tanks at Tours in July 1923 saw Ettore turn his back on such unorthodox aerodynamic excursions and he must have spent the winter of 1923 and 1924 radically re-thinking his design strategy. The 1923 French Grand Prix cars carried the Type 32 designation. The Type 33 was a 2-litre road car, with a gearbox-mounted rear axle, which was a passenger model with its mechanicals, though it never entered production The Type 34 16-cylinder aero engine suffered a similar fate. Both these projects dated from 1923 and the next number in Ettore's design register was 35, which was allocated to the new Grand Prix car and has become the best known and illustrious of the Bugatti Types.

The Type 35 Conceived

So Ettore rejected the Tanks concept. The peculiarities of the cars' roadholding must have been a factor, although he naturally did not concede this publicly. He candidly declared the decision was made solely on commercial grounds or, as he put it, 'I have abandoned the thick aerofoil in spite of its technical advantages, simply with the object of obtaining a more elegant shape, to facilitate sales.'[2] Bugatti therefore decided to follow convention, 'as far as their shape is concerned, they will be quite *normal* [my italics]' he wrote.[3] This meant that he looked, like many of his contemporaries, to his native Italy for inspiration in this respect and the wind-tunnel proven bodywork of the 805 Fiat.

---

[2] *Automobilia*, 173, 31 July 1924, quoted in *Grand Prix Bugatti*.

[3] 'Letter to Vincent Junek' in *Grand Prix Bugatti*.

The car which inspired the lines of Bugatti's Type 35; the Fiat 805.405 of 1923 which introduced the supercharger to Grand Prix racing. As the trilingual captions indicate, Carlo Salamano won the 1923 European Grand Prix, with another 805 in second place.

In a freehand sketch of the lines of his new car attached to a letter which Ettore sent to Vincent Junek on 9 April 1924, three months prior to the cars' appearance, the embryo Type 35 is shown with a conventional radiator once more in place after two years of Bugatti's Grand Prix cars having cowled fronts. The seats are depicted with the mechanic's one staggered behind the driver's, which would have resulted in a slightly narrower body than that finally adopted and culminated in a handsome, wedge-shaped tail. As on the Tank, in the interests of aerodynamic efficiency an undertray was fitted, which totally enclosed the underside of the car, with the exception of the engine's projecting air-cooled sump.

The first Type 35 to be completed, though with the shape of the radiator yet to be finalized. This car, chassis number 4323, was driven by Ettore himself at Lyon though it did not compete there. It was displayed at the 1924 Paris Motor Show, later the London one and, on 11 November, was delivered to Bugatti concessionaires, Jarrott and Letts. It had been seen at Olympia by Sir Robert Bird of Bird's Custard, who became the first private owner. In the post-war years the Bugatti went to America where it was owned by the distinguished General Motors stylist, Hank Haga.

Why then is the Type 35's body so aesthetically satisfying in a way in which the lines of the Fiat and the other cars inspired by the 805 are not? The bodywork of the P2 Alfa Romeo – a contemporary of the Type 35 – appears by contrast both flat and lifeless? Apart from its finely balanced proportions, a crucial element in the visual success of the car was that its body followed the contours of Bugatti's new gently curved horseshoe-shaped radiator, unquestionably the most beautiful ever fitted to a racing car.

This radiator is yet another ingredient of the Bugatti mystique and would appear on every Molsheim car, in one form or another, from thereon. But as will have already been apparent, the shape, in its orignal form, had first appeared on the previous year's Béchereau-bodied Type 30s that had run at Indianapolis and perfectly complemented Bugatti's *Pur Sang* trademark, first registered in Germany in 1911. This means literally 'pure blood', and was usually applied to thoroughbred horses, which echoed Ettore's passion for riding and all things equestrian. In 1914 it was refined to one word: *Pursang* and then, in 1925, the year after the Type 35's appearance bearing its lovely horseshoe-shaped radiator, was elaborated to *Le pur sang des automobiles* (the thoroughbred of cars). It comes as no surprise to find Ettore telling Junek in 1924 that his new racing car would be: 'more thoroughbred than all cars of other makes'.

Bugatti was well aware of what he achieved, as far as the Type 35's body was concerned, when he wrote: 'The shape of the coachwork seems to be very successful, something not easy to achieve, as I wanted to retain the profile of my radiator.'[4] A copy of the drawing of the original body lines, which thankfully survives, shows his pencilled amendments, adding an extra centimetre here, paring a millimetre there, in other words relying on his impeccable eye to obtain the correct proportions.

Having decided on the Type 35's appearance, Bugatti moved on to the chassis which, he pointed out, 'follows the shape of the body'.[5] This was new

---

[4] *Automobilia*, op. cit.

[5] *Ibid.*

in execution though its 2,400mm (7ft 10in) wheelbase was a perpetuation of that employed on the Type 22. The frame itself was totally enclosed for most of the car by the undertray. The members were, however, exposed at the front where they were a delicate 19mm (0.75in) deep though this progressively increased towards the centre of the car, reaching 170mm (6.75in), just behind the rear of the engine, where it was subject to the greatest stress. When considered in plan view, at the same point the side rails opened out to follow the body lines in a gentle, tapering ellipse and were enclosed, 805 Fiat-wise, within the elegant, tapering tail.

Ettore, maybe conscious that observers would draw comparisons between the Type 35 and Fiat, quickly pointed out that the contoured chassis 'was not new. I had only to draw inspiration from one of my pre-war patents for a chassis having the shape of the body and being integral with it. This is said simply so that one is not tempted to believe that I have been inspired by some other construction in designing the car.' Here the inference is that perhaps Fiat copied Bugatti rather than vice versa!

The distinctive front axle was also new and made its own considerable contribution to the architecture of the front of the car. To once again quote Ettore, never one to fail to point out his accomplishments, it was 'a mechanical masterpiece'.[6] Like the lines of the body, it was a refinement of a Fiat concept in that the front springs passed through eyes in the axle, instead of being shackled to it. Unlike the 805, where the hollow axle was in two pieces, Bugatti's 'masterpiece' was in one.

The Engine Developed

As far as the Type 35's engine was concerned, the 2-litre formula was still in force, so Bugatti effectively perpetuated the 1,991cc straight-eight, with the clean cut and distinctive architectural lines he had introduced on the Type 30, though with some important modifications. It did, however, retain two inviolate Molsheim parameters in that, at the top end of the engine, the single overhead camshaft was perpetuated, as was the familiar splash lubrication system at the other. Similarly the concept of an aluminium crankcase, with two separate 60 x 88mm fixed-head four-cylinder blocks mounted on it, was retained.

What Bugatti clearly needed was more power and he recognized that, if he was to retain the basic essentials of his straight eight, this could only be achieved by increasing crankshaft revolutions. But this meant providing it with more support. It will be recalled that the Type 30 crankshaft was made in two four-cylinder sections, a practice that went back to Ettore's wartime aero engines, and also permitted the fitment of the central double ball race. On the Type 35 Grand Prix cars, the shaft was built up from no less than twelve parts.

This was an impressive piece of precision machining for its day, as it was necessary for the sections to be interchanged with perfect accuracy and, indeed, the assembled shaft is a thing of considerable, if hidden beauty. Such a construction was adopted because, although the three double rollers of the Type 30 were perpetuated, two split single rollers were introduced between cylinders two and three and five and six.

The employment of such a built-up shaft permitted the fitment of one-piece roller bearing connecting rods 'about 50 per cent lighter than the rods on my engines of the Tours GP and Strasburg GP'[7] claimed Ettore. As a result of these ministrations, the Type 35 was capable of around 5,500rpm, or about 1,700rpm more than that advisable on the Type 30, which had a 3,800rpm rev limit.

All these modifications meant a change to the crankcase's construction and, instead of being a single casting, it was split along its centre line. Oil was directed to the shaft, courtesy of an external gallery at around 15psi, being fed from a newly located low-mounted oil pump. Lubricant was squirted at the shaft and thereafter found its way, by centrifugal force, to the rollers. It should be reiterated that such a primitive arrangement

---

[6] 'Letter to Vincent Junek', *op. cit.*

[7] *Automobilia, op. cit.*

works, if that is the word, at its best with a roller bearing crank.

At the other end of the engine, the valves were enlarged in diameter and the established two inlet and one exhaust configuration, actuated by fingers located between the valve stem and camshaft lobes, was retained. On the Type 30, the inlets were 21.5mm (0.8in) in diameter and this was increased to 23.5mm (0.9in) on the 35. The exhausts were also upped from 31mm (1.2in) to 35mm (1.3in). They incorporated one of Ettore's less successful innovations, whereby holes were introduced in the hollow stems, the idea being that the lubricant could enter and so cool the valves, water jacketing being at something of a premium around their seating! But the reality was that the orifices became blocked with carbon and the feature was discontinued. Twin triple diffuser Zenith carburettors were fitted. The magneto was dashboard mounted, which followed Grand Prix precedent, and was similarly driven from the end of the camshaft.

The engine was attached to the chassis at four points and greatly contributed to the rigidity of the front end. Bugatti also perpetuated the fitment of a separate gearbox at a time when leading designs, such as those by Fiat and Sunbeam, opted for unit construction engines and gearboxes. On the 35, Ettore relied on the strengthened Brescia unit that he had previously employed on the Type 30. But instead of it being attached to the chassis by extensions which spanned the width of the frame, these were removed leaving just the box, which was mounted on twin tubes, also contributing to bracing the somewhat flexible frame.

The gear lever was back in its conventional right-hand position and while it operated in a gate, its positions were less orthodox with first and third gear being engaged by moving the lever towards the driver and second and fourth away from him. Drive was via Bugatti's wet plate clutch and a short drive shaft. It thereafter went, via an open propeller shaft, to a rear axle which outwardly resembled that of a Type 30 unit though it was, in fact, new. The customary torque arm was fitted alongside the shaft. Driving torque was absorbed by two externally mounted radius rods located either side of the car with forward ball pivots.

Eight-Spoked Aluminium Wheels

Yet for all the Type 35's understated beauty, it outwardly resembled a relatively conventional 1924 racing car. What Bugatti needed was a striking visual feature to distinguish it from its contemporaries. Ettore found it with the memorable and distinctive eight-spoked cast aluminium wheels, which were to become as identifiable with the marque as its famous radiator. They were to feature on Grand Prix Bugattis until 1934 and their related sports models until the following year.

Over the years there has been much speculation as to their origins and one of the most ingenious is that they may have been inspired by the classical column which forms part of a 200-year-old stone fountain in the centre of Molsheim. There a lion regardant displays a representation of the town's coat of arms, consisting of a crown, surmounting . . . an eight-spoked wheel! Despite this fascinating historical by-way, I am inclined to agree with historian, Griffith Borgeson,[8] who has identified a very similar wheel, though with six spokes rather than the eight that Bugatti used. That wheel was patented in 1920, four years prior to the Type 35's announcement, by that extraordinary American artist engineer, Harry Miller. As will emerge, I am inclined to believe that this was not Miller's only contribution to the Type 35's appearance.

Back in 1920 Miller had produced two cars for Los Angeles brewer and go-getter, Eddie Maier. Called the TNT for reasons that are no longer apparent, it was powered by a purpose-built 2,720cc (166cu in) four-cylinder engine with Ballot-inspired twin overhead camshafts and four valves per cylinder. This was completely enclosed in a cast aluminium cover which merged with the bulkhead; indeed the unit appeared to be suspended from it. Two cars were built for Frank Elliot and Tom Alley to drive at the 1920 Indianapolis 500 Miles Race but

[8] Borgeson, *Bugatti by Borgeson*.

never appeared. However, surviving photographs show one completed car fitted with a conventional bonnet though the TNT then disappeared from view.

What makes this project one of particular interest is that in September 1919, Miller filed a patent (number 55070) granted on 4 May 1920, for a bonnetless racing car with a similarly enclosed power unit. It was, however, fitted with six-spoked aluminium wheels very similar in concept to those subsequently used by Bugatti on the Type 35. However, the TNT itself was fitted with conventional wire wheels and it seems likely that the aluminium ones were never built.

But where did Miller get the idea? Borgeson has opined that it may have been inspired by a flat spoked alloy wheel which formed part of the friction transmission of the rare front-wheel drive Los Angeles-built Homer Laughlin roadster of 1916. The wheel could have been cast in the Miller foundry and, in any event, after the end of World War I, Miller moved into a larger factory in the city's 2625, Long Beach Avenue – the premises having previously been occupied by the Homer Laughlin Engineers' Corporation . . .

Drawings of Miller's car, complete with its alloy wheels, duly appeared in the internationally circulated *Official Gazette*. The drawings would have been available, not only in Paris, but conveniently accessible to Bugatti at the *Bibliotèque universitaire* at Strasburg.

Perhaps the strongest pointer towards Ettore being inspired by the Miller concept is that his own patent, filed on 5 May 1924, shows what he describes as his 'cooled disc wheels'. The patent depicts the starting point of 'his' aluminium wheels, although there is no reference to the material, and the wheels only have six spokes, just like the American ones. They also incorporate a secondary row of strengthening spokes which, instead of shadowing those at the front, as on the production wheels, are set at an angle to them and therefore visually fill in the gaps in between.

None of these features found their way on to Bugatti's final wheel, apart from the incorporation of the brake drum, the spokes being slightly

A works photograph of the lovely eight-spoke cast aluminium wheels with integral brake drums which were such a distinctive feature of the Type 35. The detachable rim is secured by 24 screws. However, this example is fitted with a 710 x 90 Englebert tyre, whereas the Lyon cars were shod with . . .

. . . Dunlop Cord tyres specially made for the new wheels which gave endless trouble throughout the race.

skewed to direct cooling air towards them. A later application, of 10 June, develops the theme and refers to 'cast wheels' reinforced with iron struts, while a further patent of 18 June again makes mention of such a wheel, being used in conjunction with a type of tubeless tyre. Neither of these features reached production.

So, after thirteen years' reliance on the wire wheels for his racing cars, Bugatti switched to aluminium ones which also had a tangible advantage in that their fitment reduced the car's unsprung weight by 12–13kg (28–30lb). The wheel that was finally fitted on the Type 35 had eight, rather than six, spokes with the inners in line with the outers, so that it appeared that there was only one set. The integral finned brake drums had shrunk in steel liners, and Bugatti dispensed with the hydraulic front brakes of the previous two years, and employed the mechanical system of the type subsequently fitted to the eight-valve cars and described on page 134. These wheels had a 508mm (20in) rim diameter intended for 28 × 4 straight side tyres. They were kept in place by a detachable rim, secured by twenty-four countersunk slotted head 6mm screws. These had first appeared on Bugatti's 18 June patent fitted to both sides of the wheel and were also a feature of Miller's aluminium wheel.

There was a further innovative dimension to the Type 35 and that was its presentation. This represented something of a milestone for a European racing car and, indeed, any previous Bugatti, as 'it was so well finished, with its nickelled axles and controls, its almost show polished engine and its shapely radiator, that it was the kind of toy which an enthusiast had only to see to desire.'[9]

Beautiful Bodywork

In the years up until 1924 and the arrival of the Type 35, little attention had been paid by the racing car makers to a Grand Prix car's appearance. The bodywork had often been hurriedly and poorly painted and there was little attempt to polish engine parts, axles or running gear. In addition, races were invariably held on dusty road circuits and the cars may have been driven to the event over the public highways. Henry Segrave probably spoke for many drivers when he wrote of the condition of his car, prior to the 1922 French Grand Prix at Strasburg:

> I was not at all satisfied with the appearance of the Sunbeam. It was not only painted indifferently – to say nothing of the scratches and bruises that its surface had suffered under the usual preliminary stresses of tuning up – but it was full of oil leaks and generally filthy.[10]

As a result he had the body removed, transported over the frontier at Kehl into Germany where a local coachbuilder cleaned and repainted it. Then it was reunited with its chassis and arrived at the circuit in uncharacteristically gleaming condition. Segrave memorably added: 'My own reaction to a clean, well-painted car has always been, "Well, if it doesn't go, it *looks* good anyhow." And, in nine cases of out of ten, if a thing *looks* good it *is* good.'

This yardstick could demonstrably be applied to the Type 35 but, prior to its arrival, Bugatti was no exception to the general rule, as photographs of his pre-1924 racing cars testify, and it was a factor that was particularly noticeable when Martin de Alzaga made his unhappy journey to Indianapolis with a team of Type 30s in 1923, the year prior to the 35's arrival. He later reflected that the presentation of the American cars was far superior to that of the ones from Molsheim. 'The Bugattis were very badly finished . . . The surprise was . . . when the US cars not only performed much better than us but *were like jewels from Cartier and finished like cars for an exhibition.*'[11] (My italics.)

There was another difference, in that the all-important 500 Miles Race had been held on the Indianapolis Speedway since 1911, rather than the public roads. The Indiana bowl was a brick course while circuits, made of wooden boards and derived

---

[9] William Court, *The Power and the Glory* (Macdonald, 1966).

[10] Segrave, *The Lure of Speed*.

[11] de Alzaga Unzue, 'Indianapolis 1923'.

When Bugatti introduced his type 35 in 1924, he went to great lengths to ensure that both its bodywork and engine were impeccably presented, in stark contrast to his contemporaries. This approach already existed in America as this Miller 122 engine of 1923 shows, and may well have had made an impression on Bugatti.

from a practice initiated on small bicycle tracks, were also peculiar to the American continent with its massive reserves of timber. This meant that there was an incentive for manufacturers to spend time and effort in visually preparing their cars for a race because they would not be immediately covered in clouds of dust, as was the case in Europe. De Alzaga pointed out, 'The Packard racing cars [built by Ralph de Palma] were jewels, as were the Millers' and such observations would no doubt have been relayed to Bugatti by de Vizcaya on his return to Molsheim.

Ettore could have had an opportunity of studying Millers first-hand later in 1923 when, it will be recalled, three of the Los Angeles-built cars had come to Europe during the summer, the year prior to the Type 35's appearance. The Los Angeles cars ran at the Italian Grand Prix and a single car participated in the subsequent Spanish Grand Prix. We can only speculate whether Ettore actually saw these Millers because no Bugattis took part in either of these races. The point is an academic one because he could easily have studied photographs of them in the motoring press.

The Miller Influence?

Miller was an outstanding engineer; moreover, his engines possess an architectural ingredient that is also to be found in Bugatti's and Ettore could have only responded to their lines, though it should be noted that the American had only produced twin overhead camshaft engines from 1920. But it was the cars' presentation that had so astounded the Bugatti team at Indianapolis. The extraordinary finish lavished on his cars was prompted by Miller's perfectionist approach, his obsessive desire to save weight and a sense of showmanship which would have also struck a responsive chord in Ettore.

According to Fred Offenhauser, Miller's laconic shop foreman, 'He didn't want to see machining or file marks on metal. All his castings had to be hand-scraped and polished to a satin finish and every part was machined. It took us from 6,000 to 6,500 man hours to build a complete car in the twenties. About half that time went into the engine, clutch and transmission, the balance into frame, running gear and body. Between 700 and 1,000 hours went into beautifying, just putting the finish on each machine.'[12]

Yet another Miller initiative came in 1923 with the arrival of his 122 single-seater which became the first racing car in the world to be offered for general sale. Bugatti had, of course, already marketed his Brescia Voiturette racer, though had only produced his Grand Prix cars in penny numbers in 1922 and 1923, as was the fashion of the day. All this would change with the arrival of the Type 35 which would be the first European Grand Prix car to be offered for general sale. Work on the model was under way by April 1924, as drawings of the component parts fall between that month and November, the latter referring to some minor refinements to the design.

Ettore clearly took the approaching French Grand Prix with great seriousness. On 17 March, he had written to his English friend Llewellyn Scholte, reminding him that, in the previous year, he had made arrangements with experimental engineer Harry Ricardo to supply him with some of his Discol high octane racing fuel. In the letter Bugatti reveals that, 'this fuel has not reached me. It is still at the customs and I have been threatened with the possibility of trial for importing alcohol into France.'[13] (In a second letter of 23 April, Ettore says that he had just 'obtained the import authorisation for the fuel' but feared that it would be unusable.)

Ettore may have been inspired by Leon Cushman's success in the 1923 Brooklands 200 Miles Race when, running on alcohol, he achieved second place at an incredible 146.6kph (91.1mph) in a Brescia. Bugatti wrote: 'I continually hear that my cars which use this fuel get the best results from it' and then suggested that 'If Mr Ricardo were to have enough confidence in my discretion, I would ask him to give me details of the composition of this fuel. I would guarantee that under no circumstances would I, either now or in the future, divulge the secret he had passed to me.'

Bugatti, however, recognized that in the instance of Ricardo not wanting to impart this information, he would make the necessary arrangements, 'to import between 3,000 and 4,000 litres [660 and 880 gallons] of this fuel so that I can participate in the Lyon Grand Prix with it'. Although Ricardo was perfectly prepared to let Ettore have the specification, the latter was approached directly by Shell, which had the rights to manufacture and sell the product in France and Bugatti may have got his Discol from that source.

The first batch of Type 35s amounted to ten cars, the consecutive chassis numbers running from 4323 to 4332, and fell within the Type 30 allocation. Of these, an unprecedented five cars were completed in time for the race, which was held on 3 August. As far as the drivers were concerned, de Vizcaya and Friderich were old hands. (The latter would be driving in his last major race, as he left Molsheim that year to represent Bugatti in Nice.) Ettore's fellow countryman Meo Costantini, would take over the management of the Bugatti racing team from Ettore in 1927, and the Italian drove another car. The experienced Jean Chassagne was a new recruit, having previously driven for Sunbeam in 1922 though he would only appear at the wheel of a works Bugatti in 1924. The team was completed by the presence of Leonico Garnier, a competent amateur and Hispano-Suiza agent.

At the Grand Prix

Now to the race itself. The AC du Rhône had applied to stage the Grand Prix in 1923. This had been agreed and the 1924 event was run over part of the same circuit which had been used for the celebrated 1914 race. There would be thirty-five laps of the 23km (14-mile) circuit, making a total of 805km (500 miles).

Bugatti had not been the only manufacturer busy over the winter of 1923/1924. In his home city of Milan, Alfa Romeo was planning its debut to Grand Prix competition by producing a team of what it designated its P2 racing car. This was the work of the masterly Vittorio Jano, a former Fiat racing engineer who had been wooed to the Milan

---

[12] Griffith Borgeson, *The Miller Straight-Eight* (Profile Publications, 1967).

[13] Scholte/Bugatti Correspondence.

Ettore Bugatti proudly poses at Lyon with his masterpiece: the Type 35. The appearance of the car, with its polished radiator and distinctive cast alloy eight-spoked wheels, was a departure from previous practice and this, combined with its exquisite proportions, resulted in the most beautiful of all racing cars. This Type 35 was driven in the race by Pierre de Vizcaya who crashed on the eleventh lap.

firm by the combined efforts of Alfa Romeo driver Enzo Ferrari and proprietor Nicola Romeo.

Not surprisingly the P2's engine closely followed that of the 805 Fiat, in that it was a supercharged twin overhead camshaft straight-eight, with unit construction gearbox and torque tube final drive. This was mounted in a similarly contoured chassis to the Type 35's and cloaked in a very Fiat-like two-seater body with the obligatory pointed tail. Its 140bhp gave a formidable 70bhp per litre, compared with the Type 35's 45.

The Fiats themselves were essentially similar to the previous year's cars though output was now rated at 16bhp more, at 146. The brakes differed from those at Tours, in that the previous untrustworthy hydraulic actuation hitherto employed was replaced by a less complicated mechanical arrangement. The only other straight-eight car present was Count Zborowski's lone Miller 122 and was, unlike the Fiats, unsupercharged. The remainder of the entries, like the Type 35s, all looked very Fiat-like. They consisted of the blown six-cylinder Sunbeams, the cuff valve Schmid and Delage's three V12s which, like the Bugattis, were unsupercharged.

The Type 35s were all driven the 420km (260 miles) to the circuit, the party consisting of six cars: the five entered and a spare with Ettore himself at the wheel. They were backed up with more than 1,500kg (30 tons) of spares and the facilities provided at the circuit were impressive, even by Bugatti standards. It is best described in Ettore's own words:

> The equipment was transported in three railway wagons and two lorries with trailers ... Everything was provided: the large tent had a wooden floor, there were proper beds for forty-five people, showers, running water to each hut, plenty of electricity. Cooking was done in a wooden hut and everything was provided to feed the personnel for nearly a month[!] ... There was also a caravan for my own family.[14]

The Grand Prix, which had the added kudos of also being the European Grand Prix, was the climax of four days of racing which began with an event for motorcycles and side-cars, was followed by one for bicycles with the *Grand Prix de Tourisme* on 2 August. The authorities had gone to some lengths to improve the state of the road circuit so, for the first time, the cars were not accompanied by the clouds of dust that had been such a feature of previous Grands Prix.

According to Ettore, the Bugatti entourage

---

[14] Conway, *Grand Prix Bugatti*.

Bartholomeo Costantini 1889-1941

Although Ettore Bugatti boasted many acquaintances, he had very few friends, but one of his closest confidants was Bartholomeo Costantini, who was universally known as Meo. Born in the small Italian town of Vittoria Veneto on 2 February 1889,* he was the second of Carlo and Eugenia Costantini's children – there was already a daughter, Marianna who was a year older, and later a younger brother named Severino.

A bright child, Meo did well at school but soon displayed a strong mechanical bent and he was sent to technical college in Turin. After graduation, he took a job with the Brigita Ferrovieri Genio, a Turin railway company. Following war service in Libya, where he became a qualified pilot, Costantini entered his first major motor race, the 1914 Targa Florio. His car was the advanced Aquila Italiana, designed by the talented Giulio Cappa. Costantini did not finish the race because of engine trouble and he was similarly unlucky in the Coppa Florio of the same year. But he won his class and was fourth overall in the Parma Poggio di Berceto. He participated in the 1914 French Grand Prix but withdrew after only one lap when he was, once again, let down by his engine.

During the First World War, Costantini joined the Italian Air Service. Later promoted to the rank of Captain, he was credited with having shot down six enemy aircraft and was decorated for this achievement.

Costantini pictured at the 1925 Italian Grand Prix.

At the end of the war, Costantini joined Fiat, where Cappa had gone in 1914 as head of its technical department. The latter was working on a Bugatti, which represented Meo's introduction to the marque, and it was in this car, which had modifications to its pistons and engine, that Costantini used in the 1920 Galaratte race where he achieved third place. Later in 1921 he drove a Bugatti in the Poggio di Bercetto mountain race though, by the time that he was placed second at the Garda event in May, he was competing, probably for the first time, in a works car. Ettore had heard of Costantini's exploits and it was in 1921 that the thirty-two-year-old Bartholomeo arrived at Molsheim intending to stay for a short period though was to remain there for no less than fourteen years! As a member of the Bugatti team, he received no salary and later took up residence in the *Hostellerie du Pur Sang* though he often ate with the Bugatti family at its villa. Not only was he to prove himself a highly effective racing driver he was also able, as a qualified engineer, to make informed comments relating to the design of Bugatti cars. A strong advocate of the supercharger, Meo's was an argument to which Ettore eventually had to concede.

Costantini's most celebrated victories were in the Targa Florio – he won the gruelling race twice, in 1925 and 1926. W F Bradley had an opportunity of witnessing his first triumph and has recorded his impression of the Italian's driving, '. . . he was so completely in sympathy with his car, his estimate of speed was so precise that there was no unnecessary braking, and his gear changes were made with such wonderful accuracy of timing, that he did not give the impression of moving really fast. Time him over a given stretch, however, or, better still, get behind him in a car of equal power, and it was quickly realized that he was an uncommonly hard man to cope with.'+

It was after the celebrated World Championship year of 1926, that Costantini took over from Ettore as Automobiles Bugatti's racing manager. A strict disciplinarian, he insisted on unswerving obedience from his drivers. When one, no less than Louis Chiron, refused to stay in the hotel that Costantini had specified during the Monza Grand Prix, he was instantly dismissed and his place taken by a youthful René Dreyfus.

But as the Golden Age of the 1920s gave way to the bleaker 1930s and Bugatti's ascendancy on the motor racing circuits of Europe began to wane, Costantini must have become increasingly dispirited. He was, nevertheless, able to be a great support to Jean Bugatti with whom he had struck up a firm friendship. Then one day in 1935 Meo left Molsheim, departing, says W F Bradley, 'as quietly as he had come'.

He returned to his native Italy and his last competitive involvement came in 1940 when he managed the Alfa Romeo racing team for that year's Mille Miglia race though the Italian cars were trounced by the German BMWs.

The following year, Meo Costantini died during the night of 19/20 July; a heavy smoker, he was invariably photographed with a cigarette in his mouth, he had succumbed to lung cancer at the age of only fifty-two.

* *See* Bartholomea Costantini and Bugatti by Hans Schelegel, Pur Sang, Volume 30, Number 2, Spring 1990
+ Targa Florio by W F Bradley, Foulis, 1955

'arrived at Lyon the day before the first practice runs over the closed circuit. No work on them was necessary but the drivers washed and polished their cars from sheer pride to have them in immaculate condition for the start of practice.' This appears to have passed without incident.

Expectations were clearly high, and the optimism was noted by S C H 'Sammy' Davis, sports editor of *The Autocar*, who was to be riding mechanic in the Zborowski Miller. He recalled: 'The Bugatti encampment – the new and very fascinating aluminium-wheeled 2-litres were housed with the team in big tents – was always open to official visits from other teams, during which more lies were told about performance than we ever heard before.'[15]

Alfa Romeo's Challenge

The event, held in superb weather, was initially dominated by the Sunbeams though they suffered from the last minute fitment of new Bosch magnetos and Segrave could manage no better than fifth place. The Fiat threat did not materialize, mainly because of braking problems, and none of the cars finished. This left the field open for the new Alfa Romeos. Ascari in a bob-tailed P2 got into the lead on lap 9 and intermittently remained there until four laps from the end and looked like the winner. Then he stopped at his pit for water but was unable to re-start the car because of a cracked block, despite some frenzied efforts at push-starting it by his mechanic. However, Campari in the other P2 won, giving him and Alfa Romeo their first major international victory. Delages were second and third and another Alfa Romeo fourth.

Fiat's chairman, the great Giovanni Agnelli, was outraged. In a clear reference to the Alfa Romeo victory, he declared: 'With others stealing our best brains, we have become merely a training school . . .'[16] Fiat therefore withdrew from Grand Prix racing and, apart from a brief return in 1927, the Fiat name would never reappear, a state of affairs that continues to this day.

As will be apparent, the Type 35s did not feature in the first four cars home. The highest placed Bugatti was Chassagne in seventh place. Friderich was eighth and Garnier was flagged off an unceremonious eleventh.

So why did the cars on which Bugatti had lavished so much care and attention fail to make an impact? It seems that the problem lay with 'the special Dunlop racing tyres ordered by Bugatti and made at such short notice that they had to be flown to the Continent from Birmingham.'[17]

What happened in the race is best told by Ettore himself for, after his return to Molsheim, on 1 September, he set down an account of his experiences at Lyon, which was then circulated to his agents. De Vizcaya suffered from a flat tyre on lap 1 though Bugatti was not unduly perturbed by this because his new wheels had been designed 'to operate without air in the tyre'.

This account is confirmed by *The Motor*, which recorded:

> De Vizcaya rushes skidding up to his pit with the left back wheel burst . . . The left wheel is off and de Vizcaya is changing the right one also. The smell of burning rubber floats across to the stands, and there are whispers that the trouble, prophesied by the tyre experts owing to the fact that the Bugattis were using immature covers of a special size to fit their cast aluminium wheels, has set in.[18]

Bugatti continued: 'at the end of the third lap, Mr de Vizcaya came in again with a tread off the tyre. There was not a bit of rubber left. The adhesive that joined the carcase to the tread was soft and could be rolled into little balls, showing the bad state of vulcanization . . . I then realized that the race was lost as far as I was concerned.'

Chassagne came in on lap 4 to change his rear wheels. On lap 16 Meo Costantini's car also

---

[15] S. C. H. Davis, *Motor Racing* (Illife, 1932).

[16] Doug Nye, *Famous Racing Cars* (Patrick Stephens, 1989).

[17] Cyril Posthumus, *The Roaring Twenties* (Blandford, 1980).

[18] *The Motor*, 5 August 1924.

suffered similarly, the tread having wrapped itself around the gear lever, making it impossible for the driver to engage second and fourth gears. Martin de Vizcaya had already crashed at the *Givors* corner on the eleventh lap and his car was badly damaged. This left three cars running. Rather than risk withdrawing them, Ettore could only mount a face-saving operation and instruct his remaining drivers to keep their speed down and so be sure of finishing the race. The surviving cars therefore continued to run at speeds that were far less than their potential.

That Tyre Trouble

Why did Ettore forsake his traditional Michelins and opt instead, at a late hour, for the Dunlops which would appear to have been insufficiently cured? There is a clue contained in the circular which Bugatti issued to his agents after the race, when he stated, 'I reminded my drivers to be careful, as in tests made previously with other tyres, the throwing of a tread endangered the life of the driver. If a driver lost pieces of tread, they might wrap around the steering wheel, this had happened to Chassagne, and caused a serious accident.'[19]

Bugatti infers that this had come from a front wheel though such an incident did not occur during the race. Nevertheless this was probably sufficient reason for Ettore to change tyres, though it seems likely that he used other covers prior to the event because there are no recorded instances of such trouble on the run from Molsheim to Lyon or, indeed, during practice. The problem seems to have manifested itself, almost instantaneously, on race day.

Ettore was also quick to dispel any rumours that his new broad-spoked aluminium wheels had been in some way responsible for his difficulties. 'There are some who thought that my wheels were the cause of the tyre trouble. It is sufficient for me to say that the tyres on the front wheels never budged, so that it was not a question of the wheel, nor

[19] Conway, *Grand Prix Bugatti*.

The dashboard of one of the Lyon cars. These differed from subsequent examples in that they had a pull/push advance/retard lever for the magneto. The other instruments, left to right, are petrol gauge, clock, oil gauge, with the rev counter positioned above the steering column.

of heat.' This statement is borne out by the fact that the wheels remained essentially unchanged for the next two years. If they had been faulty, modifications would have been made to the design almost immediately after the race. But the first recorded alteration to the wheels after Lyon dates from July 1926 and is mainly related to the locking ring being modified.

So ended the Type 35's disappointing racing debut. Early in September 1924, Ettore informed Vincent Junek that he had recently driven an example from Strasburg to Paris in five and a half hours. This had included a detour, which gave him an average speed of nearly 100kph (60mph) and he found that 'all tyres will give good results except those of Dunlop. My trip . . . was made on 710 x 90 covers.'

By this time he had decided not to enter his new cars for the Italian Grand Prix at Monza in October, no doubt considering it to be an Alfa Romeo benefit, which indeed it proved to be. But the model did find its form a month after Lyon at the San Sabastian Grand Prix on 25 September. Although five cars had been entered only three Type 35s, now shod with 710 x 90 Michelins, appeared.

8 Cylinder 17.8 H.P. (2 Litres) Grand Prix Type 35

17.8 H.P. TYPE 35. GRAND PRIX

Specification

ENGINE.	8 cylinder, cast in two sets of four and in line, bore 60 m/m, stroke 88 m/m, cooled by pump circulation, overhead valves, operated by overhead cam shaft, two inlet valves and one exhaust valve to each cylinder, three ball bearing crankshaft bearings, forced feed lubrication by pressure pump, twin solex carburetter pressure fed, ignition by magneto, variable advance
CLUTCH.	Multiple (cast iron and steel) discs (Bugatti patent).
GEAR.	Gate, right hand control, four forward speeds and reverse, direct drive on top.
BACK AXLE.	Bevel drive, ratio 14 × 54.
STEERING.	Worm and helical wheel, irreversible and adjustable with ball and socket connecting rods.
FRONT AXLE.	Of circular section, with front springs passing through the axle.
SUSPENSION.	Front springs semi-elliptic, rear ¼ elliptic anchored at the rear end of chassis and extending forward to the rear axle, thus working under traction (Bugatti patent).
BRAKES.	On all four wheels, operated by pedal, and rear brakes only operated by hand.

[*Over*

Specification—cont.

EQUIPMENT.	Spare aluminium Rudge wheel complete with tyre, shock absorbers on all four wheels, grease gun lubrication and tool kit.

Bore	60 m/m
Stroke	88 m/m
Wheel base	7'10¼"
Track	3'9"
Over-all dimensions	12'1" × 4'9"
Petrol tank capacity	20 galls.
Wheels and straight edge Tyres	710 × 90

As prices are subject to fluctuation with the French Rate of Exchange, current prices will be found on a separate price sheet attached.

The Type 35, as it appeared in the Bugatti catalogue for the British market in 1926. It was listed at £1,100.

They were driven by Costantini, Chassagne and de Vizcaya, with clear challenges coming from Delage and Sunbeam.

Race day was wet, which resulted in crashes by one of the Delages, a Sunbeam and a Mercedes. Segrave was somewhat surprised to win though Costantini drove an excellent race to come in second and also put up the fastest lap at 115.3kph (71.70mph). There were only six cars left at the finish. Delages were third and fourth with the Bugattis of de Vizcaya and Chassagne fifth and sixth respectively. Thereafter the Type 35s were driven to Paris, where they were available for the use of prospective customers. The humiliations of Lyon were set aside and Bugatti circulated an account of his San Sabastian success to his dealers, concluding: 'I hope that the year 1925 will prove more favourable to my efforts.' And so it did.

The Type 35A offered owners the looks of the Grand Prix cars both outwardly and under the bonnet. In 1926 the Type 35A sold for £675 in Britain, which was £425 less than the pukka 35 which cost £1,100. This example was owned by Lord Cholmondeley.

Although wire wheels were standard on the 35A, aluminium wheels could be purchased at extra cost. Lord Cholmondeley's car has this rather unusual lining and has been equipped for road use.

A Cheaper Version

Deliveries of the Type 35 had begun in earnest in December 1924 though, at 100,000fr. (£1,136), it could only be bought by the wealthiest of drivers. In May 1925 Bugatti therefore introduced a cheaper version, designated the Type 35A. This retained the outward attributes of the Grand Prix car but was fitted with the Type 30's three bearing crankshaft in the Type 35's crankcase. Blocks of the Type 30 small valve variety were employed, the eight developing 70bhp at 4,000rpm, or about 20bhp less than the Grand Prix cars. Another departure was the fitment of coil ignition, the distributor cap initially projecting into the dashboard in much the same way that the magneto did on the Grand Prix cars. It was later driven directly off the end of the camshaft.

The mechanicals limited the model's top speed to around 140kph (90mph) and although wire wheels were fitted as standard, aluminium ones were available at extra cost. A solid front axle was fitted. The 35A cost 65,000fr. (£738) or 35,000fr. (£397) less than the Grand Prix car and the 35A, known as the *Course Imitation* (imitation racer) at Molsheim, proved to be a strong seller and decisively outsold its progenitor. No less than 124 were built between 1925 and 1927. This compared with 87 examples of the 2-litre unblown Grand Prix car built over the same period.

The same month of the Type 35A's arrival, Bugatti entered the Targa Florio for the first time.

A Type 35A engine restored by Jarrot Engines of Chalford, Stroud, Gloucester. The water pump can be seen on the left of the engine with the oil pump below it and the oil filter positioned between the exhaust manifolds.

A contemporary photograph of the inlet side of a Type 35A engine with twin Solex carburettors. The entire unit looks exceptionally neat and the plug leads are contained within the circular conduit. The drive on the bulkhead is for the revolution counter.

The extraordinary narrowness of the Type 35 engine is revealed in this head-on view of the unit. The tubes in the sump were intended to cool its contents though lubricant tended to congeal around them.

The 35A crankshaft was similar to that used in the Type 30 and also the Type 38. The shaft itself, with its circular webs, was machined from solid and was fitted with three double self-aligning ball races. The drive for the overhead camshaft is in the foreground.

Held on 3 May 1925, it was staged in Sicily, had been run since 1906 and was one of the most gruelling road races in the world. From 1919 it had been held over four laps of a 108km (67.1-mile) circuit but this was the first year of a five-lap race, making a total of 540km (335.5 miles). It proved an ideal venue for the Type 35, where sheer power was not necessary but superlative roadholding and reliability would hold sway.

Triumph at the Targa

A team of three works cars were driven by Meo Costantini, Pierre de Vizcaya and the latter's brother, Ferdinand. The main challenge came from a quartet of new 4-litre Peugeots, which were twice the capacity of the 35s, and there were factory entries from Tatra and Bianchi. Ironically it was the Peugeots that suffered from tyre trouble and when d'Auverge in the leading car crashed, his team-mate Wagner stopped to help rescue him from the resulting fire.

This allowed Costantini through to win, and he also recorded the fastest lap of 73.06kph (45.4mph) though Wagner was only five minutes behind him in second place. Another Peugeot was third and Pierre de Vizcaya fourth in a Type 35.

This gave Bugatti his first success in an international event and the marque was to dominate the Sicilian races for the remainder of the decade, the 1925 victory heralding the first of five consecutive wins there.

The Targa Florio was, of course, held on public roads but the early 1920s saw a trend in Europe towards purpose-built tracks for motor racing. The Brooklands circuit at Weybridge, Surrey had been the world's first in 1907, the American Indianapolis Speedway followed in 1909, though it was 1922 before Italy acquired its Monza Autodrome. Spain's Sitges bowl of 1923, alas, proved to be excessively dangerous and in 1924 France opened its own banked track, built at Montlhéry, near Paris. Although the first race to be staged there was held in October, on 17 May 1925, only two weeks after Bugatti's Targa Florio victory, Ettore entered a pair of cars for the *Grand Prix de l'Ouverture*, staged at the new facility to celebrate its opening.

Curiously Ettore did not field his Type 35 for this meeting but produced two single-seater cars, which carried the Type 36 designation. Although riding mechanics were no longer carried, Grand Prix cars still had to be two-seaters even if the second one was not used. However, early in 1925 the AIACR (Association Internationale des Automobile Clubs Reconnus) governing body hinted that for the following 1926/1927 season, engine capacity would be reduced to 1.5-litres, regardless of a supercharger being fitted. Also, the two seats would be reduced to a single one, as was then current in America and minimum weight would be reduced to 550kg (1,212lb).

Although these rules were subsequently modified, the Type 36 Bugatti can be considered to have been produced in response to them. Its engine was a 1,500cc unit and it thus represented the first appearance of a capacity though not the internal dimensions of the later Type 39. It was effectively a small, bored, 52 x 88mm version of the Type 35 engine but the gearbox was not located in its customary position and was incorporated in the rear axle, in the manner of the 1923 Tanks.

The suspension was also unorthodox and Bugatti maybe considered that it was tailored to the smooth Montlhéry concrete. It was limited at the front by a pair of longitudinally located quarter-elliptics which echoed Tank practice. It was even less apparent at the rear with only rubber blocks positioned between the axle and chassis. The wheelbase was accordingly slightly longer than the standard Type 35 at 2,500mm (8ft 4in), rather than 2,400mm. The theory proved itself flawed and *The Autocar* reported that 'at high speed the little car bucked and jumped to such an extent that it would have been impossible for any driver to sit on it for more than a few minutes.' So the cars did not run but would appear in suitably modified form on at least one occasion in the following year.

The next major event on the international calendar was the first Belgian Grand Prix at Spa, held on 28 June and, although Pierre de Vizcaya and Costantini were entered in works Type 35s, their cars were not, apparently, ready. This was probably not unconnected with the presence of a trio of P2 Alfa Romeos and two of them came in first and second.

In the Front Line

Ettore was there, however, for the 1925 French Grand Prix held on 26 July at Montlhéry for the

A splendid driver's eye view from the cockpit of Hamish Moffat's Type 35.

first time. Four Type 35s, chassis numbers 4571 to 4575, were entered to be driven by Pierre and Ferdinand de Vizcaya and Costantini, while forty-year-old Jules Goux (whose motor racing career stretched back to the 1907 Peugeot team) and the Italian Giulio Foresti, formerly of OM, were new recruits to the team.

Prior to the race, Bugatti was engaged in a dispute with the ACF over his use of flush fitting covers over the mechanics' empty seats. In the event, the ACF backed down and all the Type 35s participated, so equipped. The Grand Prix was marred by the death of Ascari, who had come so close to winning the previous year's event and the race was won by Robert Benoist in a supercharged Delage with another V12 in second place and a Sunbeam third. Costantini was the best placed Molsheim car in fourth place, followed, in order by Goux, Ferdinand and Pierre de Vizcaya and Foresti. Once again Ettore had shown the reliability of his cars in this longest-ever eighty-lap 1,000km (620 miles) race but he could not have failed to notice that supercharged cars had taken the first three places.

However, Ettore had some satisfaction in taking the first four places in the *Grand Prix de Tourisme*, held on the Montlhéry road circuit on 19 July, a week prior to the race proper. For this event he constructed a team of five special cars powered by 1,500cc engines – so they are technically Type 39s. Their internal dimensions differed from the previous year's 1,500s, in that they were 60 x 60mm. As petrol consumption was a key element in the event – the cars having to return 8.5km/l (24mpg) – single Solex carburettors were fitted and 125-litre (27.5-gallon) petrol tanks in place of the customary 22-gallon 100-litre ones. The bodies, designed specially for the event, which was intended for sports cars, were accordingly much higher than the Grand Prix ones, with a single left-hand door and stylish wings. The presence of headlights, windscreens and hoods perpetuated the illusion of a road car.

Afterwards, some touring bodies were removed and replaced with conventional Grand Prix ones and the 1.5-litre cars participated in the Italian Grand Prix at Monza on 6 September. They were driven by the de Vizcaya brothers, Meo Costantini, Jules Goux and Giulio Foresti. Not only did Bugatti win the *voiturette* class but Costantini came in third behind a pair of P2 Alfa Romeos. This success gave the Milan company the newly instituted racing World Championship and, having received its victory laurels, they were added to the marque's radiator badge where they have remained ever since.

Alfa Romeo did not bother to compete in the Spanish Grand Prix later in the month though the Bugatti team was once again running in 2-litre form. The race proved a triumph for the V12 Delages, which had not been at Monza, and they took the first three places, followed by Pierre and Ferdinand de Vizcaya's Type 35s.

This was the final race of the highly successful 2-litre formula, which had been in force since 1922. The following year of 1926 saw the arrival of a 1.5-litre limit which marked the beginning of a decline in the competitive element of Grand Prix racing, as the number of firms prepared to enter the fray gradually dwindled. Fiat, it will be recalled, had withdrawn in 1924 as did Alfa Romeo at the end of 1925, selling two of its all-conquering P2s to two Italian enthusiasts and a further pair of Swiss drivers. Also, after five seasons, Sunbeam departed from the Grand Prix scene late in 1925 though there were some new and, as yet unproven, Talbot–Darracqs from the same stable. More significantly, Delage was creating a new eight which did not find its ultimate triumphal form until 1927. In such circumstances, Bugatti was to enjoy his best ever racing year.

It will be recalled that, at the beginning of 1925, there had been talk of the 1.5-litre formula of 1926/1927 being for light, single-seater racers but the outcry was that this would result in cars of formidable potency that only the most accomplished drivers could control. In July 1925, the AIACR modified its rules and, while the 1.5-litre limit was retained, bodies would be those specified in 1925, in that they would be 78cm (31in) wide and, with the second, unoccupied mechanic's seat remaining. This could be cowled, which echoed Ettore's altercation at the 1925 French Grand Prix! Minimum weight was raised to 700kg (1,543lb).

At Last a Supercharger

Since the Type 35's arrival in 1924, Bugatti had yet to win a Grand Prix and he was the only front-line manufacturer to be running with an unsupercharged car. He had not initially fitted one to the model because its intricacies were probably beyond his essentially empirical approach to design and he did not want to sacrifice the reliability of his engine, which was a key ingredient in his racer's relative success.

Work on the project had begun in mid-1925 and Ettore recruited the services of a Paris-based Italian engineer named Edmund Moglia from his native Milan. He had previously been associated with the Sunbeam-Talbot-Darracq (STD) combine and has been credited with the design of the extraordinarily successful 1.5-litre *voiturette* Talbots, which had so successfully trounced Ettore's Brescias, and everything else for that matter, in the 1921/1922 era. Bugatti's employment of a fellow countryman is perhaps significant for although Ettore ran his cars under France's racing colours, he perpetuated his Italian citizenship and the supercharged Type 35 remained a wholly Italian designed car.

Factory drawings of blower rotors are dated from August 1925 and it was reported that Bugatti would enter a team of supercharged 1.5-litre cars for that year's Brooklands 200 Mile Race on 29 September. But they failed to appear and the first supercharged Bugatti did not make its public debut until eight months later.

The date was 30 May 1926 and the occasion was Bugatti's local Alsace Grand Prix, held over the 1922 French Grand Prix circuit. It was a wet day and, alas, the three Bugattis virtually constituted the entire field, the only other runners being two Sima-Violets, which soon dropped out. This state of affairs would become increasingly common throughout the remainder of the decade. There were two Type 36 single-seaters, by now employing conventional suspension and gearbox location though instead of the 1,500cc engine of the previous year, there was another variation in the shape of a 51 x 66mm, 1,092cc version. The third car, by contrast, was a conventional two-seater, although also of 1,100cc and was driven by André Dubonnet, who won the race.

One report of the Alsace event credited the blowers being belt-driven but, whatever the arrangement employed there, the conventional method of locating the Roots-type supercharger on the Bugatti's straight-eight was to position it between the right-hand side of the crankcase and chassis rail. It was driven at engine speed via a short drive shaft from two new additional gears, one of which meshed directly with the crankshaft. This meant moving the radiator slightly forward of its original position to accommodate it.

A single Zenith or Solex updraught carburettor was located at the base of the supercharger and the mixture, compressed at 10psi, reached the engine via two water-heated inlet manifolds.

Laurence Pomeroy has pointed out that less desirably: 'This layout which involves the gas being turned through six right angles during the course of its passage from the blower to the valve head, is scarcely conducive to the utmost efficiency, but on the other hand, the scheme had a neat appearance and gave reasonably good distribution as between one cylinder and another.'[20] It was surmounted by an outlet for the blower relief valve, the location being proclaimed by an accompanying hole in the bonnet. Its presence is therefore the easiest method of telling whether a Type 35 is supercharged or not, though when the model was superceded by the outwardly similar twin overhead camshaft Type 51 in 1931, the blow-off hole was accordingly lowered. Observing these respective positions is the easiest outward method of differentiating between the two models.

The supercharger was then extended to the Type 39 for the 1926/1927 formula, the resulting

---

[20] Laurence Pomeroy, *The Grand Prix Car, Volume 1*.

The radiator was moved forward to accommodate the blower drive gear which was located at the front of the crankshaft. The presence of the supercharger meant that the steering box was moved both upwards and backwards and the position of the drop arm is another method of telling whether the car is supercharged or not. This is a contemporary photograph of a 35B engine.

model being designated the 39A. It was then fitted to the 2- and 2.3-litre cars called the Types 35C and 35B respectively.

Faster! The Blown 35

The provision of a supercharger resulted in a considerable improvement in the Type 35's performance. It had developed 90bhp in its original 2-litre form and was capable of something over 160kph (100mph). The supercharged derivative, the 35C of 1927, developed about 125bhp while the ultimate 140bhp, 2.3-litre 35B of the same year was a 125kph (201mph) plus car.

However, the Type 35 ran in unsupercharged form for the first major race of 1926, the Targa Florio, held on 25 April. That year's event was particularly significant because Bugatti developed a more powerful version of the two-seater, which was designated the Type 35T (for Targa). The 2-litre engine was increased to 2,262cc by upping both the bore and stroke to 60 x 100mm. There were three cars driven by Costantini, Goux and Ferdinand Minoia, an Italian with a proven track record, who was making his debut as a works driver.

On this occasion the Bugattis were faced with a trio of V12 Delages while the new Maserati marque was making its racing debut. But once again the Molsheim cars' superior roadholding held sway. W F Bradley, covering the race for *The Autocar*, later recorded that, 'Costantini appeared to be a unit with his machine, showing comparatively little physical effort, placing it exactly where it should be, braking vigorously and firmly, never skidding wildly and accelerating with marvellous rapidity.'[21] The Delage threat never materialized; one of the cars crashed and its driver was killed. Costantini therefore won for the second year running, with Minoia second and Goux third. Privately entered Bugattis were placed fifth, tenth, eleventh and twelfth, underlying the marque's increasing popularity amongst amateur drivers.

The most prestigious race of the year, scheduled for 27 June, was the French Grand Prix. This was to be held at the new Miramas circuit, near Marseilles. Unfortunately by the time that entries closed at the end of February only two manufacturers, the Sunbeam–Talbot–Darracq combine with its new Darracq and Sima-Violet had entered. But 'The ACF had neglected to include in the regulations the normal clause to the effect that, unless a

[21] Bradley, *Targa Florio*.

Madame Junek at the wheel of a Type 39 in the 1926 Zbraslav-Jiloviste hill climb. This was one of the cars which Bugatti ran in the 1925 Touring Grand Prix which was a sports car event, hence these unusual high-sided bodies, with single passenger door and rudimentary wings and electrical equipment.

certain number of cars was entered, the race would be cancelled.'[22] With a week to go prior to the race, the Club 'declared that it would be run even if only one car entered'.

By this time STD had withdrawn and the only certainty was entries from Bugatti and Sima-Violet. In the event the two-stroke, flat-four cylinder car from Paris did not appear, leaving the field open for the three 52 x 88mm Type 39As. These 110bhp cars were to be driven by Goux, Costantini and Pierre de Vizcaya and represented the first appearance of the supercharged Bugatti, with its enlarged radiator, at a front-line Grand Prix.

The farce was extended to the event itself. This was a 310-mile (500km) race and de Vizcaya led from the outset but retired on lap 46, his tank having been filled with a fuel with a high benzol content. So regrettably had Costantini's. He also recognized his limitations and kept his revs down and de Vizcaya's retirement saw him slow even more. Goux in the third Type 39A, who seems to have been supplied with less volatile fuel, went on to 'win', while Costantini kept going and was flagged off after eighty-five laps. So Bugatti won his first Grand Prix, though the 1926 event had gone down as 'the worst fiasco in racing history'.[23] It was the hollowest of hollow victories, though Ettore can hardly be blamed for the absence of competitors.

[22] David Hodges, *The French Grand Prix* (Temple Press, 1967).

[23] Posthumus, *The Roaring Twenties*.

MASTERPIECE

The 1.5-litre Type 39A was produced for the formula that began in 1926 and was supercharged though the small radiator and brakes drums continued. This is the European Grand Prix at San Sabastian on 18 July 1926 with Ferdinando Minoia in the foreground and Jules Goux behind. The latter won . . .

. . . and is seen here with victory laurels after the event.

186

In addition to the Grand Prix there was a 1,100cc race and the entry included a pair of new supercharged four-cylinder Salmsons, which prompted the one and only meeting between Bugatti and the company's chief engineer, Émile Petit. After the main event which Goux had won at 109kph (68mph), Ettore sought out his contemporary and 'carefully told him that, on no account, was he to allow one of the Salmsons to win the 1,100cc event at a higher speed than the Grand Prix.'[24]

This placed Petit on the horns of a dilemma. He did not want to displease such a personality as Bugatti but, conversely, he was loathe to risk defeat against his long running Amilcar rival which was fielding its new six-cylinder models. The race itself attracted about thirty starters and thus became the main event of the day. 'Petit asked that the prize money be transferred to the 1,100cc event, but this was refused and Bugatti left with all the cash and the fastest average of the day.' For Casse, in one of the new Salmsons, had won at an average speed of 105kph (65mph), which was only 4.8kph (3mph) less than the Bugatti's!

San Sabastian, Spain was the venue for the next round of the championship which also had the added status of being the European Grand Prix. Molsheim entered a team of three Type 39As, their cylinder dimensions briefly standardized at 60 x 66mm and driven by Costantini, Goux and Minoia. Fortunately there was no repeat of the Miramas debacle because the new 1.5-litre Delages were ready for this event, their eight-cylinder twin overhead camshaft supercharged engines containing no less than sixty-two ball and roller bearings, and developed a formidable 110bhp per litre, compared with the 39As' 73.

Debut of the Delages

These low, wickedly purposeful cars suffered from a great design fault that was exacerbated by the searing 43°C (110°F) temperature of northern Spain. Unlike the Bugattis, the exhaust pipe of the Delages ran along the offside of the car, which was very close to where the drivers sat. This made them travelling ovens and the Delages' luckless occupants had to continually retire during the race, having been overcome by the effects of heat and physical burns. But in the latter stages of the event, nature took a hand and a fresh wind blew across the circuit.

Despite this, Goux in his Bugatti won at 113.4kph (70.5mph), with one of the Delages, which had had no less than three drivers during the course of the race, second and the Type 39As of Costantini and Minoia third and fourth.

The Delage team was present in force for the first British Grand Prix, held at Brooklands on 7 August, along with the new Talbots which, despite their impressive appearance, failed to find their form and all retired. Regrettably Bugatti did not field a team though Malcolm Campbell, who held the British concession for Grand Prix Bugattis, entered one of the San Sebastian cars, chassis number 4810. The night before the race, Campbell changed the handsome aluminium-spoked wheels for wires with Rudge-type knock-off hubs as he thought that the originals would not stand up to the pounding of the uneven Brooklands concrete. Campbell drove an outstanding race and finished second between two still overheating Delages.

The final round of the championship was the Italian Grand Prix at Monza on 5 September though both Talbot and Delage did not enter, while an expected OM team was withdrawn. The three Bugattis were only challenged by a pair of Maseratis, which soon dropped out, and an ageing Chiribiri, which caught fire. Wisely the organizers had decided to run the event in conjunction with what was to have been a supplementary 1,100cc race but it was, in truth, a Bugatti benefit. The cars were driven by Costantini, Goux and a French amateur named Louis Charaval, who raced under the pseudonym of 'Sabipa'.[25] A Molsheim victory was

[24] Chris Draper, *The Salmson Story* (David and Charles, 1974).

[25] Charavel was so called because, as a wealthy young man, his family and business associates wanted to prevent him from racing, so when he went to register with the ACF, he responded '*Sa bi pa*' which means 'I don't know' in a local dialect. He was surprised to find this entered as a name and raced thereafter under that pseudonym.

Louis Chiron pictured at Montlhéry on 2 July 1927 in the Type 35C in which he was placed second in the Formule Libre *race.*

inevitable and, although Costantini seemed destined to win after leading from the start, breaking the lap record on three ocassions, two and a half laps from the finish his engine seized. This allowed 'Sabipa' through and he went on to win. Costantini, his car sounding very rough, came in second.

This gave Bugatti the World Championship. He had won three Grands Prix, Delages had won one and Miller one – the Indianapolis 500 being included in the series. The reality was that this triumph had been gained in the face of a limited and unlucky opposition. Not surprisingly the coveted accolade was reflected by record numbers of the Type 35 and its variants being sold during the year.

The year 1927, however, was destined to be a very different story and the pattern was set by the first non-championship race of the season, held at Montlhéry on 13 March. The *Grand Prix de l'Ouverture* saw the first appearance of the revised 1.5-litre Delages, now with repositioned nearside exhaust pipes, output of the now offset straight-eight upped 5bhp to 170bhp and handsome sloped radiators. Robert Benoist ran away with the race, and the privately entered Bugatti of Lescot was second but two laps behind. Eyston and Esclasson, both in Bugattis, were third and fourth another four laps back.

The Targa Florio held on 24 April was not, of course, a championship event and it witnessed the first occasion on which the masterly Meo Costantini forsook driving for management of the Bugatti team from Ettore himself. The three cars were the new Type 35Cs, driven by Minoia, Dubonnet and, for the first time, by the talented Italian Emilio Materassi. He was destined to win, giving Bugatti his third victory in the Sicilian race, while Count Caberto Conelli in a privately entered four-cylinder supercharged Type 37A was second, though Maserati in a Maserati was third. Nevertheless, Molsheim cars occupied five of the first eight places.

A Tactical Withdrawal

It was a different story at the French Grand Prix, held at Montlhéry on 27 June. The Delage team was there in force, along with the revised Talbots. Ettore had been once again experimenting with the cylinder dimensions of his 1.5-litre, the 54 x 81mm cars being designated Type 39D by the factory. They were due to be driven by Dubonnet, Materassi and Goux but a practice session soon revealed the Delage's supremacy. Just before the race, Bugatti announced his withdrawal from the event, which produced jeers and catcalls from the 100,000 spectators and shouts of *'Il a peur'* (He's scared). This left a mere seven cars in the race; a straight fight between Delage and Talbot. Not only were the Delages victorious but the vanquished Talbot were subsequently withdrawn from the championship, leaving only Bugatti to oppose the Delage for the remainder of the championship events.

The cars from Courbevoie, Paris, went on to win every Grand Prix that they entered and this meant the Spanish, French, Italian and British races. Their only real opposition came in the Spanish Grand Prix when Materassi's 39A put up a spirited challenge to Benoist's Delage. Materassi, however, hit a wall nine laps from the finish but Count Conelli's works Bugatti was second, so preventing Delage from taking the first three places. Bugatti had given the Italian Grand Prix a miss, though the team of Materassi, Conelli and Divo was present at the British Grand Prix. But they were unable to stop Delages coming in first, second and third, so confirming, as if it was needed, their World Championship status.

This was the final year of the 1.5-litre formula. It was replaced by one which took no account of engine size, but weight was restricted to between 560 and 750kg (10.8 and 14.7cwt). However, no new cars were built for it and most organizers ignored it, staging instead *Formule Libre* (Free Formula)

Four Type 39As just about to enter the Members' Banking at Brooklands during the 1927 British Grand Prix. From left to right, Chiron, Materassi, Conelli and Campbell. Note the luckless film cameraman in the right foreground!

Emilio Materassi in a Type 39A in the 1927 British Grand Prix, held at Brooklands on 1 October. By this time, new enlarged brake drums had been fitted. He was placed fifth.

'Williams' driving to victory in the green painted 35B in the first Monaco Grand Prix of 1929.

events in which cars of any capacity could be entered. An even more complicated series of rules followed in 1929 and again, these were mostly disregarded and *Formule Libre* effectively reigned until 1933.

It would be wearisome for the reader, and the hapless author (!) to set down the scores of Bugatti successes during these six years when the sheer number of Type 35s in circulation, coupled with a very real lack of opposition, virtually ensured a marque victory, though the most significant will be mentioned. For these were the years when Bugatti's supercharged models, the Types 35B and 35C, provided the racing car *par excellence* for the wealthy amateur racing driver.

The 35B represented the ultimate expression of the Type 35, it being a supercharged version of the 2.3-litre Targa, internally known at the factory as the TC (for Targa Compressor). The cars remained much the same as in the previous year though, from 1927, the famous aluminium wheels were fitted with larger 19 x 68 tyres and later in the year the brake drums were upped in size from 270 to 330mm (10.6 to 13in).

Yet it should be recorded that, from about 1930, Alfa Romeo began to revive its racing activities and first began to challenge and finally overhaul the cars from Molsheim. Ettore had belatedly recognized the limitations of his single overhead camshaft engine and, in 1931, introduced his twin cam Type 51 which enjoyed some success, though Bugatti won no significant Grand Prix after 1933. The great racing days were over.

More Sicilian Successes

In that year of 1933 Alfa Romeo had secured its fourth Targa Florio victory but, back in 1928, Bugattis were once again victorious though Campari, in an Alfa Romeo of only 1.5 litres, broke the usual Molsheim procession with Bugattis in first, third, fourth, fifth, sixth, eighth, ninth and tenth places. A particularly praiseworthy performance came from the Czechoslovakian, Elizabeth Junek, who drove a spirited race in her yellow and black Type 35B and would have been better placed (she came fifth) had it not been for water pump trouble. Towards the end of the second lap she pulled up half a minute on the experienced Albert Divo in a works Bugatti and she effectively led the race until forced to drop back when her 35B began to boil.

With the Delage challenge no more in front-line Grand Prix races, a Bugatti was victorious in the French Grand Prix, which was won by Williams in the sole works 35C. This was the expatriate

Englishman's first win for the marque but he had an even more significant victory in the first Monaco Grand Prix, which proved an ideal Bugatti event where outright power played second fiddle to the marque's attributes of flexibility, acceleration and superlative roadholding. Driving a 35B, painted green in deference to his nationality, Williams won after a celebrated battle with Caracciola in a totally unsuitable 7-litre SSKL Mercedes-Benz, which came third. In addition, Bugattis won the 1929 French, German, Spanish and even the Australian Grand Prix in far away Phillip Island, Victoria. The Targa Florio once again fell to Bugatti, giving Molsheim its fifth consecutive win on the Sicilian circuit.

Although Ettore made his usual challenge in the 1930 Targa Florio, he was to lose for the first time since 1925 when Varzi, in an old but reworked P2 Alfa Romeo, beat Louis Chiron, in a 35B, and Conelli in a similar car. Bugattis were never again victorious in Sicily. Alfa Romeo was to win every successive Targa Florio until 1936 and Italian cars were to dominate the event until the outbreak of World War II.

A rising star, René Dreyfus, won the 1930 Monaco Grand Prix in the ex-'Williams' 35B, beating none other than the talented Louis Chiron in a works 35C, into second place. It was a feat that was to earn him a place in the Bugatti team between 1933 and 1935. The blue cars from Alsace won the

The heyday of the Type 35. Of the 23 cars entered in the 1930 Monaco Grand Prix, no less than 14 were Bugattis. Borzacchini in a Maserati (24) makes a good start though would only last until lap 10 with 'Williams' in a 35C on his left though he dropped out on lap 29 with mechanical trouble. In the second row are Zanelli (14), Chiron (18) and Bouriat (16). The race was won by René Dreyfus who beat the works cars in a 35B from the Friderich stable which can just be seen behind Bugattis number 42 and 26.

Hans Stuber in his own 35C leads another Grand Prix Bugatti at the 1930 Monaco Grand Prix, staged on 6 April. Stuber was placed sixth behind no less than five Bugattis.

1930 French Grand Prix, there being no less than seventeen Molsheim cars present, Bugattis being placed first, third, fourth, fifth and seventh.

The relatively unimportant Tunis Grand Prix, of 28 March 1931 was also a Bugatti victory, with Varzi winning in his own car, but it was significant for the appearance of the new Type 51 Grand Prix model, outwardly similar to its predecessor though powered by a new twin overhead camshaft engine. Ettore had, at long last, adopted the layout of his competitors which permitted him to employ a more efficient hemispherical cylinder head – although the configuration had already been foreshadowed on his Type 50 passenger car of the previous year. Ironically, the inspiration for this layout came not from Europe but America in the shape of the omnipresent spirit of Harry Arminius Miller.

Twin Cam Consensus

In 1925 Miller had produced a revolutionary front-wheel drive version of his 122 racer, the first practical racing car to successfully employ the configuration. The concept was perpetuated for the 91 (1.5-litre) formula, America and Europe being so allied at this time, for 1926. That year, the American racing driver Leon Duray ordered one of the last front-wheel drive cars to be built and fitted it with an enlarged, specially designed 229mm (9in) supercharger and an intercooler.

In the qualifying trials for the 1928 Indianapolis 500 his average speed of 196.96kph (122.39mph) stood for nine years and, in the race proper, he enjoyed a record-breaking 321km (200 miles) until his overstressed engine failed. Two weeks later, on 14 June 1928, Duray achieved a 238.44kph (148.17mph) closed circuit world record at Packard's still unfinished proving ground at Utica, Michigan, which stood until 1934.

For 1929, Laon Duray was backed by the Packard Electrical Company of Warren, Ohio. It should be made clear that this firm, which made electrical cables for cars, had no direct connection with the respected car company of the same name. It had been established prior to the Packard brothers starting to build horseless carriages in 1899 and remained in Ohio after the automotive operations were transferred to the Motor City of Detroit in 1903. At the time of its involvement with Duray, Packard Electrical was independent but, in 1932, was purchased by General Motors.

This new financial support permitted Duray to

The 5-litre chassis 471, and Bugatti's own car, pictured in 1979 after restoration by its then owner, Nigel Arnold-Forster.

The 471's, engine. It differed from that of the later 5-liter cars in having a five- rather than three-baring crankshaft and cruder lubrication system!

A Type 23 of 1925 with open four-seater wood-panelled touring body.

The Type 23's engine. Note the cast-aluminium bulkhead with the dynamo driven from an extension of the comshaft, a layout that was inherited by the Type 35A

A 1926 Type 23 with, desirably, original coachwork. It is by W. Short of Winchester, and the car was originally owned by a Southampton butcher.

A 1924 Type 30, which still retains its original hydraulic brakes.

Prescott on a summer's afternoon. A 1925 former works Type 35 on the strting line at a Bugatti Owners' Club meeting in 1975.

A 1926 Type 37 equipped for the road.

In 1926 the Type 35's engine was enlarged to 2.3 liters, of which this is one, and that model designated the 35T.

A 1926 Type 37, as restored by Jarrot Engines.

A close-up of the Type 37's clutch and tubularly suspended gearbox.

The 1927 Type 43 owned for many years by the late Bob Roberts who bought it in 1947.

The Type 43's 2.3-litre supercharged engine was essentially a detuned version of the 35B unit.

For the Type 35 owner with everything: the Type 52 electrically powered miniature for his son.

Talking Bugattis! BOC council member A. F. Rivers Fletcher (left), with the famous Czech lady racing driver, Elizabeth Junek, and the great authority of the marque, Hugh Conway, at Prescott.

The magnificent lines of the Type 55's Jean Bugatti-designed roadster body. This is a 1932 example. ▶

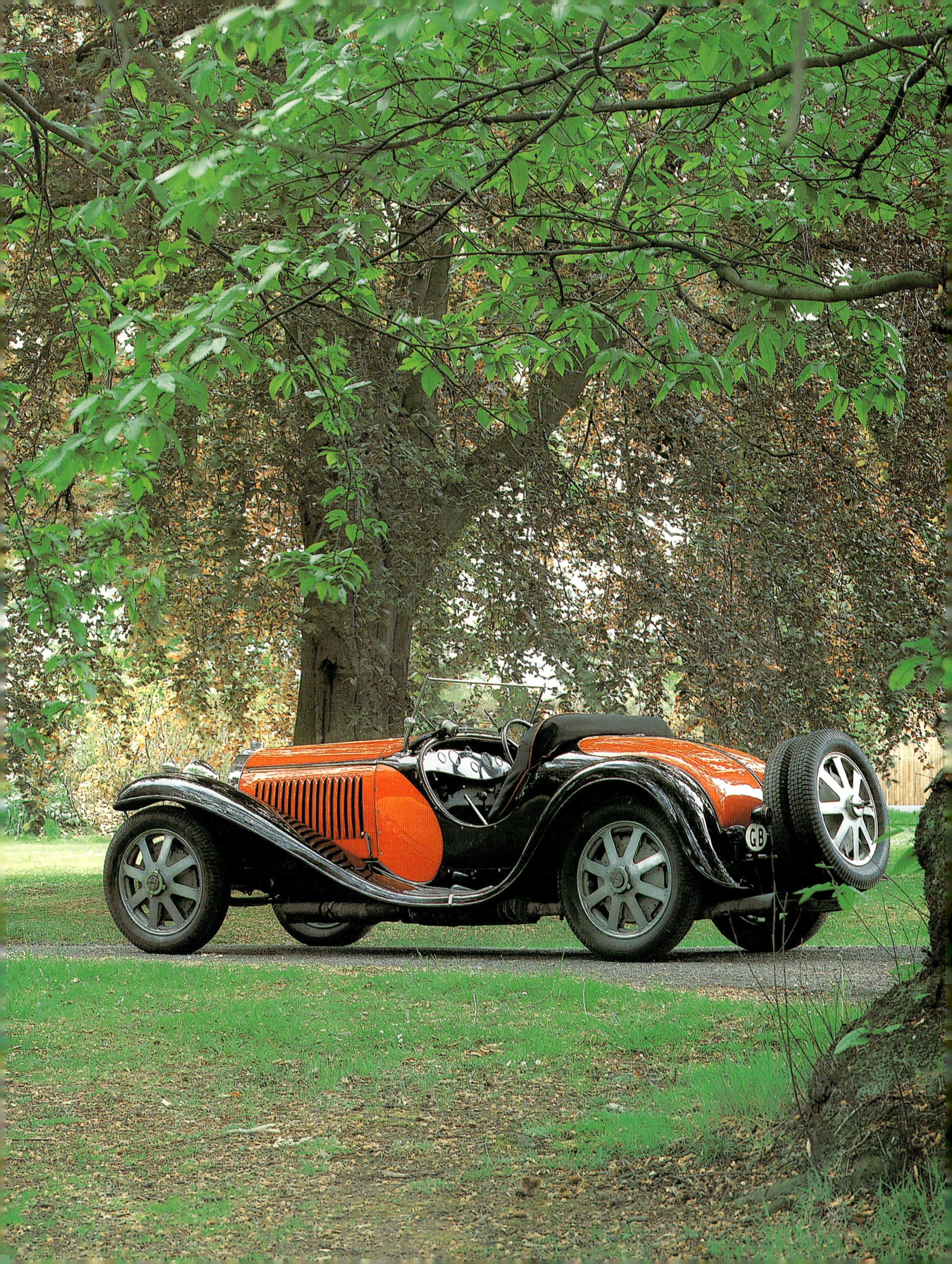
GB

The 1934 Type 59 imported to Britain by C. E. C. Martin in 1935 and today owned by E. A. Stafford-East.

The Type 59's twin overhead camshaft 3.3-litre supercharged eight-cylinder engine.

A 1936 Type 57 with drophead coupé coachwork by James Young of Bromley.

The original owner of this 1935 Type 57 was Colonel Godfrey Giles with the Bertelli coachwork designed by his brother Eric. It foreshadowed the body that he also designed on his brother's later Corsica-bodied 57S.

The impeccable lines of a 1937 Atalante coupe.

Underbonnet sculpture. The engine of the Atalante coupe.

1938 57S drophead coupe by Vanvooren.

The 57S's supremely elegant V-shaped radiator.

The driving compartment, with distinctive Bugatti steering wheel, of the 1938 Vanvooren-bodied 57S.

The 1938 Type 57SC Atlantic coupe when in the ownership of its original owner, Richard Pope of Ascot, Berkshire.

This is a 1939 Type 57C with drophead coupe coachwork by Gangloff.

The unconventional Colombo-designed Type 251 Bugatti Grand Prix car of 1956; with centrally located transverse engine and all-round non-independent suspension.

A 1991 Bugatti, EB110, pictured at the model's launch in Paris on Saturday 14 September, the day before Ettore's 110th birthday.

buy a further two new Millers 91s, the only front-wheel drive example to be built in 1929 and a rear-drive one.

Duray retained his 1928 car and all three single-seaters were finished in distinctive purple livery, with the chassis rails and wheels painted in yellow, as opposed to the previous year's black. The cars carried the Packard Cable Special name on their bulkheads. Funnily enough Duray preferred his original car to the new ones and others, namely Ralph Hepburn and Anthony Gulotta, usually drove the later Millers. Duray qualified for the 1929 Indianapolis 500 Miles Race at 191.63kph (119.08mph) though, in the event proper, he led for a time but later retired with carburettor problems.

It was during the summer of 1929 that Duray decided to take all three cars to Europe. He was accompanied by the French-born Jean Marcenac, formerly mechanic to two deceased Indianapolis stars, Frank Lockhart and Ray Keech, who had died in 1928 and 1929 respectively. Marcenac had originally gone to America in 1920 to fit front brakes to Ralph de Palma's Ballot in the Elgin road race and decided to stay! The principal reason for Duray's trip was to embark on a series of record-breaking runs at Montlhéry and also 'to create publicity for the new Cord car. It was being heavily promoted in Europe where great stress was placed on its "front-drive based on Miller patents".'[26]

What happened then, according to W F Bradley, is that Duray successfully broke a spate of records at the Paris Autodrome, then participated in the Monza Grand Prix, whereupon the two Packard Cable Specials were snapped up by Ettore Bugatti. A financially embarrassed Leon Duray was pleased to swap his brace of front-wheel drive Millers for three new Type 43A Bugattis and Ettore had a twin overhead camshaft layout that he could then adopt. In some respects, this has been the accepted wisdom from thereon. But the reality, as will emerge, was probably the complete reverse of what Bradley chronicled. Griffith Borgeson has already pointed out that it was Duray who got the best of the deal.

[26] Borgeson, *The Golden Age of the American Racing Car*.

An American in Paris

Once established in Paris, Duray made a considerable impact on the French capital and Bradley, who had no liking for the man, has recorded that:

> His commanding figure, rendered more impressive by the black overalls and the black skull cap he always wore, his confident swagger, his cool contempt of all rivals, his remarkable powers of expectoration and his pungent remarks which lost nothing in the translation, made him a temporary idol for the Parisians who had never seen a Westerner otherwise than on the screen.[27]

Soon after his arrival in France, in the summer of 1929, Duray immediately set out for the Montlhéry circuit, near Paris. His first sortie was to attack the 1.5-litre standing start record on the local Arpajon road where he attained about 272kph (138mph). Back at Montlhéry on 19 August, he broke the world's International Class F five-hour record at 220.82kph (137.2mph) and the ten-hour one at 217.78kph (135.33mph). He subsequently improved on the five-hour one and averaged an impressive 224.04kph (139.22mph).

Not surprisingly these successes, achieved as they were with a supercharged engine of a mere 1.5-litre capacity, caused something of a sensation in France. *Paris Match* was moved to comment: 'To achieve such speeds with a 1,500cc car . . . proves clearly that the Americans are far in advance of us. We have no cars of similar displacement capable of rivalling the speed of these Millers.'[28]

Duray went on to compete at Monza in the following month and Bradley says that, as a result of his European trip, the American 'had not enough funds to pay his fare home. In his extremity he appealed to Bugatti, who saw in these Packard Cable Specials a chassis quite inferior to anything he had ever produced, but with an engine which appeared to have qualities. The outcome was a deal under which the Millers went to Molsheim and

[27] Bradley, *Ettore Bugatti*.

[28] Borgeson, *op cit*.

three 2,300cc [Type 43A] Bugatti were dispatched to California.'[29]

A Bargain for Duray

This is almost certainly a travesty of what actually occurred. It seems far more likely that it was Bugatti who approached Duray following his record-breaking achievements, rather than the other way round and the fact that such a barter ensued, suggests that it may have been Ettore who was strapped for cash! I pointed out earlier that he had probably developed a considerable regard for Miller's cars from at least 1923.

At this time a new Type 43 sold for 130,000fr. which was then the equivalent of $5,094 or some £1,048. Duray therefore got $15,282 (£3,144) worth of top line Bugattis for two obsolete Millers, given that they had cost $15,000 (£3,092) new, which Packard may even have paid for in the first place! For according to Miller historian, Mark Dees, 'Leon knew that he would have to completely rebuild the cars to meet the new Indianapolis formula, [the stock block "junk" one of 1930] and, since they were front-drives, they would have only been competitive there and at Altoona [Pennsylvania], as the rest of the board speedways were being demolished.'[30]

On the other hand Bugatti got, for little cost, a twin overhead camshaft layout which he could copy without the embarrassment of aping one of his European competitors. Having said that, Miller had in turn been inspired by a French racing car, the Grand Prix Ballot, when he had built his first twin overhead camshaft engine for the TNT racing car, which we have already encountered, back in 1919.

First Duray had to compete in the Monza Grand Prix which was held in mid-September. Bradley has recounted that Vincenzo Florio, founder of the Targa Florio, who had taken over the management of the Monza track, 'was attracted by the "Asso Americano" and invited him to the autumn races'. Duray therefore entered his front-wheel drive Millers in the event which was held, regardless of the AIACR, to take the place of the European Grand Prix. It consisted of preliminary races for 1.5-litre and 3-litre and *Formule Libre* cars, the winners then competing in the final.

Duray achieved a record-breaking speed of 190.12kph (118.14mph) during the third of his practice sessions which augured well for the race proper. The 1.5-litre heat took place on 15 September and, despite making a slow start because of the car's gearing and its centrifugal supercharger, after ten laps he was in the lead, having overtaken Arcangeli and Nuvolari in privately entered Talbots. Then on lap 12, Duray characteristically withdrew, having lost oil pressure. He ran the newer of the two cars in the 3-litre event, claiming that it had been bored out to 1,505cc, so excluding him from the 1.5-litre race, but it suffered from transmission problems and he retired in third place, leaving the race to one of the ageing but still formidable P2 Alfa Romeos.

The Millers were then handed over to Bugatti and Duray returned to America with his trio of Bugattis which he subsequently sold. After arrival at Molsheim, the Miller engines were removed from their chassis and underwent a series of dynamometer tests. These began on 30 November 1929 though, on that occasion, a 'rod seized' with the unit running on alcohol fuel and it recorded a mere 160bhp. Later on 11 January 1930, using a fuel consisting of 60 per cent petrol, 35 per cent benzol and 5 per cent ethyl fuel, one of the 1.5-litre twin overhead camshaft straight-eights attained a remarkable 208bhp at 6,400rpm, before the centrifugal supercharger broke up. Nevertheless this was the equivalent of 138bhp per litre, which was infinitely superior to the Type 39A's 73.

These figures told their own story and Bugatti wasted little time in transferring the Miller 91's head and valve gear to the bottom end of the Type 35B unit, so creating the Type 51. Therefore the hemispherical head, twin overhead camshafts, driven via spur and bevel gears from the front of the engine and two valves per cylinder at an

[29] Bradley, *Ettore Bugatti*.

[30] Dees, *The Miller Dynasty*.

Although the Type 51 closely resembled the 35B, there were detail differences and most are shown here. Because of its twin overhead camshaft engine, the supercharger relief valve was lower than that fitted to the single cam engine. Similarly a new pattern of aluminium wheel was introduced with fixed-well base rims.

The Type 51 was the first Grand Prix Bugatti to be powered by the Miller-inspired twin overhead camshaft engine though the lower half was essentially that of the 35B. This is actually a rare 1.5-litre 51A, restored by Nigel Arnold-Forster though externally it resembles the larger capacity car.

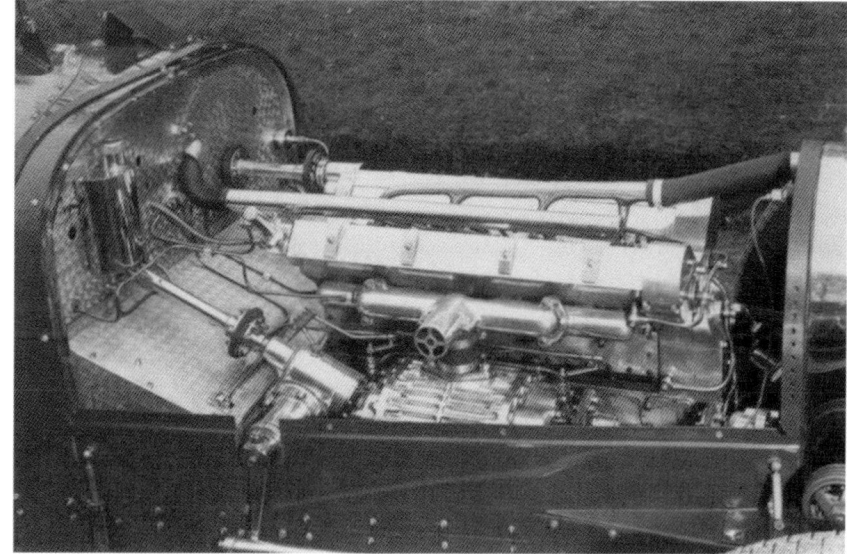

The cockpit of the same car. It is essentially similar to that of the Type 35 though the magneto, driven off the exhaust camshaft, is accordingly offset to the left. The lever also directly activates the instrument's advance/retard mechanism.

The French Grand Prix, 1931, was held on 21 June, at Montlhéry. Chiron shared the driving in this 10-hour race with Achille Varzi and they won with Alfa Romeos in second and third places.

A Memorable Monaco

The works cars made their first appearance at the 1931 Monaco Grand Prix, on 19 April, always a good Bugatti circuit, and there were no less than eleven of all types present. Chiron won and though Fagioli in a Maserati was second, Varzi was third in another Type 51. The model was also successful two months later in the French Grand Prix at Montlhéry when a Type 51, driven by Chiron and Varzi, managed to win, coming in ahead of the Italians, in the shape of the increasingly threatening presence of a Monza Alfa Romeo and Maserati.

It was to prove to be Bugatti's last genuine French Grand Prix victory. At the 1932 race held at Rheims, the best that Bugatti could manage was fourth place, the race being won by Nuvolari in the new 2.6-litre P3 *monoposto* (single-seater) Alfa Romeo, essayed by the talented Vittorio Jano, which overnight rendered the Type 51 and all two-seater racers for that matter obsolete. That year, the P3 also won the front-line German, Italian and Monza Grand Prix and Bugatti was relegated to victories in the second string Dieppe, Tunis, Oran and Czechoslovakian events.

included angle of 96 degrees, were directly inspired by the car from Los Angeles. The result: not only breathing but water jacketing was a great improvement on Ettore's single camshaft, three-valve Type 35 head.

In fact, according to Dees, Bugatti, 'copied the top ends of his Millers so faithfully that the spark plug caps and valve gear of the Type 51 . . . will interchange with that of a late 91'. The 51's 60 x 100mm dimensions were essentially those of the 35B and its capacity was therefore of 2,262cc though there was a shorter 88mm stroke version resulting in 1,991cc. The outcome of these ministrations saw the Type 51's 2.3-litre engine develop 160bhp, though 185 has been recorded, compared with 140 for the 35B. The 51 was also available in rarer 60 x 66mm, 1,493cc 51A form.

Minor modifications were also made to the rest of the car. The fuel tank was enlarged and twin filler caps, a feature of some late 35Bs were standardized. There were new well base aluminium wheels which were no longer fitted with a locking rim. Otherwise the car closely resembled its predecessor.

A fine study of a Type 51 at speed. This is English driver Tim Rose-Richards in 51145, delivered in February 1933. It was in fact a 35B, 4963–214T, converted by the factory. Note the twin filler caps, another Type 51 identifying feature. He was placed fourth.

Varzi receives admiring glances in the streets of Monaco in April 1933. He won the event...

...after a celebrated tussle with Tazio Nuvolari in a Monza Alfa Romeo. Here the Bugatti is ahead at the Station hairpin.

No chapter on Bugattis in Grand Prix racing would be complete without a picture of Ettore himself, pictured at a race meeting, making a point to his son, Jean, with his back to the camera.

Then at the very peak of their invincibility, Alfa Romeo announced at the beginning of the 1933 racing season that it was withdrawing its all-conquering *monopostos* from competition. This was because of the firm's bankruptcy although, later in the year, the Mussolini regime nationalized it. As a consequence, Scuderia Ferrari, which ran the Alfa Romeo racing team, was forced to fall back on the older, 1931 two-seater Monzas for the first half of the 1933 season and it was not until August of that year that the unassailable P3s reappeared.

The new 750kg formula was due to begin in 1934 so, in any event, 1933 was the Type 51's last competitive year. The Monaco Grand Prix not only proved to be the marque's last win in the Principality but it also witnessed a memorable battle between two Italian arch rivals: Achille Varzi in a Type 51 Bugatti and Tazio Nuvolari at the wheel of a Monza Alfa Romeo. The personalities of the respective drivers are best summed up by W F Bradley, who has recorded that 'Nuvolari was daring, fiery, impetuous, a mass of nerves which in his earliest days he had not always been able to control. Varzi was of a more sullen disposition, with a cool, calculating daring and a reluctance to forget a defeat.'[31] The two were old adversaries and had previously met at the 1931 Targa Florio, which Nuvolari had won and Varzi wanted his revenge.

The 1933 Monaco event marked a milestone in Grand Prix racing history in that, for the first time, grid positions were related to practice times, rather than by a draw as had hitherto been the case. As a result of his impressive pre-race performance, Varzi was in pole position with Nuvolari directly behind him, in the second row.

Varzi made an excellent start and initially took the lead, but by the end of the third lap, Nuvolari was only a second behind and the two cars were locked in a struggle which endured throughout the race. Nuvolari got ahead on lap 4, Varzi regained it on the 13th but lost it on the 17th but the Bugatti was in front on the 19th. And so it went on until the penultimate 99th lap. Nuvolari was just ahead on the 98th. Then on the climb to the *Casino*, 'Varzi held on to third, stretching his engine to the limit and risking all at over 7,000rpm to take the lead.'[32] Nuvolari held on but a piston failed, leaving Varzi to win, the duel having lasted for 97 of the 100 laps, making it unique in Grand Prix history.

It was motor racing at its very best and I cannot think of a better note on which to conclude this chapter.

---

[31] Bradley, *Ettore Bugatti*.

[32] David Hodges, *The Monaco Grand Prix* (Temple Press, 1964).

8
Four, Eight and Sixteen Cylinders

'Bugatti was proud of his eye. He loved engines that had straight lines with flat and polished surfaces, behind which manifolds and accessories lay hidden. The general effect obtained gave, even the uninitiated, an impression of mechanical beauty.'

Jean-Albert Grégoire[1]

As will have been all too apparent, Bugatti was fully engaged in Grand Prix racing in the years between 1924 and 1933. But during this period Molsheim was also involved in the creation and production of a variety of other models. These included a cheaper range of racers and one of the world's fastest sports cars while its twin cam replacement is often considered to be one of the most beautiful. Then there were well-mannered tourers, such eccentricities as 16-cylinder and four-wheel drive Grand Prix cars and even a toy racer for children. This chapter deals with these single and twin overhead camshaft cars which invariably came up to the usual Bugatti standards though there were the occasional duds. Even Ettore had his peccadilloes.

It will be recalled that Bugatti had introduced his four-cylinder 16-valve model in 1920. This very Edwardian car lasted until 1926 though, in the previous year, its Brescia sports derivative was effectively replaced by a much more up-to-date model in the shape of the Type 35-based 1.5-litre four-cylinder Type 37. This offered potential customers the looks of a Grand Prix Bugatti but without the cost or complication of its eight-cylinder roller bearing engine. In fact it was almost half the price. In 1927 a pukka Type 35 would cost £1,100 in Britain while a Type 37 was £550 or about half the price of that year's native 1.5-litre Aston Martin.

The new four-cylinder model resembled its

Most Type 37s were fitted with the factory's Grand Prix bodywork but some were imported to Britain in chassis form and bodied by local coachbuilders. This one has a Compton body. The 37 used a similar chassis to that of the 35 though was fitted with a solid rather than a hollow front axle and, of course, 1.5-litre four-cylinder plain bearing engine, offering owners the looks of a Grand Prix car though without the complications.

[1] Grégoire, *Best Wheel Forward*.

A Type 37 engine receiving admiring glances from visitors at the Lewes Speed Trials in the early 1930s. Note the Type 43 on the car's left.

eight-cylinder relation though was fitted with wire wheels as standard, in place of the Type 35's aluminium ones. Interestingly, nowadays the Bugatti eight-spoked wheels are almost universally revered though this was not always the case, particularly in Britain. When *The Autocar* announced Type 37's specifications, it commented that 'On the four-cylinder model wire wheels have replaced the aluminium wheels which, however satisfactory in service, gave the cars a somewhat *curious appearance* [my italics] not appealing to every user.'[2]

The Type 37 was fitted with the small Lyon-type radiator throughout its six-year manufacturing life. Its 69 x 100mm four-cylinder engine shared the same clean-cut architectural lines of the Type 35 and was, in effect, half the stillborn eight-cylinder Type 28 which had appeared at the 1921 Paris and London Motor Shows.

Its 69 x 100mm dimensions, giving a capacity of 1,496cc, was the same as its 16-valve predecessor. The layout thereafter followed what had become established Bugatti practice with the cast iron cylinder block mounted on an aluminium crankcase which was split along its centre line. The lower part contained the sump, complete with cooling tubes, and also incorporated four substantial bearers, the whole weight of the engine being therefore carried by them.

The crankshaft was a one-piece unit with circular webs. It ran in plain bearings, there were five in all, as did the connecting rods. Initially the lubrication system followed Bugatti's familiar but haphazard jet arrangement. Oil was squirted at about 15psi, the oil pump being driven off the front of the crankshaft via a scroll gear. However, from 1928, a fully pressurized system was introduced, the crankshaft then being drilled to permit direct oiling, which ran at 45psi, of the main bearings and big ends.

The single overhead camshaft was shaft driven, via bevel gears, from the front of the engine, while Bugatti's familiar two inlet and one exhaust valves configuration was perpetuated. These were, incidentally, larger than those used on the Type 35, inlets and exhausts being of 26.5 and 36mm (1.04 and 1.41in) respectively. The water pump, located on the right-hand side of the engine, was driven via the cam drive. Nearby, a single Solex carburettor was fitted while coil ignition was standardized, its

[2] *The Autocar*, 18 December 1925.

A Type 37 engine in which the dynamo was driven from the front of the crankshaft while the distributor for the coil ignition system was attached to the end of the camshaft. The unit was, in effect, half Bugatti's still-born Type 28 3-litre. This particular engine is number 8 and about the first ten examples had their cambox plugs on the exhaust side. Thereafter they reverted to the inlet one. Another sensitive restoration by Jarrot Engines.

distributor being driven off the end of the camshaft.

When bench-tested at Molsheim and running on petrol, this four-cylinder engine developed a conservative 49bhp at 4,200rpm though was capable of approximately 60bhp at 4,700rpm. Top speed was in excess of 136kph (85mph) with road equipment and good 16kph (10mph) more in stripped form. For the Type 37 was offered complete with headlights, the dynamo being driven directly off the end of the crankshaft and therefore running at engine speed.

The first example, chassis number 37101, came to Britain and was delivered to Malcolm Campbell on 18 November 1925 and was one of only three built in that year. The 37 was an immediate success and 120 were produced in 1926, sales no doubt being assisted by Bugatti winning that year's World Championship. In 1927 output amounted to 87 cars and 45 in 1928. Thereafter production began to tail off. Sixteen cars were delivered in 1929, six in 1930, eight in 1931 and the last two in June 1933. The final Type 37 chassis number was 37388.

The Supercharged 37

The Type 37 was also available in 80bhp supercharged form, popularly known as the Type 37A and 37C, for compressor, at the factory. The first car, chassis number 37269, was delivered to Turin in June 1927. The model was fitted with the smaller supercharger used on the Type 39 with a casing length of 135mm rather than the 185mm of the Type 35B. To make room for the blower, the steering box had to be moved back in the chassis while the water pump was transferred from the right to the left-hand side of the engine for the same reason.

The 37A dispensed with the coil ignition of the

Supremo Montanari in his Type 37, during the 1926 Targa Florio. He was placed twelfth, the event having been won by Costantini in a works 35T with Minoia second and Goux third, all in similar cars.

unblown car and, instead, was fitted with a dashboard-mounted magneto in the manner of the eight-cylinder cars. The model was externally similar to its unsupercharged predecessor, apart from the presence of the blower relief hole in the bonnet. Although wire wheels were fitted, Grand Prix aluminium ones were available at extra cost. The 37A remained in production until 1930. Inevitably, the supercharged 37 sold less well than the cheaper unblown car, though production amounted to a respectable seventy-seven examples, making a total of 290 Type 37s of both types.

The Type 37 was a perfectly usable road car but was also a versatile performer in *voiturette*, road and circuit races as well as hill climbing (although it was invariably overshadowed by its Type 35 contemporary). Nevertheless it did not prevent Cacciari from undertaking an 8hr 50min drive in the 1926 Targa Florio to finish eleventh, an indication that the 37 inherited the established Bugatti virtue of reliability. In the following year, Count Conelli was placed second in such a car.

The 37A supercharged model proved to be more popular in competitive events than its unblown predecessor. Its racing debut was the *Coupe de la Commission Sportive* in 1927, which Jules Goux won in an example, while Count Conelli was fifth in another. A youthful René Dreyfus was placed eighth in another in the 1928 Targa Florio and in Britain, Chris Staniland achieved an impressive 196.44kph (122.07mph) in a supercharged Type 37 at Brooklands in 1929.

Practically all the Type 37s were fitted with the usual open two-seater Grand Prix bodywork. However, a handful of cars were converted to boxy coupes. This was the Million–Guiet built Toutalu which was patented by de Vizcaya. It retained the Bugatti bonnet though, from thereon, the remainder of the body was replaced by a rather cramped coupe, the interior being re-upholstered and the gear lever and handbrake mounted inboard. The doors were fitted with sliding windows and the wire wheels were partially covered by discs. The conversion was completed by the presence of a rather Voisin-like mascot. An example was owned in 1932 by the Brazilian ambassador to France, Count de Souza Dantas.

A Touring Bugatti

In essence the unblown Type 37's four-cylinder engine also powered a new touring car, designated the Type 40. This appeared in June 1926, seven months after the 37's arrival. The gearbox was also similar to that used in the 37 though the gear lever and handbrake were centrally mounted. Initially the Type 40 used the longest of the 16-valve chassis, with its small drummed four-wheel brakes and 2,550mm (8ft 4in) wheelbase. But after about 150 examples had been built, this frame was replaced by a more robust one, which was similar to its eight-cylinder Type 38 contemporary, but shortened to a 2,714mm (8ft 9in) wheelbase.

Although the Type 40 was first available in coachbuilt form, from mid-1926 it was offered with

Édouard Charavel who raced under the pseudonym 'Sabipa', pictured in a Type 37A in a race at St Brieuc in northern France.

A handful of Type 37s were offered in coupe form, the conversion being undertaken by Million-Guiet, complete with disc wheels and the spare relegated to the rear. Here Mlle Dalbaicin displays the car at the Championnat Automobile des Artistes *in 1928.*

a factory Grand Sport open body, the first occasion on which a roadgoing Bugatti was fitted with standardized coachwork. This had a single left-hand door, seating for 'three and a half' and a stylish pointed tail. It was subsequently and successfully scaled up and used in the following year on the newly introduced eight-cylinder Type 43.

The Type 40 represented excellent value for money. It sold complete for £465 in Britain, which was about £100 cheaper than an open 12/50 Alvis. It had respectable road manners and was good for 112kph (70mph) plus. The first chassis number was 40101, while the last one dated from May 1931, the final allocation being 40880. The Type 40 was a popular model, a total of 775 examples being built.

Although many Type 40s were fitted with the factory Grand Sport bodywork, coachbuilt examples were also produced. More significantly, in 1928 a chassis was the recipient of the first body to be designed by Ettore's eldest son, Jean, a remarkable achievement for a nineteen-year-old. It was a two-door, drophead coupe with a separate trunk at the rear, in his father's familiar *fiacre* style with its echoes of horse-drawn vehicles.

The central portion was finished in yellow, the remainder of the car being painted in black, showing Jean's use of contrasting colour to great effect.

The car, chassis number 40623, was designed for his twenty-one-year-old sister Lidia. This disc-wheeled example was also fitted with a supercharger, courtesy of the Type 37A, and later had

A frontal view of a Type 40 at Molsheim. It initially used the Brescia chassis though later its own stronger frame.

The Type 40 was offered with this factory Grand Sport bodywork which was similar to that later used on the Type 43. This English example has been fitted with a non-standard driver's door.

A famous Type 40, chassis number 40623, with fiacre drophead coachwork, designed by Jean Bugatti for his sisters Lidia and L'Ébé though the latter never learnt to drive. Apparently this 1928 car was bodied with the help of the factory workers and without the knowledge of Ettore! It was also supercharged with a 37A unit, note the blower relief hole. Lidia drove the car for many years and, after her marriage, displayed it in the hall of her château. She retained it until 1972 and it has been in Britain ever since. It is pictured at the Chateau of St Jean near the factory which provided an appropriate background for many factory photographs.

Another fiacre-*bodied Type 40, left, and a Type 44 with similar coachwork ahead.*

large Type 46 brakes. Its body was subsequently offered in fixed head coupe form. The Type 40 enjoyed some popularity with Parisians as a fashionable town car. Jean was on his way.

In 1931 the Type 40 was succeeded by its 40A derivative, with enlarged 1,627cc engine, achieved by upping the engine's bore from 69 to 72mm. This followed the 3-litre eight-cylinder Type 44, which used two Type 40 cylinder blocks. The Type 44 was replaced in 1930 by the 3.3-litre Type 49 and it therefore had twin sparking plugs per cylinder, demanding a twin spark plug distributor.

The Type 40A's factory open bodywork, which is credited to Jean Bugatti, was very transatlantic in concept and resembled that of the existing eight-cylinder Type 43A of 1929. It was therefore a two-seater with dickey, or what the Americans called a rumple seat and a fashionable separate compartment for golf clubs. The first Type 40A was chassis number 40900 and the last car, 40930, was not delivered until October 1933. A mere thirty-two or so Type 40As were produced and, after it was discontinued, Bugatti built no more four-cylinder cars.

Across the Sahara

It may well have been due to the fact that the Type 40 was a relatively light, uncomplicated model, that early in 1928, Ettore Bugatti was approached to supply four or five chassis for Lieutenant Frédéric Loiseau, who wanted to lead an expedition of ordinary French touring cars to Africa to drive the 16,093km (10,000 miles) from the colony of Algiers, south across the Sahara Desert, to Lake Chad. The destination was subsequently changed to a shorter journey of a mere 12,874km (8,000 miles) to Oran on the Ivory Coast and back.

The party departed on 29 January 1929 but, after reaching Gao in the Sahara, Loiseau lost his colleagues who preferred hunting to the perils of desert motoring! He pressed on alone with a friend and reached Bassam on the Ivory Coast. The return journey was made via Bamako and Timbuctu, Gao and then virtually non-stop back to Algiers on 4 March.

The Type 40, with its pick-up body, performed perfectly throughout and it seems did not even require topping up with water. It was only on its return to Paris that it had the misfortune to run a big end but nevertheless provided Bugatti with valuable publicity, Loiseau writing a book about his experiences. Afterwards the car, chassis number 40811, was returned to Molsheim and today this Type 40 can be seen at the *Musée national*.

The model's durable qualities were also underlined when two lady drivers, the first ever to enter the event, drove an open two-seater example in the 1930 Le Mans 24-hour race. This was the marque's first appearance at the Sarthe circuit since two privately entered Brescias had run there, and were placed 10th and 22nd, in the inaugural 1923 event.

The Type 40 of seven years later was entered by Mme Odette Siko and her co-driver was Mme Mareuse. Although eight cars retired, the Bugatti kept going and came in seventh, ahead of two front-wheel drive Tractas. The doughty female duo entered the 1931 race though, on lap 41, had the misfortune to be disqualified when their pit refuelled their Bugatti too early. Madame Siko was back again in 1932 though, by this time, she had switched to a 1750 Alfa Romeo, shared with former Bugatti driver 'Sabipa', and came in a very creditable fourth.

A Type 30 Replacement

So much for the four-cylinder Bugattis but what of the mainstream eight-cylinder cars? Bugatti had introduced his first production straight-eight, the 2-litre Type 30, in 1922. In effect, its 60 x 88mm, 1,991cc three bearing crankshaft engine was carried over for its Type 38 replacement which appeared in 1926. But whereas, at its longest, the Type 30 had a 2,850mm (9ft 3in) wheelbase, the 38's was elongated to 3,122mm (10ft 2in). This put the chassis weight up to 860kg (16.9cwt), which was also an encouragement for coachbuilders to produce relatively heavy bodies for it.

A new gearbox was designed for the Type 38 though it was still separate from the engine. However, Molsheim bowed to convention with the gearchange. First was now engaged by moving the lever towards the engine, instead of away from it as had previously been the case and, in the usual way, top was back instead of forward!

The first Type 38, chassis number 380101, was delivered to a banker and friend of Ettore's, Leo d'Erlanger, in March 1926. In an effort to pep up the model's performance, in July 1927, at chassis number 38435, Molsheim introduced the supercharged 38A. This used the same Type 39 blower

A Type 38 with French touring coachwork. The model was produced between 1926 and 1929, latterly in supercharged form and its engine was used in the Type 35A.

as had been employed on the 37A though the three bearing crankshaft was always the limiting factor. The Type 38 was not a particularly successful model and bespoke coachwork by Lavocat and Marsaud and Vanvooren in France, Murphy in America and the British Compton company did little to enhance its elongated lines. Curiously, it was particularly popular in Brazil, many cars going to Matarazzo, the Bugatti agent there. The model lasted until 1929 and was discontinued at chassis number 38487. A total of 387 Type 38s were built and, of these, around fifty were supercharged.

If the Type 38 was a controversial model, the Type 43, which arrived in mid-1927, was not. This is one of the truly great Bugattis, a 160kph (100mph) sports car offered, from the outset, with stylish, functional works Grand Sport open bodywork, scaled up from that used on the Type 40 of the previous year. It was, inevitably, expensive. British customers paid £1,365 for a Type 43 in 1928 so perhaps, not surprisingly, only fifteen were imported while the car was current. The first of them, chassis number 3156 went, once again, to Leo d'Erlanger.

A Super Sports Car

The model's radiator, steering, gearbox and axles were all related to the Type 38 though the 43's chassis radically differed and was shorter and contoured, just like the 35/51 Grand Prix cars, though it had a 2,972mm (9ft 7in) wheelbase. At the heart of the Type 43 was its 2.3-litre supercharged straight-eight engine, courtesy of the Type 35B Grand Prix car. Developing a slightly reduced 120bhp at 5,000rpm, it was virtually identical to it though had a conventional sump in place of the tubed variety.

The car's impressive performance was underlined by W F Bradley when, accompanied by Ettore on a drive in an early Type 43, he reported that 'with the standard gear ratio, it will run up to 112mph ... Most of the run was made on the greasy, winding lane of the Vosges ... and even under such conditions the suspension was good.'

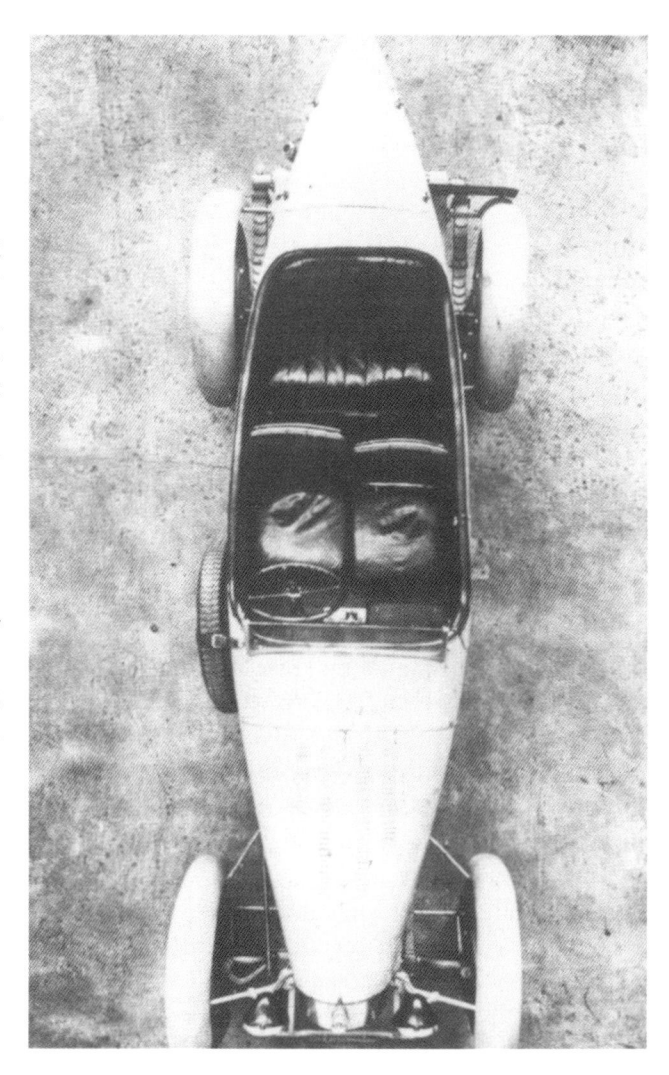

The Type 43's Grand Sport bodywork viewed from above. Like the Type 35 and 51 Grand Prix cars, its chassis had curved long irons which are reflected in the body contours. The front seats are upholstered in unpleated leather while the rear one is ribbed. The small luggage compartment is located in the tail while the reversed quarter elliptic springs are also apparent. The leather piping around the top of the seating compartment provides a pleasing finishing touch.

Bradley was also impressed by the car's flexibility and reported that 'Bugatti, during a trial run we made with him in Alsace, stopped the engine, engaged top gear, and then started away by pressing the starter and accelerating without gear changes to a speed of 90mph.'[3]

[3] *The Autocar*, 18 March 1927.

A 1927 Type 43 raising the dust at the Middlesex Automobile Club's Amersham Hill Climb in 1930. This particular car was registered JTX 575 in 1950 and survives in Britain.

We have no better illustration of the Type 43's performance and, indeed the driving style of eighteen-year-old Jean Bugatti, than an account by Lieutenant Frédéric Loiseau, who was at Molsheim in 1927 to talk to Ettore about his impending expedition across the Sahara. It was 6.30 in the evening, he needed to get to Paris and Jean offered to take him in a Type 43 but told him: 'Do you mind if we don't stop on the way . . . I must be in Paris at half past ten – they're expecting me at Maxim's.' Loiseau later recalled: 'I was tickled to bits. I love speed. But this! My driver played with the steering wheel like Paderewski with a piano. Only a piano doesn't run at 160kph.' The pair arrived in the rue Royale four hours and five minutes later, at 10.35. ' "Not too shaken?" asked my driver, ". . . Meanwhile let's have a drink. What about some crayfish and a little cold chicken?" '[4] This meant that Jean had *averaged* 107kph (67mph) over the 534km (331 miles) between Molsheim and the French capital.

The first Type 43, chassis number 43150, was delivered to works driver Pierre de Vizcaya, on 16 June 1927 although the later 43155 had been dispatched to *Agenzia Generale* in Turin, four months earlier, in March. Inevitably the model was produced in relatively low numbers and, in 1930, came the supplementary Type 43A with a works open two-seater body with dickey of the sort to be introduced on the Type 40A in the following year.

In reality only a few 43As were built. Deliveries fell off badly after 1931, the combined effects of the world depression and the fact that Bugatti had switched to twin overhead camshafts, no doubt playing their parts in this. The final car, chassis number 43310, was completed in February 1935. This made a total of 160 examples of the Type 43s produced.

For all its performance, running a Type 43 represented a formidable challenge to a prospective owner because of the Grand Prix engine's roller bearing crankshaft. The problem was that the rollers themselves were subjected to wear, rather than their tracks on the journals, or the one-piece connecting rods. On the Grand Prix cars the cranks were re-rollered between races.

Although the 43 would hardly be subjected to the same stresses, an expensive overhaul about every 8,000 or so kilometres (around 5,000 miles) became an all too familiar aspect of Type 43 ownership. Then the lubricating oil had to be warmed up to prevent the rollers skidding, while the passages in the crankshaft could become carboned up, so preventing the flow. If engine maintenance was ignored, the roller cage could seize, taking the rod

---

[4] Frédéric Loiseau, 'Interdit au moins de cinq ans' translated in *Bugantics*, vol. 27, no. 3 (autumn/winter 1964).

Malcolm Campbell's much-photographed fire in the 1928 Tourist Trophy race at Ards, Northern Ireland. The Type 43's hood frame can be seen as the cars ran with their hood raised. The rear of the car ignited after an experimental rubber fuel tank fractured and leaked fuel.

This Type 43, registered YT 8241, belonged to Earl Howe. It then passed to Denis Evans, centre, of the famous Bellevue Garage at London's Wandsworth Common and he used it at Brooklands and for hill climbing. A plug change was clearly a family occasion, with sister Doreen who was also to race, left, Mrs Evans, Denis, father Bertie Graham Evans and racing driver, Kenneth Evans, right. A non-standard exhaust manifold has been fitted. This car, alas, has disappeared.

Louis Dutilleux, the Bugatti test driver, in the Type 43 in which he nearly won the 1928 Boillot Cup at Boulogne. He was leading the field when one of his wheels collapsed. But who is the gentleman with the cine camera?

with it which could then appear through the side of the crankcase.

A Disappointing Mille Miglia

Although Type 43s were campaigned by private owners in races and trials, in 1928 the factory entered a team of three cars in the second Mille Miglia, held on 31 March and 1 April 1928. The entry was initiated by Count Aymo Maggi, one of the event's instigators, who had begun his racing career in 1922 on a Chiribiri but had later switched to a Brescia Bugatti.

His connections with Molsheim resulted in Ettore granting him the Bugatti franchise for Italy and Maggi opened a showroom and service department in Milan's Porta Roma. Count Giovanni 'Johnnie' Lurani has recalled that Maggi's Bugatti agency was 'not merely a commercial office but a meeting place for technicians, supporters, beginners and would-be drivers of the blue meteors . . .

the agency finished by becoming practically a club, business taking second place'.[5]

By this time Bugatti had won the Targo Florio race on three separate occasions and great things were expected from the trio of Type 43s. But rather than relying on French drivers, as he had in the Targa, Bugatti team manager Meo Costantini opted instead for Italian ones on the basis that they would best know the roads. They were Pietro Bordino, Count Gastone Brilli Peri and Tazio Nuvolari.

These were early days in the career of Nuvolari, arguably the world's greatest racing driver, who had been placed in a creditable tenth place in the previous year's Mille Miglia at the wheel of an uncompetitive works Bianchi. Also in 1927, along with his trusted mechanic Decimo Compagnoni and drivers Achille Varzi, Cesare Pastone and others, Nuvolari established a Type 35 racing stable in his native Mantua. He had gone on to wrest victories in the 1927 Rome and Garda Grand

[5] Count Giovanni Lurani, *Nuvolari* (Cassell, 1959).

Léon Duray with one of the type 43As he exchanged for three Miller racing cars. Introduced in 1929, the 43A differed from the 43 in having this American-inspired two-seater bodywork and rear 'rumple' seat. The body mouldings on this Jean Bugatti-designed body are another transatlantic influence.

Prix from Count Maggi, so was well acquainted with the marque.

In the 1928 Mille Miglia, the Bugattis were challenged by a cross section of the products of the Italian motor industry, in the shape of entries from Alfa Romeo, Lancia and OM. The three Type 43s began in fine style and were the first three cars to arrive at Bologna. But from thereafter the challenge faded. All the cars suffered from a variety of maladies, mainly relating to time-consuming plug troubles, overheating and braking problems though all finished.

Brilli Peri was the highest placed, coming sixth, and was second in the 3-litre class, with Bordino seventh in the same section. The fact that the cars had completed the course was, according to Count Lurani, 'solely due to the stoicism and tenacity of their drivers'.[6] The event was won by Campari in an Alfa Romeo and the make was to dominate the event until 1939. Bugatti, by contrast, stayed away.

Later in 1928 another, though mostly unofficial trio of Type 43s, entered the revived Tourist Trophy race, on 18 August, having that year been transferred from the Isle of Man to the Ards circuit in Northern Ireland. Leo d'Erlanger's car had works support and was driven by the Molsheim tester Louis Dutilleux, with further entries from Malcolm Campbell and Viscount Curzon, who became the fifth Earl Howe in 1929.

Although Curzon put up the fastest lap, at 70.15mph (112.89kph), he later withdrew with fuel problems. Dutilleux finished ninth though suffered in the latter stages from fuel pump trouble. Campbell's car had the misfortune to memorably and spectacularly catch fire. It had been fitted with an experimental aircraft-type flexible fuel tank of British manufacture which leaked on to a full length undertray. The formidable blaze saw the flames lick halfway across the track and there was plenty of smoke.

As the fire raged, 'the Bugatti's aluminium wheels melted right away one by one, and when the car was resting on its drums the back axle casing melted and the prop shaft and bevel wheel with its bearing fell out.'[7] But it was not, by all accounts, insured and Campbell lost over £1,000 as a result of the inferno. Fortunately there were no

[6] Count Giovanni Lurani, 'Mille Miglia' (Automobile Year, 1981).

[7] Richard Hough, *Tourist Trophy* (Hutchinson, 1957).

casualties. However, the car (43171) was not totally destroyed and probably, though partially, survives.

Earl Howe was back with his Type 43 in 1929. There was another private entry from J F Field while the works entered no less than four cars which were driven by Albert Divo, Williams, Count Conelli and Baron d'Erlanger. The bodies of these works cars differed from the production ones, in that the pointed tails were replaced by flat-backed ones leaving the petrol tanks exposed. The event was hardly a Bugatti triumph. Both Conelli and Williams put up the fastest times in Class D but none of the Type 43s finished, while d'Erlanger was disqualified from participating because he had been delayed in Paris, had missed the boat from Liverpool to Belfast, and had been unable to practise.

An Exercise in Eccentricity

This rather dismal racing record reveals that the Type 43 did not begin to match its Type 35B first cousin for consistency and reliability but, late in 1928, Molsheim announced what it described as the Super Grand Sport Type 43A for 1929, which would have probably been positively disastrous in competition, had it been built. It should not, incidentally, be confused with the Type 43A which entered production in 1931. This version was a 3.8-litre model, powered by an engine with no less than sixteen cylinders and would have sold for an impressive 400,000fr. (£3,225). Perhaps fortuitously, it never entered production.

The Type 43A of 1929 had its origins in the 16-cylinder racing car, designated the Type 45, that Bugatti designed in 1928, apparently to compete in that year's French Grand Prix. It was proposed to run the event with no limits to engine size while the cars were required to weigh between 550 and 600kg (10.8 to 11.8cwt). But as only six entries had been received by the closing date, these requirements were scrapped and replaced by a race for sports cars. Despite this, Williams in a Type 35C Grand Prix model but fitted with wings and headlamps to present the semblance of a road car, won the Grand Prix!

Bugatti was therefore left with his 16-cylinder model which was revealed to the press in April 1929. He was, of course, no stranger to the concept. His U16 wartime aero unit consisted of two eight-cylinder units geared together, one mounted alongside the other, on a common crankcase. This configuration was perpetuated for a stillborn Type 34 aero engine of 1925. However, its design differed radically from the earlier U16 in that, instead

A contemporary photograph of the 16-cylinder 3.8-litre engine, as announced in 1929. The blocks were single castings, each with its own supercharger. This is a dry sump unit.

of the eight consisting of two in-line fours, each block of cylinders was a single casting. This new arrangement meant that the nine bearing crankshaft could be attached directly to the block. It was a concept that would first be applied to the eight-cylinder engine of the mighty Royale car and then the 16-cylinder Type 45.

Bugatti's idea to revive the configuration may have been sparked by Fiat. It will be recalled that Italy's largest car maker had withdrawn from Grand Prix racing in 1924 but briefly returned to the fray with the 1.5-litre twelve-cylinder Type 806 in 1927. Its engine consisted of two 742cc six-cylinder blocks mounted on a common crankcase, the crankshafts of which were geared together. It briefly appeared at the Monza Grand Prix, in September 1927, with a gaggle of Bugattis and a lone Alfa Romeo. Although it won, Fiat never raced again.

The 16-cylinder unit had a capacity of 3,801cc. Each 60 x 84mm eight-cylinder block was fitted with its own rear-mounted Zenith carburettored supercharger, the inlet manifolds running between the middle of the blocks. Each block also possessed its own nine bearing crankshaft, eight of which were rollers though the central one was plain to permit oil to reach the mains and the plain bearing connecting rods.

The cranks were geared together and the camshafts and superchargers were all driven from the rear of the unit by a train consisting of no less than nineteen gears! One of these also drove the front-mounted, centrally located water pump by a long shaft that ran between the blocks. In view of the innovative bearing construction, the sump was little more than an oil pan and there was a total of three oil pumps. One was needed for lubricating the crank, and its bearings, at 100psi, another for the valve gear, while the final one was a scavenger unit. Drive was taken from a central pinion, via the usual Bugatti wet plate clutch, to a separate gearbox.

Because the 16-cylinder engine was a remarkably compact unit, it was fitted in a chassis very similar in concept to that used on the Type 35 though with a longer 2,600mm (8ft 5in) wheelbase. Because of this, there was no room to take the twin exhaust pipes below the chassis rails, as on the 35,

This example of the extraordinary 16-cylinder Type 45 Grand Prix car, chassis 47156, was owned by the Bugatti factory until its purchase by the Schlumpf brothers in the 1960s. It is seen here after its rebuild at Molsheim and prior to its delivery to Malmerspach.

so for the first time on a Grand Prix Bugatti, they were run along the body sides. Also unlike the 35, where the rear quarter-elliptic springs followed the contours of the chassis members, those of the Type 45 were mounted outboard and therefore projected beyond the bodywork. The customary aluminium spoked wheels were fitted.

The Type 45 was only competitively active in 1930 though it never appeared on the race track, where what had proved to be its engine's relatively fragile rear-drive train would have probably succumbed but in short, hill climb events. During that year, Louis Chiron drove it in no less than three including the Austrian Klausen and at Frieberg in Germany. The car made fastest time of day on all occasions but it was not run again.

The Type 45 also spawned a short 66mm stroke 2,986cc Type 47 Grand Sport derivative with a projected price of 250,000fr. (£2,016). Intriguingly, the publicity leaflet produced for the model makes reference to an optional, enlarged 4-litre version. The 47's chassis differed from that of the 45, in that its side members were an unprecedented 228mm (9in) deep and the wheelbase was longer at 2,750mm (9ft).

As on the Grand Prix version, external exhaust pipes featured while the projected factory bodywork, which was 20,000fr. (£161) extra, was a simple two-door, four-seater tourer as the model was intended to run at long distance sports car events, such as the Le Mans 24-hour race. It is the Type 45's engine that would have also powered the Type 43A of 1929 and we should perhaps be grateful that none of these 16-cylinder Bugattis ever entered production.

The chassis of only two 16-cylinder cars, one of each type, were completed though only the Type 45 was bodied and both were retained by the factory. In 1963 they passed into the hands of the Schlumpf brothers and are just two of the historic Bugattis which can be seen at Mulhouse. A third 16-cylinder engine, built up from spare parts by Belgian Bugatti dealer, Jean de Dobbeleer, in the late 1950s and mounted in a Type 35 chassis, survives in Britain where this Type 47-based car greatly enlivens the vintage racing scene.

Despite the fact that neither the Type 45 and 47 entered production, the concept probably inspired Maserati to produce its own 4-litre 16-cylinder V4 model, which appeared five months after the announcement of Bugatti's 16, in September 1929. It consisted of two of its 26B twin overhead camshaft blocks geared together though only two examples were built. Also two years later, in 1931, Alfa Romeo introduced its own 3,504cc, twin-engined Type A, a model which paved the way for the formidable Type B *Monoposto* of the following year. Unlike the 16-cylinder Bugatti, both these multi-cylinder cars were run, with some success, in Grand Prix events.

The Popular Type 44

In 1927 at about the same time that the Type 45 was being conceived, Bugatti introduced his most popular touring model of the inter-war years. This was the 3-litre Type 44 which was the largest capacity Bugatti of its day. It sold 1,095 examples in the little over three years between October 1927 and November 1930 and chassis numbers 44251 and 44134–5.

The model shared the same 3,122mm (10ft 2in) wheelbase of its Type 38 predecessor though its performance was transformed by the fitment of its 2,991cc eight-cylinder engine which resembled that of the stillborn Type 28 of 1921. It therefore consisted of two 69 x 100mm cylinder blocks, of the variety used on the Type 40, mounted on an aluminium crankcase. But instead of the camshaft drive being located at the front of the engine, it was neatly sandwiched, Type 28-wise, between the two cylinder blocks. From there it drove the oil and water pumps, which were mounted in tandem and positioned on the left-hand side of the engine between the exhaust manifolds. Instead of fitting the customary French carburettor, for the first time Bugatti opted for a single Schebler unit of American origin which was mounted on a water-heated inlet manifold.

As refinement rather than performance was the order of the day, the Type 44 had the distinction of

A Bugatti catalogue which draws the parallel with horse-drawn vehicles of the past and a fiacre-*styled Type 44.*

being the first eight-cylinder Bugatti to reach production with a plain rather than roller bearing engine. It used no less than nine of them which were initially jet-lubricated though in 1928, from engine number 292, the system was converted to a fully pressurized layout which meant that the oil pump was relocated from the centre of the engine to sump level. Bugatti took the opportunity to redesign the crankshaft so, instead of it being split into two halves (effectively two four-cylinder shafts), it was replaced by a better balanced 2–4–2 single-piece configuration. The separate gearbox with its central gear lever was courtesy of the Type 38, as were the axles and brakes.

The 44's arrival coincided with an expansion of Molsheim's body building facilities which were enlarged to reflect the increasing popularity of saloon bodies on sporting cars in France. The Type 44 therefore became the first Bugatti to be offered with a range of stylish, mostly closed coachwork. These included two- and four-door saloons, *fiacre* two-door coupes and a two-seater roadster in the manner of that employed on the Type 40 and 43A. It was also fitted with bespoke coachwork, with examples being bodied by such establishments as Labourdette and Vanvooren in France, and Harrington and Corsica in Britain.

The Type 44 was particularly popular in closed Weymann bodied form and a road test of such a car was published in the July 1928 edition of *Motor Sport*. After the editor had deprecated 'the fitting of a saloon body on a sports car' he was soon won over by the Bugatti's road manners:

> In traffic, the car proved a delight to handle as either the top gear crawl or the second gear 'buzz and jump' method could be indulged at will. It was soon realized that all the drawbacks of the more 'racing' Bugattis were absent, giving place to real docility, smoothness and silence at low speeds and at the same time not excluding such desirable *traits* as splendid acceleration and deceleration.

At Brooklands the car averaged an impressive 128kph (80mph) before the plugs 'cooked'. The Type 44 sold for a respectable chassis price of £550, excellent value for a straight-eight, that *Motor Sport* considered to be 'equivalent to that of the dull British medium powered tourer'.[8]

A 3.3-Litre Single Cam

The Type 44 was replaced in 1930 by the mechanically similar 3.3-litre Type 49 model, the first chassis number being 49111. This was destined to be the last of the single overhead camshaft Bugattis and is also considered as the best of Molsheim's touring models. The increase in capacity, to 3,257cc, was achieved by upping the bore size from

[8] *Motor Sport*, July 1928.

A 1929 Type 44 with fashionable Weymann body by a British coachbuilder. That year a Type 44 chassis sold for £550 in Britain which was excellent value for an eight-cylinder car.

69 to 72mm, output rising by 5hp to 85bhp. Top speed remained at about the same as the Type 44's though it was even more flexible, underlining the 49's dual role as a tractable tourer and well mannered town carriage.

Other mechanical departures from previous practice included the introduction of two sparking plugs per cylinder, the single block also being used in the four-cylinder Type 40A. The 49's engine was fitted with a cooling fan and the radiator was also enlarged. Other differences between it and the Type 44 included the grinding of the gearbox gears, while its gate change was replaced by a more fashionable ball unit. A helical bevel drive was also introduced, all these improvements being intended to quieten the drive train, such undesirable sounds being more apparent with the increasingly popularity of closed bodywork. But unlike the 44, which was only available in one chassis length, its Type 49 successor was offered in additional 3,222mm (10ft 5in) long wheelbase form.

The Type 49 could be had with wire wheels though it was distinguished by new purpose-designed aluminium ones. The later sported diagonal cooling fins around the periphery to direct air to the brake drums, and resembled smaller versions of those fitted to the Royale. The Type 49 lasted until September 1932, and chassis number 49118–9. A total of 470 examples were built and, of these, no less than 127 were bodied by Gangloff, mostly in two-door drophead coupe form.

Bugatti had been moving progressively up market as far as his passenger cars were concerned and, at the 1929 Paris and London Motor Shows, he introduced his eight-cylinder 5.3-litre Type 46 which can be considered as a scaled-down version

A Type 49 fitted with Gangloff drophead coupe coachwork on home territory.

A 1932 Type 49 with a sports saloon body, of which there were a number, by James Young of Bromley. The distinctive cast aluminium wheels were usually though not exclusively fitted to this model. This one has lost its enjoliveur *used to cover the wheel securing bolts.*

of the gargantuan 12.8-litre Royale (Type 41) considered in the next chapter. The Type 46 was the most luxurious Bugatti, the Royale excepted, of its day, its chassis selling in Britain for £975 in 1931, which was £425 more than the Type 44. It was about the same price as the similar capacity though more pedestrian 35hp Daimler six, with the L head straight-eight 833 Packard chassis £200 less and, although a respected American make, it lacked the mechanical refinement of the overhead camshaft Bugatti.

The 46's engine radically differed from the eight-cylinder Type 44 and 49 because of its resemblance to the Royale and it even retained the mammoth's 130mm stroke and one, as opposed to two-piece cylinder block. This and a bore size of 81mm produced a capacity of 5,359cc and the engine developed 140bhp at 3,500rpm. Its layout also followed Royale, and 16-cylinder Type 45/47 precedent, in that the nine bearing, circular webbed crankshaft was attached directly to it. The only trouble with such a construction was that, to grind or replace the twenty-four valves in this fixed head unit, the engine had to be removed from the car,

The other side of a Type 49 cast alloy wheel, revealing the integral brake drums and cooling fins. They also appeared, with minor differences, on the Types 46 and 50.

then the cylinder block, crankshaft, connecting rods dismantled to gain access to them!

The single overhead camshaft was driven from the front of the engine while a single Smiths' multi-jet carburettor was employed, fed from a pair of Autopulse electric pumps. A dry-sump lubrication system was fitted, fed from a dashboard-mounted tank.

Rather than the gearbox being located in its traditional position, the Type 46's three-speed unit was incorporated in the rear axle and actuated by a centrally located ball change, which also perpetuated Type 28 and Royale practice. The Type 46 also boasted its own optional finned, aluminium wheels.

A Refined Approach

The model had its own chassis with a wheelbase of 3,500mm (11ft 4in) which was therefore demonstrably longer than the 44/49 one. Although Molsheim offered its own range of open and closed bodywork, many customers preferred to order their own. In Britain, despite its 5,359cc, Type 46 was popularly known as 'the 5-litre' and Freestone and Webb, H J Mulliner and James Young all produced mostly closed bodies for the model. A chauffeur-driven saloon provided appropriate transport for Lieutenant-Colonel Wyndham Sorel, manager of Bugatti's London-based sales outlet. In France, Gangloff bodied twenty examples with such Parisian *carrossiers* as Saoutchik, Figoni and Cartier bodying Type 46s.

As far as performance was concerned, the model could be wound up to around 144kph (90mph) and it benefited from the customary Bugatti flexibility. Top gear could therefore be engaged at around 16 to 24kph (10 to 15mph) though with only three speeds; on the debit side, acceleration was less spectacular than earlier models and braking left something to be desired.

The Type 46 was destined for a two and a half

A Bugatti in the Jazz Age. The 5.3-litre Type 46 could also be fitted with wire wheels. This example has Corsica coachwork, decorated in what we today refer to as the Art Deco style.

An early Type 46 engine with single sparking plugs though twins were later standardized. The carburettor is a Smith-Bariquan multi-jet unit.

year manufacturing life though, from 1930, it was to some extent challenged by its less successful though potent, supercharged, twin overhead camshaft Type 50 derivative. Therefore from that year the 46, which developed 140bhp, was offered with an option of a supercharger, though the resulting 160bhp Type 46S did not compare with the popularity of the basic model and accounted for a mere eighteen of the 400 or so Type 46s built. Production of both models ceased in 1932. The last unblown car was chassis number 46579 and the final 46S, built in June of that year, 46588.

This was a respectable enough figure, particularly at a time when the market for luxury cars had been badly hit, from 1929 onwards, by the onset of the world depression. Nevertheless a Type

The 4.9-litre Type 50 was the first Bugatti to benefit from a twin overhead camshaft engine. A fast expensive model, it was built in relatively small numbers. This is a magnificent example of a coupe with imitation canework by Weymann.

46 was displayed on the Bugatti stand at the 1933 London Motor Show and was still being listed in Britain in 1934, which was two years after it had ceased production. So this model, which was reputed to be Ettore's favourite, was a slow seller in its latter years. Having said that, by this time the last of the Type 50s had been built and this amounted to a mere sixty-five cars produced in four years.

Logically, the Type 50 should never have been built. It is difficult to think of a worse time, from an economic standpoint, for Bugatti to produce yet another luxury model. It retailed in Britain for a chassis price of £1,725 which was about the same as a Rolls-Royce Phantom 11. Perhaps, not surprisingly, only one example was sold by Bugatti's London depot. Colonel Sorel strongly disapproved of it, and chassis 50118 went to that great Molsheim enthusiast, Sir Robert Bird Bt., chairman of the Birmingham-based Bird's Custard.

It seems far more likely that the Type 50's 4.9-litre eight-cylinder engine was created, not for what proved to be a rather unhappy road car, but because Bugatti needed a big twin overhead camshaft supercharged eight in the increasingly competitive field of *Formule Libre* racing.

Straight-Eight Luxury

The fitment of a twin overhead camshaft Miller-inspired cylinder head to the big eight preceded its arrival of the Type 51 Grand Prix car though the 50's valves were at an included angle of 90 degrees, instead of the racer's 96. The twin overhead camshafts were driven from the crankshaft, via a drive shaft through five spur gears, an idler and a final pair of bevels, Like its single cam predecessors, the Type 50's cylinder head was a fixed one, a practice that was perpetuated on all the Bugattis produced throughout the 1930s.

Apart from its new head, the engine was similar to that of the Type 46, having a nine plain bearing crankshaft supported by the cylinder block and dry-sump lubrication. It was, however, supercharged from the outset, a pointer perhaps to its ultimate, Grand Prix status. In view of the potency of the new head, the engine capacity was reduced from 5,359cc to 4,972cc, effected by the bore size being upped from 81 to 86mm and the stroke reduced from 130 to 107mm. It developed no less than 225bhp at 4,000rpm.

The Type 50's chassis was effectively carried over from the Type 46 with its rear-mounted, three-speed gearbox. Unlike its predecessor which could be fitted with optional wire wheels, finned aluminium ones were standardized, though they differed slightly in pattern to those of the earlier car. On the Type 50, the 46's wheelbase was somewhat reduced at 3,100mm, (10ft 1in) rather than 3,500mm (11ft 4in). This latter dimension, however, was revived for the detuned 200bhp 50T, for Touring, version which effectively replaced it in March 1933.

The first Type 50, chassis number 50115, was delivered to Bugatti Paris agent and family friend, Dominique Lamberjack, on 11 October 1930. Ironically, the first chassis, number 50112, was completed a year later, on 21 October 1931. The final type 50 chassis, number 50176, left Molsheim in December 1934.

The model was pricey and fast, too much so for some and it has consequently acquired something of a chequered reputation. However, the 50 inherited the traditional flexibility of the Bugatti touring cars, being able to accelerate from 8 (5mph) to 180kph (15mph) in top gear. This made it the fastest Bugatti road car of its day and, as such, naturally attracted the attention of some of the leading Parisian *carrossiers*.

The reality was that the majority of the Type 50s were bodied by the factory. It was with this model that the precocious stylistic talents of Jean Bugatti who, incredibly, had only celebrated his twentieth birthday in 1929, began to reach maturity. Prior to the model's arrival, his influences had essentially been two-fold: the *fiacre* style of his early work, which dated from 1928, echoed his father's fascination for horse-drawn vehicles of the previous century while the next stage, which manifested itself from about 1930, saw Jean clearly coming under the influence of the American custom coach-builders with the two-seater with dickey roadster bodies of the Types 43A and 40A. A coupe body

A two-door Type 50 saloon with French coachwork.

which Jean executed for works driver Louis Chiron on a Type 50 in 1932 clearly reveals another transatlantic influence and is his interpretation of the contemporary L-29 Cord cabriolet.

When Type 50 production got underway in earnest in 1931, Jean initially perpetuated the *fiacre* style he had previously employed. However, a new element arrived in the shape of coupes with mildly sloping, low windscreens which reflected contemporary French thinking, though their cabriolet-like rear folding roofs were unusual for such bodies. These, in turn, paved the way for the daring Profilée two-door saloon, in which many of the transatlantic influences were banished. The result was a daring, visually extravagant coachwork which dramatically complemented the Type 50's formidable performance.

With the Profilée, the rake of the windscreen was considerably increased and the cabin abruptly cut off at the rear, the body being completed by the integrated, visual presence of the boot outline. At the same time, the front wings were raised, accentuated and extended so that they also did double duty as running boards. Less happily, the bottoms of the doors had to be angled to clear them.

The impact of these audacious low lines was enhanced by the use of contrasting body paints. The wing colour was extended in a dramatic, curvilinear sweep to flood the upper portion of the body which then tapered down the bonnet contours to end at the radiator. This sweep panel device was an American technique, that had been first used by the New York-based Le Baron styling house, and successfully employed on the Double Cowl Phaeton bodies it had produced for the 1929 Model J Duesenberg, a line that was available until 1934.

The impeccably proportioned frontal view of a Type 50 fitted with Jean Bugatti's Profilée *saloon coachwork. Note the mouldings on the bonnet sides and around the periphery of the front wings.*

A three-quarter rear view of the same car, highlighted by body moulding and two-tone paintwork. The wheels are now retained on knock-off hubs rather than bolts. As the style evolved . . .

. . . the rear of the body was extended to incorporate the boot and the rear wings were reshaped. This is the car exhibited at the 1932 Paris Show.

The Profilée body style was to evolve and mature by the time that the Type 50 ceased production in 1934. It was perpetuated on the longer wheelbase 50T from 1933 onwards, by which time the front and rear wing lines had merged into a single, controlled sweep, the boot had been eliminated and the rear roof line extended.

Despite the Type 50 being a road car, its engine was clearly destined for greater things and the model made its first competition appearance in the 1931 Mille Miglia, held on 11 April. According to Count Lurani,[9] Achille Varzi, who had just signed for Bugatti as a works driver entered as an independent, as he was 'anxious to avenge his 1930 defeat by Nuvolari' and had chosen a Type 50 which he had been given by the firm. It was perhaps just as well this was a semi-official entry because Varzi had the misfortune to retire with engine trouble a few minutes after the start of the event.

The Type 50 at Le Mans

However, the Type 50 did form the basis of a Bugatti team that entered the 1931 Le Mans 24-hour race. It will be recalled that no works cars had run in the event during the 1920s, probably on account of the fact that Molsheim's racing and sports cars were of a relatively low capacity and Ettore tended to avoid competing in events which he did not think he could win. The decade had been dominated by the 3, 4.5 and 6.5-litre Bentley victories at the circuit and these may have produced the legendary Bugatti quip that 'Monsieur Bentley produces the fastest lorries in the world!'

But the Type 50 was a 4.9-litre car and a team of three, with simple four-seater black Weymann touring bodies by Vanvooren, were entered for the event. W F Bradley had informed us that, at this time, 'Ettore had fallen out with Dunlop . . . and declared impetuously that he would find French tyres which were just as good.'[10] So he opted instead for Michelins and the trio of Type 50s, managed by Jean Bugatti, were driven by Chiron and Varzi, with Albert Divo and Guy Bouriat in another car and Count Carlo Conelli and Maurice Rost in the third.

Tragically, tyre trouble manifested itself prior to the event but the works Bugattis 'continued to use them even when tread-throwing in practice confirmed their weakness at racing speeds'.[11] Bradley

---

[9] Lurani, 'Mille Miglia'.

[10] Bradley, *Ettore Bugatti*.

[11] David Hodges, *The Le Mans 24-Hour Race* (Temple Press, 1963).

The Type 50, driven by Maurice Rost, that crashed at Le Mans in 1931. As a result the team of three cars was withdrawn.

says that Varzi described the cars as 'dangerous . . . and . . . the other men agreed with him'.[12] It was too late to make a change and Rost has recalled that for the race 'we were under orders not to exceed 4,000rpm, along the straight, i.e. 200kph [125mph] since our ratio gave 50kph [30mph] per 1,000 revs. This was good when you consider that, with its 210kg [463lb] of ballast, the Bugatti weighed about 2,032kg [2 tons].'[13] Rost has divulged that the Type 50's 4.9-litre engines were boosted to 275bhp for the 24-hour event.

The main challenge came from a team of works Alfa Romeos and, says Rost, 'according to our orders Louis Chiron was to take the lead ahead of Albert Divo and me.' Then after about two and a half hours, Chiron stopped to change a tyre and he later threw a tread trying to pass Birkin's Alfa Romeo and dropped back to ninth place. After Chiron's difficulties, Jean Bugatti ordered the three Bugattis to slow to about 185kph (115mph) down the straight. But just as Rost, who had worked himself up to third place, was approaching the curve prior to *Mulsanne*, he felt 'a tremendous blow at the rear of the car. [It] slewed across the road and I could do nothing to right it. I rolled myself up into a ball and at the first impact was thrown out of the car.'

His left-hand rear tyre had thrown a tread and, as a result, the wheel locked and the car left the road, killing one spectator and injuring several others. While Rost was being rushed to the Delageneire Clinic, Chiron lost another tread so Jean Bugatti telephoned his father, who ordered that the cars be withdrawn. Thankfully Rost recovered from his injuries after two months 'but I only realized the gravity of the accident when [Pierre] Marco, at the Molsheim factory, lifted up a tarpaulin and showed me the remains of my poor car. There and then I became frightened for the first time – a kind of delayed action fear – and I have not raced since.'[14] It was little consolation to Rost that the curve on the Le Mans circuit, where he had crashed, was named after him.

Bradley says that after the accident Bugatti 'reproached himself for allowing the cars to start in an unsatisfactory state. He had always been proud of the fact that no accident had ever occurred which could be directly attributed to faulty mechanism.' His anguish was compounded by the fact that there were insinuations that the tragedy had been caused by the cars' aluminium wheels, and not faulty tyres, which revived memories of the Lyon debacle seven years previously.

As a consequence, Ettore declared himself against night racing with touring equipment and

[12] Bradley, *Ettore Bugatti*.

[13] Georges Fraichard, *The Le Mans Story* (The Bodley Head, 1955).

[14] Fraichard, *op. cit.*

proposed the creation of a new race, spread over five days, with the cars being locked in a garage every night. He offered a prize of 500,000fr. (£2,865) which he said he would forfeit if a Bugatti won, though the suggestion was not taken up. He also settled his differences with Dunlop.

Fast and Fateful

Three months after the Le Mans incident, the Grand Prix car which was powered by the Type 50's 4.9-litre engine, made its appearance. It was designated the Type 54, and in his article for *The Autocar* of 11 September 1931, W F Bradley reported that: 'The new racing Bugattis possess the distinction of having been designed, built and put on the road in thirteen days.' As the car was largely created from stock parts this seems more than likely.

Th model was created to meet the increasingly strident Italian challenge in *Formule Libre* events where Bugatti's 2.3-litre Type 35B, which was being replaced by the newly introduced, twin cam Type 51 of a similar capacity, was being increasingly outclassed. Ironically, it had probably been the arrival of Ettore's stillborn 16-cylinder Type 45 in 1929 that had prompted Maserati's 4-litre V4 of the same year and it was to follow this with the 4.9-litre V5 in 1932, also with sixteen cylinders. The V4 had a limited and chequered competition career. It managed sixth place in its 1929 Monza Grand Prix debut though proved itself better for straight-line record breaking. It did, nevertheless, score a rare victory in the relatively new Tripoli Grand Prix of 1930 and was always threatening.

Alfa Romeo's twin 1,750cc-engined, 3.5-litre Type A single-seater first appeared in May 1931 at the Italian Grand Prix. There, it was said to have been fast but something of a handful to the extent that even the great Tazio Nuvolari was relieved when Vittorio Jano withdrew the car, even though it was well placed at the time. But at the Acerbo Cup Contest at Pescara in August, Campari had a fairly easy win in a Type A, ahead of Chiron's Type 51 Bugatti, a victory that may have prompted Molsheim's panicky response.

The Type 54 was effectively the Type 50's 4.9 twin overhead camshaft engine, with its output boosted from an original 215 to 300bhp, and separate, three-speed gearbox mounted in the redundant 16-cylinder Type 47 chassis with 2,750mm (9ft 3in) wheelbase. It was fitted with the usual broad-spoked aluminium wheels though with non-detachable rims, as featured on its Type 51 contemporary. But the greater power produced by the 4.9-litre engine was reflected by the fitment of enlarged 400mm integral brake drums. The 51, by contrast, used 330mm ones.

The Type 54 used the redundant chassis of the 16-cylinder Type 47; note the depth of the side member. External exhaust pipe also differed from previous GP practice.

Two cars were completed in time for the model to make its competitive debut at the Monza Grand Prix on 6 September 1931. First there were two eliminating heats, one of cars of under 3 litres and the other for racers over that figure. The latter event was billed as 'The Battle of the Giants' and the single line of racers consisted of Varzi and Chiron in Type 54s, Nuvolari and Campari in Type A Alfa Romeos and Ernesto Maserati in a V4 Maserati. Things looked good for the new Bugattis as Varzi and Chiron took first and second places, though in the race proper, Chiron retired when a tyre failed and left him without brakes. Varzi in the other Type 54 came third despite tyre troubles behind Fagioli's 2.8-litre Maserati, which won, and a Monza Alfa Romeo.

A total of thirteen chassis, from numbers 54201 to 54213, were eventually allotted to the series. But the Type 54 soon acquired a deserved reputation for poor roadholding, the result perhaps of the rushed union of its unrelated engine and chassis. The brakes were also unable to cope with the 300bhp developed by the supercharged twin cam eight. It was no doubt in view of these all too obvious deficiencies that the first five cars, 54201 to 54205, were sold off by the works during the first four months of 1932. Two went to French drivers: Marcel Lehoux and Jean-Pierre Wimille and a similar number to Britain, where Kaye Don and Earl Howe each took deliveries of 4.9s. Bugatti's Zurich-based Swiss agent Bucar, was the recipient of another for Prince Lobkowits. The model's difficulties were underlined by Wimille who soon returned his car to Molsheim because he was dissatisfied by its performance. It was then dismantled.

Further Type 54s were then built and campaigned by the factory over the next two years. Varzi and Divo each drove 4.9s in the 1932 French Grand Prix with private entries from Howe, Lehoux and Don though the last named was a non-starter. The race did not last long for Varzi, who dropped out with gearbox trouble on lap 12, and the same trouble afflicted Lehoux on his Type 54. Divo withdrew on lap 52 with a split fuel tank.

The highest placed Bugatti was Chiron in a Type 51, which came fourth behind a trio of P3 Alfa Romeos. Howe, meanwhile, also suffered from gearbox maladies; he only had the use of first and top gears. He shared the driving with Hugh Hamilton, and came ninth. About the only other noteworthy victory for a Type 54 during the remainder of the year came when Williams won the La Baule Grand Prix.

Because of the cars' unhappy reputation, during the winter of 1932/1933, modifications were made to the works Type 54s and to the cars belonging to Don and an example, number 54209: the property of a Polish aristocrat, the Paris-domiciled Count Stanislas Czaykowski. These changes included moving the engine back in the chassis, which had the effect of improving its roadholding though the brakes were left unaltered.

Count Czaykowski, who had previously successfully raced a Type 51 Bugatti, was to give the Type 54 its only major international accolade when, on 5 May 1933, he wrested the world hour record from George Eyston who had gained it in the previous year at Montlhéry in an 8-litre sleeve valve Panhard Levassor. The Pole averaged 213.8kph (132.87mph) for 293km (182 miles). His record was, alas, to be short-lived, because in 1934 Eyston regained it and achieved a marginally better 214.05kph (133.01mph).

Czaykowski's triumph was achieved at the Avus circuit, located in the Grünewald district of Berlin. Built in 1921, it consisted of two parallel roads, only 7.6m (25ft) apart, and linked by curves at either end. It was accordingly a very fast track and Czaykowski's run, after two days of practice, took place between five and six o'clock in the morning when the air was cool and fresh and well before the day's racing began.

This venue was to be scene of the last great Bugatti racing triumph for, after the Czaykowski's successful record run, it was prepared for the 1933 Avus–Rennen, held that year in place of the German Grand Prix. This was attended by Adolf Hitler, who had been elected German chancellor just over three months previously. The opening race was dominated by a battle between two Bugattis, one driven by a German driver, Ernst Gunther Burggaller, and the other by a Frenchman, Pierre Veyron. A tremendous tussle ensued and Veyron finally won, a mere three-tenths of a second ahead of the German.

The main event of the day was a fifteen-lap event for cars of an unlimited category and, of the eleven starters, no less than three were Type 54s. Czaykowski was in his record-breaking car, while works machines were driven by Varzi and Williams. The opposition consisted of Nuvolari, Borzacchini and Chiron, all driving Alfa Romeos. German honour was upheld by von Brauchitsch in a streamlined Mercedes-Benz.

The Type 54s soon dominated the event though Williams dropped out when his car caught fire. Czaykowski, who was familiar with the circuit following his earlier record run, managed to just keep ahead of Varzi. Both cars increased their

This is 'Williams's' works Type 54 pictured at the 1933 Avusrennen on 21 May. He retired on lap 7 when his car caught fire.

There were no less than three Type 54s in the 1933 Avusrennen and this is the rear of the other works car driven by Achille Varzi who won the race. From this viewpoint, it looks similar to the Type 59. Note the twin fillers and that of the scuttle-mounted oil tank.

Varzi photographed after his win at Avus in 1933. Note the divided front axle, designed to introduce a modicum of independence.

performances so that, by the end of lap 12, they both managed just over 214kph (133mph) and Barre Lyndon has recorded that, 'The spectators had eyes for no other machines now. The two cars, never more than a few yards apart, were moving down the straights at a pace hardly slower than the fastest speeds ever recorded at Avus, and it roused the excited crowd to cheers which sounded above the volleying of the howling exhausts.'[15]

On the penultimate lap, Varzi managed to get ahead of Czaykowski and, on the fifteenth and final circuit, the Count attempted to regain the lead but was unable to do so and Varzi won but only by a tenth of a second. Both cars averaged 208.08kph (129.3mph). We can only speculate at Hitler's thoughts at this double French victory, for he stayed to the end of the second race. The following year, however, would see a revival of German racing supremacy with annual state subsidies of RM450,000 (£33,088) provided respectively for the Mercedes-Benz and Auto Union teams.

Success at Brooklands

Czaykowski entered his 54 in the 1933 French Grand Prix in June but retired on the ninth lap with the inevitable gearbox trouble. Two months later, he brought his car to Britain and, on 1 July, won the British Empire Trophy race at 198.87kph (123.58mph) with Kaye Don, who was coincidentally also of Polish extraction, second in his red-painted Type 54. Don called his 4.9, 'Tiger Two', to complement his ownership of the ex-Henry Segrave world land speed record Sunbeam Tiger.

Don had earlier run it in the International Trophy meeting at Brooklands, held on 6 May, on the day after Czaykowski's Avus triumph. Then Don had the misfortune to have the car's back axle fail, eight laps from the end, while he was in second place. In addition to the modifications made by the factory intended to improve the 54's roadholding, Don had attempted to rectify his car's poor braking and had had the mechanical system replaced by a Lockheed hydraulic one. It had been fitted by Louis Coatalen himself, who had forsaken his stewardship of the STD combine for a post as the head of Lockheed's French division.

Two months after his Brooklands success, Czaykowski was in Italy for the Monza Grand Prix. This was staged on the afternoon of 10 September, to follow on from the Italian Grand Prix held earlier in the day. The Monza consisted of three heats and a final and the Count won the first at 182.6kph (113.47mph), ahead of two Monza Alfa Romeos. The second heat was tragically marred by the deaths of Campari and Borzacchini. Both died after their respective Alfa Romeo and Maserati cars had skidded on a patch of sanded oil that had been spilt by Trossi's new Scuderia Ferrari single-seater

[15] Barre Lyndon, *Circuit Dust* (John Miles, 1934).

Duesenberg. The latter had thrown a connecting rod in the previous race in its driver's attempt to catch the flying Czaykowski.

In view of this double fatality, the final race was accordingly reduced from twenty-two to fourteen laps. Czaykowski in the Type 54 soon put up the fastest lap, at 187.98kph (116.81mph), but then tragedy struck for a third time when he crashed on the eighth lap, having skidded on the same patch of oil on the South Curve that had earlier claimed the lives of the two Italian drivers. He spun off the track, the Bugatti exploded and he was killed. The event was won by Lehoux in a Type 51.

Czaykowski's car was, of course, his own – though the works Type 54 only appeared occasionally in second-stream events during the 1933 season. Bugatti was represented by privateers at that year's French Grand Prix because his new Type 59 Grand Prix car, although entered, was not ready and was withdrawn just before the race. The 54 in which Varzi had triumphed at Avus was used by Dreyfus in the Marseilles Grand Prix at Miramas in August but, while in the lead, he had the misfortune to lose a rear wheel during the closing stages of the race. However, an example was also victorious, for the second year running, at La Baule.

The Type 54's career, which began at Monza in 1931, also ended there two years later, for this was its final appearance in a major international event. Although Count Czaykowski's death there could not be directly apportioned to his car, it had been the second fatality of a driver at the wheel of a Type 54. The first had been in 1932 when Prince Lobkowitz, supplied with the first example (54201) by Bucar of Zurich in April 1932, had crashed fatally at Avus that month.

The model had, in truth, been the unhappiest of Bugatti Grand Prix cars, lacking the precision of its Type 35 and 51 predecessors. While the latter two models possessed an enviable record for reliability, the 54 was also deficient in that respect. Its transmission, usually its gearbox but sometimes the rear axle, was unable to cope with the 300bhp generated by its engine. This was a reflection of an economic chill at Molsheim at the time of its 1931 creation, a state of affairs which would continue to dog the Bugatti racing programme throughout the decade.

Four-Wheel Drive

The Type 54 did not represent Bugatti's only attempt to break the Italian domination of *Formule Libre*. In addition to the Type 50's 4.9-litre engine being used in the 54, it was also employed to power an extraordinary four-wheel drive Bugatti which was a contemporary of the big 4.9, as it first appeared in 1932. Not only was it unusual for a Molsheim product because it was not designed by Ettore Bugatti, but it also represented the first

The chassis of the type 53 four-wheel-driven car was based on that of the moribund 16-cylinder Type 45 sports model. This example was restored by Uwe Hucke. Note the rare Bugatti oil can!

serious attempt by a manufacturer to employ four-wheel drive – a configuration which had hitherto been only used on commercial vehicles – on a racing car.

The idea had originated with the talented former Fiat engineer, Giulio Cesare Cappa, who under managing director Guido Fornaca, had overall responsibility for the fabled Fiat Grand Prix cars of the early post-war years. After Fiat withdrew from competition in 1924, Cappa departed and set up an engineering consultancy in Turin. One of his first assignments was for the local Itala company for whom he designed its 2-litre Type 61 of 1924 powered by an aluminium six-cylinder pushrod engine. Far more interesting, when viewed in the light of future events, was an advanced front-wheel drive single-seater Grand Prix car which Cappa created for the 1.5-litre formula of 1926 and 1927 and was also available in 1,100cc form for *voiturette* racing, the concepts being designated the Types 15 and 11 respectively.

Adventurously the engine was a supercharged aluminium V12 unit and, consequently, its minute 53 x 55mm dimensions, with 45mm bore for the smaller version, were not rivalled until the appearance of the controversial 1.5-litre V16 BRM of twenty-four years later. Cappa's unorthodox approach did not, however, stop with the power unit for it drove the car's front wheels. The front suspension was therefore independent and consisted of an upper transverse spring with lower swinging arms of a similar conception, separately enclosed in streamlined steel fairings. A similar independent layout featured at the rear. In the interests of weight saving, Cappa used a wooden chassis. As a result the car only weighed 520kg (10.25cwt).

Despite this Itala appearing in 1926, Cappa does not seem to have been influenced by Harry Miller's front-wheel drive racers of the previous year. The design had been completed by the time that the American car had made its first appearance at that year's Indianapolis 500 Miles Race. It was, however, a contemporary of the pioneering front-wheel drive Grégoire-designed Tracta sports car from France which arrived in the same year. It is more likely to have been conceived at about the same time as the front-wheel drive Alvis hill climb car, which appeared in Britain two months prior to the Miller, in March 1925.

Unfortunately the 1926 Itala was never to run in a race, though a connection between Cappa and Bugatti dates from this time because Ettore's confidant and team manager, Meo Costantini, drove the sole example at Monza. He found that the 1,100cc car was capable of about 168kph (105mph) and considered that it performed well, apart from a tendency for it to skid on corners.[16] It seems that, despite its impressive specifications, the car was low on power, the V12 having been rumoured to have developed as little as 50bhp. Also, Itala was in financial difficulties at the time, and would have disappeared completely by 1934. Thankfully, this historic car has survived and can today be seen in the Turin Motor Museum.

Just when Cappa decided to extend the concept of his front-wheel drive design and develop it into a four-wheel drive one, is not known. We do have it on the authority of his young associate, Antonio Pichetto, that he had been working on the project in 1930, and that 'the design was offered to Bugatti and Jean Bugatti became enthusiastic.'[17]

Once Bugatti was committed to the concept, Pichetto was seconded to Molsheim to see the car into production. This was probably in the autumn of that year because the original Bugatti drawing of the car is dated November 1930. The Italian engineer initially stayed at the Hotel Heim in the town where he often dined with René Dreyfus who was also lodging there. The latter has recalled that, on such occasions Pichetto 'could talk about little else, he thought the Type 53 was a great car. I thought it was a monster.'[18] For his part, Antonio was to remain at Molsheim after the Type 53 had been completed and he stayed with Ettore until the latter's death in 1947.

[16] Cyril Posthumus, 'The 1926 1100cc 12 Cylinder Itala' in *Motor*, 1 November 1969.

[17] *Bugantics*, vol. 27, no. 1 (spring 1964).

[18] René Dreyfus with Beverly Rae Kimes, *My Two Lives* (Azbex, 1983).

For Once Independent Suspension

As in the case of the Type 54, the four-wheel drive car incorporated as many existing Bugatti parts as possible. Therefore the chassis had originally been conceived for the redundant 16-cylinder car though it used the shorter 2,600mm (8ft 5in) wheelbase of the Type 45 version, as opposed to the Type 47 one used on the 54. The car was powered by its 300bhp 4.9-litre supercharged engine, which was conventionally mounted and thereafter drove the separately located, purpose-designed, three-speed gearbox in the usual way.

The box, which contained its own differential, also incorporated the primary, intermediary and driven shafts which stepped the drive sideways to two outputs positioned on the left, one at the front and the other at the rear of the casing. The latter drove an offset differential in the rear axle by a short propeller shaft. Power was transmitted to the front wheels, via a drive shaft, that ran to the left of the engine having passed through its rear mounting, which incorporated a steady bearing. It was joined to the forward-mounted differential, contained within a split aluminium housing, by a universal joint.

As the differential was offset, the drive shafts were of varying lengths and drove the front wheels through universal joints though they were not of the constant velocity type. Suspension was by transverse leaf springs located above and below the shafts, in the manner of Cappa's still-born Itala racer, so the Type 53 remains the only Bugatti to be fitted with independent front suspension. The customary rear, reversed quarter-elliptic springs were perpetuated.

The Type 53 photographed at Shelsley Walsh in 1932, prior to Jean Bugatti's spectacular crash.

The Type 53's engine was essentially similar to that of the Type 54 Grand Prix car and that, in turn, to the Type 50; a 4.9-litre twin overhead camshaft supercharged eight.

The body was rudimentary, with the familiar contours of a Bugatti radiator concealed by a wire meshed cowl which also masked the front drive mechanism. Unlike the contemporary Bugatti Grand Prix cars, no tail was fitted, so the twin-capped petrol tank was untidily exposed to view, perhaps to save weight. The finned, cast aluminium wheels were similar to though not the same as those fitted to the Type 50.

The concept was, alas, far from successful as a Grand Prix car, though it proved to be better as a hill climb machine. During his two years at Molsheim, René Dreyfus was responsible for testing the Type 53. While it became apparent that the car

Starting the Type 53's engine, with the dry sump reservoir in the foreground.

possessed some advantages, namely that of its ability to stop and start easily on a wet road, Dreyfus soon found that 'the car was impossible to drift; whatever direction you aimed it, that was the direction you went. Accuracy and a tendency not to hesitate were called for. I could handle that. The strength of Hercules was also called for. That I couldn't.'[19]

The Type 53 made its first appearance at a hill climb, early in 1932, when Albert Divo, who was physically the strongest of the Bugatti drivers, drove in the Mont Ventoux event. There, he put up second fastest time of the day, at 15min 21sec, only 9 seconds behind Carraciola in an Alfa Romeo. In March, when the 53 was unveiled to the press, it was Louis Chiron's turn to hill climb the four-wheel drive and he broke the record at La Turbie.

In the following month the 53 made its one and only, albeit tentative, Grand Prix appearance at the Monaco event. It was no doubt thought that the race, with its elongated, twisting circuit through the streets of the Principality, would be an ideal venue for such a car. But in practice Divo was witnessed, 'cornering gingerly as if he expected it to turn over'[20] and according to Dreyfus, who was driving an independent Type 51 at the same event, after these preliminaries, Divo was left with 'a car that overheated, brakes that worked none too well and a driver who was dead tired after only a few laps.'[21] Consequently the 53 was not run and Divo drove a Type 51 instead. As it happened, the first Bugatti home belonged to a privateer, namely the indefatigable Earl Howe in his new Type 51, who came fourth behind two Alfa Romeos and a Maserati.

A second Type 53 was completed during 1932, which was when Chiron and Varzi competed in the Austrian Klausen Pass hill climb and were to finish second and third respectively behind Carraciola in a new P3 Alfa Romeo. They were to also compete at a similar event at Freiburg in Germany.

Sensation at Shelsley

The most celebrated climb of all for the Type 53 did not take place on Continental Europe but the unlikely Worcestershire surrounding of the Shelsley Walsh hill climb, which is Britain's oldest, having been established in 1905. The Type 53's driver was none other than Jean Bugatti who brought one of the cars to the Open meeting held on 25 June 1932. Jean and a lady friend drove in the latest Type 55 Bugatti, an open black and red two-seater, with the Type 53 transported in its own van accompanied by two Molsheim mechanics.

Jean's host was Britain's leading motoring peer and Bugatti driver, Earl Howe, at his family seat, Penn House at Amersham, Buckinghamshire. The lunch party held on the day prior to practice was completed by the arrival of the eldest son of the Marquis of Camden, Lord Brecknock, a Talbot director, who arrived in a Speed 20 Alvis, on loan from the rival Coventry company.

Practice took place on Friday 24 June, prior to the event proper. Fate ordained that few individuals would witness Jean's first ascent though, in the words of Austen May, 'the car put up the most startling climb ever seen on the hill up to that date and returned a time so remarkable that the figure has never been revealed, the secret remaining locked in Leslie Wilson's bosom [secretary of the organizing Midland Automobile Club] from whence wild horses will not drag it.'[22]

Similarly, Alec Rivers Fletcher has written more recently that 'Jean himself and Lord Howe also knew but agreed to keep quiet. In later years, I raised the matter with Earl Howe but when I did so, he merely changed the subject!'[23]

Jean was less lucky on his next run because, 'on a further trial spin young Bugatti unfortunately misjudged the difficult Kennel bend, his terrifically fast getaway landing him there at a very high speed. The car went into a tremendous broadside skid, pushing its tail through the iron railings on

---

[19] Dreyfus with Kimes, *op. cit.*

[20] Hodges, *The Monaco Grand Prix*.

[21] Dreyfus with Kimes, *op. cit.*

[22] CAN May, *Shelsley Walsh* (Foulis, 1946).

[23] AF Rivers Fletcher, *Mostly Motor Racing* (Foulis, 1986).

the one side of the road, and then hit the bank on the other side head-on.'[24] As a result, the 53's front wheels were splayed out and the centre of the axle beam became detached. Yet another version of the same story has it that Jean had previously agreed to wave at Kennel, which is the first corner at Shelsley, to a young lady with a rose pinned to her coat, did so but misjudged his speed and crashed.

Fortunately Jean was unhurt but managed to convince the stewards to accept a substitute entry and ran the Type 55 on the Saturday. Rivers has recorded that he made a 'very fast but rather wild ascent . . . I stood with his girlfriend watching the start, she asked me many questions but she was so excited and spoke so quickly that my school-French was quite useless.'[25] But was she the same young lady to whom Jean had agreed to wave on the previous day?

Fastest time of the day did go to a Bugatti. Earl Howe achieved this with an ascent of 44 seconds, for which he was awarded the Shelsley International Cup and 100 guineas. He drove the Type 51 in which he had achieved his creditable fourth place at Monaco two months previously, untouched apart from being checked over by Thomas, his mechanic, who fitted a new set of Dunlop tyres.

Then there was the matter of the damaged Type 53 and Howe insisted that his young guest telephone his father to inform him of the incident. According to May, 'holding the receiver in one hand, young Bugatti gesticulated wildly with the other, while he poured out a flood of voluble French into the mouthpiece, becoming more and more excited as the conversation progressed.'[26]

The damaged car was returned to Molsheim and, from thereon, the Type 53 was only occasionally seen. Chiron drove one in the 1933 Parma Poggio hill climb in Italy though the result is not known. Later, on 29 March 1934, René Dreyfus took the wheel of a Type 53 at La Turbie. There he managed fastest time of the day, with a climb of 3min 45⅖sec and exceeded 100kph (60mph) for the first occasion on the hill. But afterwards, 'I was exhausted . . . My shoulders felt as if they had been broken, my arms were dishrag limp with fatigue, my nerves were frayed.'[27]

The Type 53 made its final appearance at the second-string Château-Thierry hill climb in May 1935, with an ascent of 30⅖ seconds. Interestingly the car had the same registration number as the one carried by Jean Bugatti at Shelsley, so that example may have been rebuilt after its crash. Thereafter the Type 53 faded from view.

It is thought that only three examples of this four-wheel drive design were built and one, number 53002, survives in the south of France while the frame of a second, with its unique front-drive mechanism though minus the remainder of its mechanicals and body, can today be seen at the Mulhouse museum.

Forward Thinking?

History has cast the Type 53 a failure, largely because it was not fitted with constant velocity joints, which would have been needed to cope with the tremendous torque developed by its engine, particularly on corners. But the reality was that front-wheel, let alone front-drive technology, was in its infancy at the time of the 53's creation and very few cars were fitted with them.

This certainly applied to the ex-Léon Duray front-wheel drive Millers stored at Molsheim while Pichetto was working there on the four-wheel drive design. These were fitted with their maker's customary De Dion front axles. The make's historian, Mark Dees, has pointed out that the front-wheel drive Millers only needed to turn sharply, where constant velocity joints would have been an advantage, into the Indianapolis pits:

> The springs permitted a little amplitude for vertical movement (1½ in each way from rest), so Miller and Goossen could get away with fabric inner

[24] May, *op. cit.*

[25] Rivers Fletcher, *op. cit.*

[26] May, *op. cit.*

[27] Dreyfus with Kimes, *My Two Lives.*

couplings and standard Spicer outer universals of the Hooke type. Miller front drives used sliding block inner joints, but never constant velocity joints as original equipment until 1935, as far as we know.[28]

The joint that Miller finally adopted was the Rzeppa unit which was patented by Ford engineer Alfred Hans Rzeppa that year.

It is also worth recording that at about the same time Bugatti was developing 'his' four-wheel drive, Miller was engaged in a similar exercise. This was a logical development for a front-wheel drive pioneer and resulted in two cars. These were seen for the first time after the Type 53 Bugatti had made its appearance, at the Indianapolis 500 Miles Race of March 1932. They were powered by a purpose-designed 308 Cid V8 engine which, by coincidence, had an almost identical capacity as the 53, being 4,954cc, rather than the Bugatti's 4,972cc.[29]

There is evidence from front-wheel drive pioneer, Jean-Albert Grégoire, to indicate that Bugatti was told that his Type 53 needed such joints. Grégoire had been responsible for conceiving the innovative Tracta which appeared in France in 1926 and ran regularly at Le Mans between 1927 and 1934. It used the Tracta constant velocity joint, developed by his associate, Pierre Fenaille. According to Grégoire, although he encountered Bugatti looking 'more red faced and jovial than ever',[30] he knew that Ettore was experiencing problems with his four-wheel drive racer.

'He said to me: "It doesn't work, your front wheel drive!" and added, "the moment one wants to take a corner with such a machine, one can hardly hold the steering wheel and the car tries to leave the road!" ' Grégoire then enquired what sort of transmission his car had and Bugatti 'indicated that it was by means of simple carden joints. It was useless trying to explain the cause of the trouble and the necessity of using homokinetic [constant velocity] joints; he was not prepared to admit the obvious. As a result of his experience he condemned front wheel drive for ever.'

Perhaps the last word on the model should go to English *Bugattiste*, Geoffrey St John, who drove Uwe Hucke's rebuild car, (53002) very appropriately at Shelsley Walsh in 1985. St John has recorded that, to his surprise, he found that he actually liked it:

It was infinitely better than I imagined . . . the Type 53 handled so well . . . it was certainly quicker than the Type 51 would have been. You could really hoof it through the corners and throw it about just like any Grand Prix Bugatti. In my view the absence of constant velocity joints didn't seem to matter when the wheels are slipping and sliding. I can imagine that it might become tiresome to drive, as we have been led to believe on a long climb like Mount Ventoux . . . but its 'point and squirt' character was ideal for short bursts.[31]

By coincidence, St John's run took place in the wet, and Dreyfus has said it performed better in the rain. Nevertheless, his account suggests that the car was not quite as bad as it was made out to be. And significantly it was not until 1960, no less than twenty-eight years after the Type 53 had appeared, that the resources of Harry Ferguson helped create the first successful four-wheel drive racing car.

Keeping Up Appearances

Just as the Type 53 had made maximum use of available parts at Molsheim, the same approach applied to perhaps the most visually memorable of Bugatti's sports cars, the Type 55, which replaced the Type 43 late in 1931. Just as its predecessor had employed the 35B Grand Prix engine, so the 55 was powered by the Type 51's 2.3-litre twin overhead camshaft unit. This was almost identical to the pukka racing eight though the compression ratio was lowered by the fitment of a 9mm (0.3in) compression plate, rather than the racer's 6mm

---

[28] Dees, *The Miller Dynasty*.

[29] *Ibid*.

[30] Grégoire, *Best Wheel Forward*.

[31] 'Quattro' in *Classic and Sportscar*, June 1990.

The artist and his work. Jean Bugatti pictured at Shelsley Walsh in 1932 with a Type 55 when he attended the meeting with the Type 53 four-wheel-drive car.

(0.2in), so that it would run on petrol. Another difference was that an AC mechanical petrol pump was driven off the right-hand camshaft and the supercharger drive was modified so that it also turned the Scintilla dynamo which projected beneath the radiator.

The 55's separate gearbox, with its central ball-type change, was courtesy of the Type 49 and was mounted in a full width casting to strengthen the frame. This was, once again, the 16-cylinder Type 47 one with its distinctive full width rear end and quarter-elliptic springs projecting beyond the bodywork. It therefore shared the same 2,750mm (9ft) wheelbase as the Type 54 Grand Prix car but had a narrower track of 1,250mm (4ft 1in) instead of 1,350mm (4ft 4in). The specifications were completed by the fitment of the Type 51's handsome eight-spoked aluminium wheels.

To these formidable mechanicals, Jean Bugatti contributed coachwork of supreme elegance. This took the form of an open two-seater body, along with a coupe derivative. But the roadster's lines were by far and away the most impressive. It was also the most popular of the Type 55's body styles and, of the thirty-eight cars built, it has been estimated[32] that sixteen were so endowed, seven were factory coupes, eleven cars were bodied by outside coachbuilders, while the origins of the coachwork of a further four are, so far, unidentified.

It seem probable that a synthesis of the Type 43A and 40A's bodywork represented Jean's starting point for the Type 55 body. This is because the former's chassis dimensions were far closer to the 55 than its Type 43 predecessor. Also, the 40A was fitted with full length wings while its sports opposite number used cycle type ones. It has already been recorded that Jean had clearly been influenced by contemporary transatlantic practice when he came to design these bodies. This was echoed in the body mouldings, which tapered along the sides and centre of the bonnet and boot, a feature which had been pioneered on the 1926 Auburn. Their presence permitted the introduction of a second, contrasting body colour.

On the 55 roadster, Jean expanded these mouldings to great effect. The wings and sections of bodywork around the driving compartment were painted in one colour and the bonnet and part of the scuttle in a secondary colour contained within a curvilinear outline. As already mentioned, this so-called sweep panel technique was another American initiative.

A different colour combination featured on the coupe and there the principal one was used for the

---

[32] Hugh Conway, 'Type 55 – What and Where!' in *Bugantics*, vol. 49, no. 4 (winter 1986).

wings, roof, top of the bonnet and boot, leaving only the sides and portion of the doors for the contrasting hue. These closed 55s offered their owners weather protection though at the expense of noise from the engine's straight cut gears and heat from the exhaust pipe.

These essentially decorative features could only enhance the magnificent proportions of the Molsheim-built roadster. At the front of the car, the horseshoe-shaped radiator – the 55 was the last Bugatti road car to retain a honeycomb core – was in perfect visual harmony with the superbly balanced wing line, which ran the length of the car. There were no doors, so the cockpit sides were deeply curved to permit entry, while the boot profile was similar to that employed on the earlier 40/43As.

If a chronological interpretation is correct, the lines of the open Type 55 were the inspiration for perhaps the most celebrated Molsheim body of all. It graced the first example of Bugatti's gargantuan Royale to be sold and Jean created a superb, open two-seater body by successfully scaling up the exemplary lines of the Type 55 roadster, despite the fact that the 12.8-litre model was about half as long again as the sports car. The Royale was delivered to Parisian couturier, Armand Esders, in April 1932 and therefore postdated the 55's arrival by six months.

The first Type 55, chassis 55201, was completed in time for the 1931 Paris Motor Show and that particular car was subsequently purchased by the Duke of Tremoille, while the second example went to Bugatti team manager Meo Costantini. The model lasted until 1935, the final chassis, 55238, being delivered on 30 July 1935.

Contemporary road tests enthused about the Type 55's 177kph (110mph) or so top speed, its acceleration and roadholding. When *Motor Sport* evaluated a 55 roadster in July 1932, it pointed out that, although it was capable of about the same speed as the Type 43, the latter 'took about 45 seconds to reach 80mph [130kph] from a standing start, while the Type 55 reaches this figure in about

Arguably the finest of Jean Bugatti's body designs, the type 55 was introduced in 1931. Under the bonnet was the mildly detuned version of the Type 51's 2.3-litre twin overhead camshaft supercharged engine.

Peter Hampton pictured in his Type 55, beside the Byfleet Banking at Brooklands with his wife Lola.

A coupe version of the Type 55 which did at least have opening doors!

20 seconds from 15mph [25kph] ... the performance of the new car is still about twice as good as the earlier one.' Roadholding was up to the usual Bugatti standards, 'cornering, of course is effortless, and if a bend is taken too fast, the car seems to correct the error without help from the driver.'[33]

Despite this, the road tester's 'first thought was the comfort of the front seat, and our second the possibility of being catapulted over the low-cut side of the body. Familiarity soon bred confidence, and we found it possible to make intelligible notes when the car was doing 90mph [144kph].' The magazine succeeded in recording a top speed of 180kph (112mph) although, 'at speeds above about 80 the sound of the blower and its gears rises above audibility, and one glides along as in a fast aeroplane, with no sound but the rushing of wind.'

The car clearly went very well, despite the gearchange requiring 'considerable practice before one gets the knack of it' while the brakes also gave *Motor Sport* concern. The writer candidly observed that, the 'figure of 57 feet [17.3 metres]

[33] *Motor Sport*, July 1932.

from 37.5mph [60kph] does not seem in keeping with the rest of the car's performance' which were put down to either the linings not being properly bedded in or the use of 'unsuitable' material.[34]

Nice but Noisy

Despite its sensational looks and performance, the Type 55, with a mere thirty-eight examples produced over four years, was a slow seller. Maybe this was not surprising, bearing in mind that it was launched in the very teeth of the world depression and cost no less than £1,350 when it arrived in Britain in 1932. Yet this is only part of the story. The truth was that, fabulous as the 55 appeared, it was an unrefined product with the tracks of its lovely cast aluminium wheels planted firmly in the 1920s. The model's lack of doors must have frightened off plenty of prospective customers and their passengers and one perceptive *Bugattiste* has described the 55 as 'one of the draughtiest cars ever for travelling in'.[35] Significantly, practically all the surviving 55s fitted with other than factory roadster coachwork, have doors.

Its mechanicals were also extraordinarily noisy, a deafening combination of the engine's roller bearing crankshaft, with its limited running life between costly overhauls, and the noisy straight cut gears.

By contast, the contemporary 3C Alfa Romeo from Ettore's native Milan, essayed by Vittorio Jano, was considerably more modern in concept. It mirrored the marque's growing stature, at Bugatti's expense, on the race track. The car from Milan featured a detachable cylinder head, plain ten bearing crankshaft, dry plate clutch and unit construction gearbox. Having said that, its mechanical specifications were remarkably similar, the 8C having a capacity of 2,336cc, compared with the Bug's 2,262cc, and possessing a twin overhead camshaft supercharged eight-cylinder engine which de-

veloped 142bhp, compared with the 55's 130. In its short chassis form, clad in superlative open two-seater coachwork by Touring or Zagato, the 8C was also lighter than the 55, weighing 1,000kg (19.6cwt), compared with the 1,180kg (23.25cwt) of the Bugatti. It was a reminder that, with the exception of its aluminium bonnet, the roadster's body was made from heavy gauge steel, a reflection that the 55 was a more cost-conscious creation than might be readily apparent.

Not surprisingly, with these impressive specifications, the Alfa Romeo's sales were considerably greater than the Bugatti's, a total of 188 being built in short and long chassis form between 1931 and 1934. This was despite the fact that the Italian car was also £400 more expensive than the Molsheim product, selling in Britain for £1,750 in 1933.

The 2.3 Alfa Romeo's sales can only have been helped by an example winning the 1931 Le Mans race and, in 1932, despite Ettore's protestations after the previous year's event, two Type 55s were entered by the works, or with factory support. Because of the Le Mans regulations, they were not run with the delectable factory coachwork but were fitted with simple open four-seater bodies. One car was driven by Bouriat and Chiron and the other by Czaykowski and Friderich. The latter's performance was promising and the car survived until the 186th lap on the Sunday morning. However, it retired while in third place behind two 8C Alfa Romeos and that model went on to win for the second consecutive year. The other Type 55 only endured until lap 23 and withdrew when it was in eighth position.

Thinking Small

Bugatti's cars had invariably appealed to wealthy racing enthusiasts, while his road models were often purchased by the rich and fashion conscious and, in 1927, Ettore introduced a charming creation for such a clientele, or for the Type 35 owner with a young family. It was a miniature version of the Grand Prix car, intended for children, and Ettore may well have been inspired by André

---

[34] *Motor Sport*, July 1932.

[35] Ronald Barker, *Bugatti* (Ballantine, 1971). [An excellent book.]

FOUR, EIGHT AND SIXTEEN CYLINDERS

Two Type 55s were run at Le Mans in 1932 though bereft of their lovely roadster coachwork. This was one, driven by Bouriat and Chiron, dropped out on lap 23 when in eighth position. The other 55 was . . .

. . . driven by Czaykowski and Friderich though withdrew on lap 180 while lying third. The Alfa Romeo is being driven by M Siko and Sabipa, both former Bugatti drivers, who came in fourth.

240

As a Bugatti driver and British importer of Bugatti GP cars, it was wholly appropriate that Malcolm Campbell should introduce his son, Donald to the Type 52, Bugatti's electrically powered children's car. This is one of the early examples, which is exactly half the size of the Type 35. Later the wheelbase was extended from 1.2 to 1.3m.

Citroën's *Citroënette* pedal car, which appeared in 1925 and was scaled down from his newly introduced 5CV model. The intention was that a youngster's first words would be '*Maman, Papa, Citroën*'. It proved to be immensely popular throughout France and, over the next decade, Citroën would produce an impressive half-million toys of various types.

Rather than produce a common-place pedal car, Bugatti opted, instead, for a more sophisticated electrically powered one and the first example was created, in 1927, for his five-year-old-son, Roland. W F Bradley saw the miniature car on one of his regular visits to Molsheim and was to later recall that 'Baby Roland first drove it up and down the corridors of his home, then was also allowed in the grounds, where he was taught to take bends, *à la* Costantini, and to avoid such foreseen obstacles as his father's hat thrown directly in his path.'[36]

A photograph of Roland at the wheel of his toy Bugatti, alongside his older brother Jean in a Type 43, appeared in the motoring press and, Bradley says, the little car drew so much admiration and so many requests for Molsheim to produce duplicates that it was decided to put it into production. As a result, the model was titled the Baby Grand Prix Type 52, its number being no doubt publicized to echo Type 35 practice. It was launched at a 1927 motor show held, appropriately, at Ettore's Milan birthplace.

With a wheelbase of 1,200mm (3ft 9in), and a 625mm (2ft 6in) track, it was almost half the size of the genuine article, the dimensions of which were 2,400mm and 1,200mm (7ft 10in and 3ft 9in) respectively. Similarly the 35's 710 x 90 tyres were scaled down and the detachable, cast alloy eight-spoked wheels fitted with specially made 335 x 45mm Dunlop Cord covers. There was even a spare wheel, strapped in its traditional position, on the left-hand side of the scuttle. Cable-operated four-wheel brakes, with wooden shoes, were fitted.

It was intended for children of six to eight years old and this 68kg (150lb) miniature could accelerate to about 19kph (12mph). This was achieved by the child operating a throttle-like pedal which accelerated a rear-mounted 12 volt electric motor that drove the right-hand rear wheel, leaving the left free. Reverse was actuated by a switch which changed the polarity of the motor. Power came from a 12 volt accumulator, located under the bonnet.

Some way into production, the Baby Bugatti's

---

[36] Bradley, *Ettore Bugatti*.

dimensions were enlarged, the wheelbase being extended to 1,350mm (4ft 4in), which slightly upset the aesthetics but increased leg room. These later cars can be easily identified by their longer bonnets, containing twenty-one louvres on each side, in place of the original's sixteen.

The model remained in production for approximately three years and, during this period, about 250 examples were built. It became a desirable accoutrement for well-to-do families at such fashionable Riviera resorts as Nice and Deauville and Ernest Friderich, Bugatti's Nice representative, even staged the Baby Grand Prix at the town of Barcellonette, north of Cannes. Similar events were also held in the French capital and, in 1931, no less than eleven Type 52s lined up at Paris' Buffalo Stadium for a children's race, where the winner received the customary bouquet of flowers. These little cars survive in respectable numbers and two of them can be seen at the *Musée national* at Mulhouse.

At the Turin Motor Museum is a further example which once belonged to the Agnelli family, czars of the Fiat empire. It is pleasing to reflect that Giovanni Agnelli, the present head of Fiat who was born the year before Roland, was introduced to the world of motoring at the wheel of a Baby Bugatti.

The Type 52 was not Molsheim's only electrically powered vehicle. As photographs of Ettore in the early 1930s testify, he had acquired, what is politely referred to as a middle-aged spread, despite getting some exercise by bicycling around his factory. In 1931 he designed his own open two-seater carriage, which was allotted the Type 56 number, and created for the sole purpose of comfortably and quietly transporting him around his estate. It was, however, also registered for the road. The chassis was of wood and metal, while single transverse springs at the front and vertical quarter-elliptic springs featured at the rear. The driver sat on the right of the bench-type seat and speed was regulated by a single lever, positioned on the same side. The steering was operated by centrally mounted tiller and the front wheel pivoted in the manner of early cars and eighteenth-century carriages. This curious device was electrically powered in a similar way to the Baby Bugatti though the rear-mounted motor was a larger 36 volt, 1hp unit. Energy was provided by no less than six 100amp batteries, wired in series and located beneath the seat, Only foot-operated, rear wheel drum brakes were deemed necessary though there was a hand lever for parking. The wire wheels were shod by 26 x 3.5in tyres.

Ettore's electrically propelled carriage photographed at Molsheim. Designated the Type 56, it was offered for public sale and exhibited alongside the Royale at the 1931 Paris Show. No less than four survive.

The 56 weighed about 360kg (800lb) and was capable of around 25 to 28kph (16 to 18mph). There were some thoughts of marketing the device even though it resembled a turn-of-the-century American 'horseless buggy'. The vehicle appeared on the Bugatti stand at the 1931 Paris Salon, and a specification sheet was prepared. There might have been at least ten examples in all. Of the four survivors, one belonging to that great enthusiast, Uwe Hucke, which he purchased from the estate of Roland Bugatti, bears the 56106 number. Ettore's personal vehicle found its way to America but is now at Mulhouse, the Schlumpf brothers having acquired it in 1962 when they bought a collection of Bugattis belonging to John Shakespeare of Centralia, Illinois.

9

The Royale Rebuffed

'Good car, eh? Good car.'
Ettore Bugatti's comment, in English, to W F Bradley, while at the wheel of the prototype Royale.[1]

The date was Sunday 15 July 1928 and the occasion the third German Grand Prix, held for the second successive year at the recently opened and spectacular Nürburgring circuit located in the thickly forested Eifel Mountains, about 200km (124 miles) north-west of Molsheim. It was a sweltering day, the temperature reached 40°C (94°F), and the vast majority of the 90,000 spectators were no doubt there to support a trio of 7-litre works blown SS Mercedes-Benz that were being challenged by four works supercharged Bugattis. All the cars were fitted with wings and lights, as required by the Automobilklub von Deutschland.

The Grand Prix was attended by many dignatories from the world of motor racing who had travelled to Germany from all over Europe. Vicomte de Rohan represented the ACF French regulating body, from Italy came Renzo Castagneto, one of the so-called four musketeers who had conceived the Mille Miglia race of the previous year, along with Arturo Mercanti, instigator of the Monza Autodrome. British interests were upheld by the Royal Automobile Club's Colonel Lindsay Lloyd. A future West German chancellor, Dr Conrad Adenauer, Mayor of the nearby city of Cologne, who had played a crucial role in the creation of the extraordinary circuit – its construction in 1925/1927 had helped to alleviate local unemployment – headed the list of visiting politicians.

Ettore Bugatti was also present to see how his racers performed: there were no less than thirteen private entrants, in addition to the quartet of factory cars. But there can be little doubt that it was the dramatic nature of Ettore's arrival that created the greatest impact. Fortunately we have a first-hand account of the occasion which also presents a revealing insight to the status Bugatti enjoyed at this time as the manufacturer of the world's most successful racing cars. It came from Henry Ralph Stanley 'Tim' Birkin, who had just co-driven a 4.5-litre Bentley to fifth place at Le Mans in the previous month and was using the same car at Nürburging.

He recorded that, what he describes as the 'chief diversion', occurred just before the start of the Grand Prix:

> A commotion was observed among the great crowd, as if someone was trying to get through and finding it hard. Gradually, the throngs parted, as down the lane of inquisitive Germans came a wonderful . . . car, and in it reclined, like a Roman Emperor in Rome's most apolaustic days, the creator in person, none other than the great, the sublime Ettore Bugatti. It was indeed to be a big event, when the Golden Bug attended it.[2]

The car which Birkin so graphically described was the prototype of Ettore's most extraordinary

---

[1] Bradley, *Ettore Bugatti*.

[2] Sir Henry Birkin, *Full Throttle* (Foulis, 1932).

creation, the massive Royale, conceived for the crowned heads of Europe though no monarch ever owned one. A mere six were built but such was the lack of demand that, of these, only three were sold to private customers. Paradoxically, at the time of writing, the Bugatti Royale is one of the most sought-after and expensive cars in the world. As such it is still the subject of more folklore, myth and exaggerated rumour than any other Bugatti, or any other car for that matter, sixty-four years after it first appeared.

We can only speculate as to Ettore's thoughts as he left the Eifels in his majestic seven-seater tourer after the close of that summer day's racing in 1928. He had the satisfaction of seeing his cars win two classes in an event which saw only ten entrants survive from a field of forty-one. The smallest capacity 750 to 1,500cc Class 1 was won by Simons in a privately entered car, while works driver Brilli-Peri was victorious in the 1,500 to 3,000cc Class 2. The Italian Count was also placed fourth overall behind the trio Mercedes-Benz which were over three times their engine capacity. But Ettore would have been saddened by the death, on the fifth lap, of the Czech banker Vincent Junek, who had just taken over the wheel of his Bugatti from his wife, and had crashed his at *Breidscheid*. The car overturned and he was crushed to death. As a result, Elizabeth Junek never raced again.

It should also be recorded that Birkin, in his lone Bentley, drove a model race to come in eighth and was the first unsupercharged car to finish, for which he was much praised in the German press.

In many respects, Bugatti's presence at the Grand Prix in his Royale perfectly encapsulated two divergent elements of his personality. For if the exquisitely proportioned Type 35 racing car represented one facet of his artistic mastery, then the dramatic gesture of the extraordinary Royale reflected his more flamboyant, extrovert mood.

Pre-War Thoughts

Although Bugatti probably did not begin serious design work on the Royale until 1925, he had harboured thoughts of producing a luxurious eight-cylinder model from at least 1913. In the spring of that year, he confided his thoughts to his friend and aviator, the wealthy Dr Gabriel Espanet.

Ettore obviously thought that Espanet might be a prospective customer for his new car and what has now become a celebrated letter, dated 11 April 1913, is addressed to him at the Hôtel Renaissance in Monte Carlo. In the key sentence of the communication, Bugatti stated: 'As for the eight-cylinder, it is on the drawing board, but not yet in production. [It] ... will have an engine with 100mm bore, but the stroke has not yet been decided.'[3]

Bugatti was, of course, no stranger to large cars. His 1903 Paris to Madrid car had been of 12.9cc capacity and one of the models he designed for Mathis was of a similar capacity. These were, of course, fours but the 1913 statement suggests that Ettore might have been intending to produce a straight-eight of 10 or so litres capacity, by effectively coupling two of his four-cylinder 5-litre Garros type engines, which had a 100mm bore, in tandem. This would have followed on from a similar exercise he had undertaken when he produced a crude eight by joining two of his 1.3-litre engines together in 1912.

Ettore went on:

> It will be larger than a Rolls-Royce, but lighter; [the original typed text has *HISPANO-SUIZA* but another hand has crossed this out and substituted *Rolls-Royce*] with a closed body it will reach a speed of 150kph [93mph] and I hope to make it quite silent. When the first [car] is on test, I intend taking it on a long journey, and we shall not fail to come and see you to have an opinion of it.

The deleted reference to Hispano-Suiza is intriguing because that respected Spanish company, which opened a Paris factory in 1912, is best remembered for its mainstream Alfonso XIII model of 1911, though this had a T head four-cylinder engine of a mere 3.6 litres. However, Ettore may well have had in mind the shadowy 11,150cc six-

---

[3] Conway, Bugatti, *'Le pur-sang des automobiles'*.

cylinder 60/75hp, which came closest in engine capacity to his projected eight.

The latter was first specified in 1907 and by 1909 had a chassis price of £1,000, which was slightly more than a contemporary Rolls-Royce. That Derby-based firm, at the time of Bugatti's letter was building its worldwide reputation with just one model, Henry Royce's meticulously refined 7.4-litre six-cylinder, seven-year-old 40/50hp, which we know today as the Silver Ghost. Just one of its requirements was that it had to be capable of carrying sufficiently large closed coachwork to convey a wealthy British family and its retinue in comfort and understated splendour to winter in Cannes or Monte Carlo. In 1913 such a car would have been capable of a respectable 112kph or so (70mph) and would have cost about £1,400.

There is no doubt that Bugatti was thinking of producing an expensive car, for he goes on to tell Espanet that its production would be 'very limited' and the model would consequently be 'extremely dear, but will bear no comparison with any other of its kind'. This suggests a car of an unprecedented price which certainly foreshadows the Royale. Manufacture would be 'faultless' and, prior to delivery, each car would be tested for at least 1,000 kilometres (620 miles).

In the final reference to his projected model, with characteristic aplomb, Bugatti wrote: 'If I succeed in getting what I am striving for, it will undoubtedly be a car and a piece of machinery beyond all criticism.'

We do not know how far the project advanced in the eight months that elapsed between Ettore writing this letter and Dr Espanet and his wife spending the Christmas of 1913 with the Bugattis at Molsheim. It is possible that Ettore was diverted from his intentions by Espanet's friend, Roland Garros, who, it will be recalled, wanted a purpose-designed aircraft in which to cross the Atlantic. As discussed on page 94, he would have been responsible for the airframe and Bugatti its two engines.

It is possible that, as a result of Garros's overtures and the outbreak of World War I in August 1914, the projected 10-litre straight-eight car engine may well have evolved into the 14.5-litre aero unit that Ettore completed during the war. Its 120 x 160mm dimensions confirm that it consisted of two enlarged big bored fours of the Garros type, though employing four, rather than three valves per cylinder.

Perhaps the closest indication of Ettore's thoughts, as far as the design of his pre-1914 luxury car is concerned, were revealed after World War I. It is more than likely that he literally scaled his ideas down and the outcome was the 3-litre straight-eight Type 28 prototype, which appeared at the 1921 Paris and London Motor Shows. Aimed at the expensive end of the market, some of its features, namely the concept of its rear axle located two-speed gearbox and details of its camshaft drive, intended to quieten its valve gear, were later enlarged for the Royale.

The Type 28 never entered production but, instead, Bugatti created a car more closely tailored to the economic climate; his first production straight-eight, the cheaper 2-litre Type 30 of 1922. There Ettore's plans for his large luxury model may, for the time being, have rested, had it not been for the French government for, without its intervention, there would probably never have been a Royale.

Aero Engine Origins

In 1923 the French administration decided to encourage selected specialist manufacturers to produce a new generation of aero engines. Bugatti was one of them and a contract was placed for six units. The existence of this commission is confirmed by a reference to the project in the April 1924 issue of the Bulletin of the STA, the *Section technique de l'Aéronautique militaire*. This described a U16 unit which perpetuated the configuration of Bugatti's wartime one, with a 125mm bore while the earlier sixteen's 160mm stroke, that had first appeared on Ettore's Paris to Madrid car back in 1903, was continued. It was expected to develop 600bhp at 2,100rpm and the new engine, with its cast iron blocks, was likely to weigh 600kg (1,320lb) though the anticipated 19bhp per litre could hardly be considered sensational.

Designated the Type 34, this sixteen was of a larger capacity than the wartime engine and was of 35.4 litres rather than 24.3. It similarly perpetuated its three valves per cylinder, though featured the clean, angular lines that had first appeared on the Type 28 of 1921. Hugh Conway has written that the new engine seems to have had a pressurized lubrication system, so Ettore had probably belatedly taken note of Charles King's transatlantic modifications to the first U16, despite his original protestations!

Whereas the parallel eights of the wartime engine consisted of two four-cylinder blocks mounted in tandem, the Type 34's design differed radically in that, instead, they were single massive eight-cylinder castings that extended downwards to support their individual crankshafts in nine main bearings which, in the case of this Bugatti sixteen, were also water cooled. Such a layout was not uncommon on cheap engines, the Model T Ford is a prime example, and Griffith Borgeson[4] has pointed out that the Frontenac which won the 1921 Indianapolis 500 race, was powered by a straight-eight with such a block/main bearing construction. This dry-sump unit was the subject of a widely publicized paper by its creators, Carl van Ranst and Louis Chevrolet, which appeared in the 1921 *Transactions* of the Society of Automotive Engineers.

It is not known whether any Type 34 engines were completed, though at least four blocks, sufficient to build two U16s, were produced. What is certain is that the design was not approved by the French Air Ministry, maybe on the grounds of its relatively low anticipated power output. It is not clear when this decision was taken, or what stage the construction of the U16 had reached. But the only surviving drawing of the unit, a cross-sectional one, is dated 2 September 1925, which suggests that the design had been completed by then, so work on the project probably ceased in the latter part of that year.

Bugatti was therefore left with a batch of substantial straight-eight blocks, each of 17.7 litres capacity, and for which the French government may well have paid. It was then that he probably thought of reviving his long cherished ambition to produce a costly straight-eight model, using one of the 1,397mm (4ft 7in) long cylinder castings as his starting point and proportionally relating the rest of the car to it. But a model with a capacity of close on 18 litres was probably too great, even for Ettore to contemplate, so for the prototype engine, the 125mm bore was retained but the stroke reduced by 10mm to 150mm, resulting in 14,760cc.

Some modifications would have to be made to the casting, namely the removal of the semi-circular propeller shaft lugs and paring down the bore length to clear the shorter connecting rods required by the reduced stroke. A new crankcase would be required, but there is no apparent reason why the Royale engine could not evolve from the left-hand of the Type 34's twin blocks. A comparision of the respective drawings of the two engines confirm the striking similarity between them.

By embarking on such a huge car, Ettore was committing Automobiles Bugatti to a tremendous amount of expenditure. Apart from the engine, there could be no posssible interchange of existing components, as with his current models, because the largest capacity car he was then producing was of a mere 2 litres. Everything else would have to be specially designed for the Type 41. However, the production cars would have 125 x 130m engines, giving 12,763cc.

In October 1925, Bugatti made his regular visit to the Paris Salon and there he purchased a left-hand drive Packard seven-seater touring Second Series Eight. He is said to have bought this particular model of what was then America's most prestigious marque 'to see what it was made of'.[5] His interest was probably aroused by the fact that it was powered by Packard's first straight-eight engine, a side valve unit of 357cid (5,850cc). There were also, of course, distant wartime connections with the Detroit company as it briefly produced the LUSAC-21 biplane, intended to be powered by Ettore's controversial U16 aero engine.

---

[4] Borgeson, *Bugatti on Borgeson*.

[5] L'Ébé Bugatti, *The Bugatti Story*.

The Golden Bugatti Arrives

The Eight had first appeared in 1923 and was available in two chassis lengths. Bugatti bought the longer of the two, which was designated the Model 143 to indicate, in inches, its 11ft 11in (3,606mm) wheelbase. The well-proportioned, understated lines of the factory coachwork, which Packard listed as the Type 245, was the work of the company's body engineer, Archer L Knapp.[6] Although Ettore and his family used the Packard, probably some time in the first half of 1926, its body was destined for greater things, as it was later removed and fitted to the chassis of the prototype Royale.

The first that the English-speaking world knew of the project was in *The Autocar* of 11 June 1926, which reported on what W F Bradley described as the 'Golden Bugatti' for reasons that were not explained. Perhaps Ettore was contemplating so plating the brightwork, maybe as an indication of its superiority to the Rolls-Royce Silver Ghost, as mentioned in Ettore's 1913 letter to Dr Espanet. The original Silver Ghost of 1907 was so called because of its relative silence and silver-plated fittings. When the Bugatti was completed in 1927, and Bradley once again went to Molsheim to report on its progress, he made no further reference to the 'Golden Bug'.

Back in mid-1926, Bradley gave an account of the car's engine with precise details and weights of its parts, whereas those facts relating to the chassis are vaguer and refer to its dimensions, for instance, as 'slightly more than 15ft [4,572mm] and a track of about 5ft 6in [1,676mm].' This seems to suggest that the engine was rather more advanced than the frame.

There is no doubt that at least one straight-eight engine was ready by this time because Bradley says that: 'Examined externally, it is difficult to understand how the engine is built, for it appears to be a huge block of aluminium, with carburettors [only one was fitted to the production cars] added on one side and exhaust pipes on the opposite side.'

The carburettor that Ettore eventually created specifically for the Royale was a sidedraught unit and a remarkable cocktail of French and American influences. It inherited its barrel throttle and valve assembly from Bugatti's own Zenith-inspired unit, as fitted to the Type 28. The design of the twin accelerator pumps was courtesy of the American Schebler carburettor, of the type Bugatti later used on the Types 44 and 49.[7]

The engine is recorded as having a 125 x 150mm bore and stroke, giving a capacity of 14,726cc. It was claimed to develop an optimistic '300bhp at 1,700rpm' though the true figure was probably nearer the 200 mark. Described as being no less than 1,219mm (4ft 7in) long, its massive eight cylinders were cast in one piece, weighing 107kg (238lb), from which the crankshaft that ran in nine plain bearings was suspended, *à la* Type 34. This massive circular webbed component, with bolted on balance weights, was made in two pieces in the usual Bugatti manner and weighed 100kg (220lb).

Because of the cylinder construction, the crankcase was a simple oil-tight aluminium box – a dry-sump lubrication system had been adopted, 'there being two scavenging pumps and one feed pump from the tank on the forward face of the dashboard'. The Royale was therefore the first Bugatti car to be fitted with a pressurized lubrication system.

The enclosed camshaft drive was at the front of the engine and was driven, by bevel gears, from the forward end of the crankshaft. As usual the magneto was on the right and the water pump on the opposite side. To keep noise to a minimum, Ettore scaled up an idea which had first appeared on the Type 28 prototype. Although Bradley does not mention it, at the upper end of the drive shaft, 'were two spiral bevels toothed at opposing angles and meshing with concentric bevels on the camshaft; one of the latter was positively geared to the shaft, whereas the other was one tooth out and drove through a cone friction shaft, the object being to damp camshaft torsionals and so eliminate backlash chatter.'[8]

---

[6] Rae Kimes, ed., *Packard, a History of the Motor Car and the Company*.

[7] Henry Posner, 'The Royale carburettor', in *Bugantics*, vol. 52, no. 4 (winter 1989).

[8] Barker, *Bugatti*.

The engine of 41141 looking refreshingly original, complete with two plugs per cylinder and Bugatti-designed carburettor. Note the bonnet louvres which are designed to trap oncoming air which was then let out at the rear.

The camshaft itself was made in two parts and operated the three valves per cylinder via the usual hinged fingers. The other end of the shaft drove the coil's distributor as the Royale was fitted with double ignition, in the Rolls-Royce manner. There were two sparking plugs per cylinder. Bradley quotes the complete engine as weighing 770lb which is 349kg. It was attached to the frame by three hangers each side, the securing bolts passing through the block which was an inheritance from the eight's aero engine parentage.

There was no flywheel, the drive was conveyed through the usual Bugatti wet plate clutch, which used a light oil and paraffin mix, and behind it was a drum for the transmission handbrake. The propeller shaft thereafter ran to a three-speed gearbox, mounted, Type 28-wise, in the rear axle's differential assembly. It was similarly actuated by a centrally located gear lever.

So much for the mechanicals. The 'Golden Bug's' chassis side members deepened towards the centre in the manner of the Type 35 and were extended to a depth of 254mm (10in), just below the rear engine mounting. There were no less than seven tubular cross members together with a rein- forcing cruciform just in front of the rear axle. The hollow front axle was also scaled up from the racing cars, with the half-elliptic front springs also passing through its integral eyes. At the rear were Bugatti's customary reversed quarter-elliptic springs though, uniquely, on the Type 41 there was a secondary set, 'which only come into play when a heavy load is carried, being ahead of the axle and beneath the frame members'.[9] The petrol tank could be better described as a reservoir which contained 173 litres (38 gallons) of petrol.

For the brakes, Bugatti forsook a mechanical servo system of the type that Marc Birkigt had sagaciously created for his superlative 6.6-litre H6 Hispano-Suiza of 1919. It was an arrangement which was taken up and refined by Henry Royce and fitted to the Silver Ghost in 1924, its penultimate year of manufacture. Ettore ignored such a system and retained his usual cable-operated brakes to arrest the progress of this 3,200kg (3.1 ton) car.

The 455mm (18in) diameter drums were integral with the specially cast aluminium wheels, though these differed from the Type 35's in that they consisted of an enlarged centre section with turbine-like cooling fins extended to the rim to direct air to the brake drums. The specially made Rapson 1,000 x 180mm tyres were secured by a detachable rim, in the manner of the racing cars, secured by set screws.

Bugatti no doubt thought that such an editorial trailer as *The Autocar*'s would excite great European interest in his new car because it was stated that 'there is no intention of exhibiting this car at either the Paris or London Motor Shows, for it is recognized that it will not appeal to the ordinary purchaser, and those who are interested in it will have other opportunities of examining it.'[10] History was to decide otherwise.

One item to which Bradley does not refer was the car's mascot. Up until the arrival of the Royale and, indeed, after it, no Bugatti was fitted with a

---

[9] *The Autocar*, 11 June 1926.

[10] *Ibid.*

The Bugatti family in the prototype Royale, chassis no. 41100, fitted with Packard bodywork, pictured at Molsheim during the winter of 1926/27. Jean is at the wheel and Lidia and L'Ébé are in the back with young Roland, all well protected against the cold. The pavilion in the background was used by Bugatti for distilling plum brandy.

mascot as its original equipment. But the Type 41 was no ordinary car and, from the very outset, the prototype was fitted with an elephant mascot which has thereafter become inexorably entwined with the Royale legend.

The Elephant Mascot

It will be recalled that Ettore's younger brother Rembrandt, was a talented *animalier* (animal sculptor). In 1903 he had produced a bronze of an elephant standing on its hind legs with the front two placed on what appears to be a tree stump. Later in 1904 Rembrandt scaled the design down as a car mascot when the piece was described as *Éléphant faisant le Beau* (elephant begging).[11] At some unspecified date, Rembrandt presented Ettore with a version for use as a seal with the initials

[11] Harvey, *The Bronzes of Rembrandt Bugatti*.

'EB' on its base. For this version, the tree trunk was omitted. Rembrandt was to pursue his interest in elephants and was to execute a number of such bronzes after he moved to Antwerp and had access to its famous, well-stocked zoo in 1907.

With his new massive car in the offing and, maybe wanting to underline its size, strength and dependability, Ettore approached the Hébrard foundry in Paris. Hébrard had cast most of Rembrandt's work, and Ettore commissioned them to produce a small batch of copies of the elephant seal, which the firm duly executed. There is a characteristic twist to this story because Galerie A A Hébrard was located in the fashionable street that ran between la Madeleine and the Place de La Concorde. Its address was, 8 rue Royale . . .

By early 1927, the first car was completed and being run on the road. In *The Autocar* of 18 March, Bradley once again wrote about what he described as 'this super car' and described a run with Ettore at the wheel. However, no photographs of it were

A close-up of the elephant mascot, the work of Rembrandt Bugatti, fitted to some of the Royales.

allowed but the magazine's distinguished artist, Frederick Gordon Crosby, produced an impression of a saloon version at speed, suitably dwarfing another lowly, anonymous vehicle.

For his description of the mechanicals, Bradley relied on his article of the previous year. A few changes are, however, discernible. The tyre, and therefore the wheel size, had shrunk slightly from 1,000 x 180mm to 980 x 170mm and we have the first inkling of the number of cars that Ettore was contemplating, for Bradley says that, 'M. Ettore Bugatti has such confidence in his new model, the design of which represents ten years' work, [this is stretching it a bit but may relate to the first U16 aero engine of 1917] that he is putting through a batch of twenty-five.' For the first time, a chassis price is quoted for this, 'The World's Most Expensive and Largest Car', of 500,000fr. (£4,098). This compared with £1,900 for the contemporary Phantom 1 Rolls-Royce of a mere 7.6 litres capacity.

By early 1927 Bugatti and his massive car were probably a familiar sight in Alsace but, in April 1927, he took the 14.2-litre leviathan on what was probably its first long journey and this was across the Alps to his native Italy. Once there, he stopped at Turin and the Stablimenti Farina coachbuilding establishment. Later he drove on to Rome where he had an audience with *Il Duce*, the country's fascist dictator, Benito Mussolini. During a long interview, Ettore discussed plans for establishing a new branch of his factory in Turin. There was also talk of Ettore designing a fleet of pursuit boats, each powered by eight of the big car's engines apiece, and designed to make the Brest to New York crossing in a mere 50 hours. Nothing more was heard of either of these projects.

The car was still unnamed during Ettore's Italian sortie. It was simply described as the 'Super Bugatti', which echoed Bradley's reference of the previous month.

The Royale Revealed

The next occasion that the public had an opportunity of witnessing Ettore with his huge Bugatti was three months later at the Spanish Grand Prix, held at San Sabastian on 25 July. There it was inspected by King Alfonso XIII, with whom Ettore was already acquainted. Possibly as a result of this meeting, Bugatti would proclaim in his 1927 catalogue that 'His Majesty the King Alfonso XIII will receive this year the first example of this privileged construction.' It was no doubt on account of this apparent commitment by the Spanish monarch that the Type 41 at last received a name; what Ettore probably considered to be the highly appropriate one of 'Royale'. It is spelt as such because the French word for car is *voiture*, which is feminine, so it was proclaimed as *La Bugatti Royale*. But, conversely, *chassis* is masculine and is

Putting the prototype Royale through its paces. Members of the Bugatti family, though without young Roland, out for a winter drive. Note that the elephant mascot was fitted to this car from the very outset. This car is reputed to have had a longer chassis than the other Royales.

therefore *Le Chassis Royal* while the engine is similarly *Le Moteur Royal*.

Paradoxically, King Alfonso decided against buying a Royale though not, as has been often stated, because of his subsequent exile. In fact he fled his country in April 1931, nearly four years after his encounter with the Packard-bodied Royale. By this time the King had purchased a Duesenberg Model J, with a blue five-seater four-door touring body by Hibbard and Darrin of Paris. Its chassis price was $8,500 which was the equivalent of 216,920fr. or about half of that of a Royale.

One monarch whose name is often associated with the Royale was King Carol of Rumania but David Scott-Moncrieff, that celebrated *Purveyor of Horseless Carriages to the Nobility and Gentry*, has shed some light on the procedures necessary to 'sell' a car to such a distinguished customer. He had, by all accounts, 'a very nice taste in cars but to actually sell him one was not a question of the king just writing out a cheque for it, but of the payments coming through the "usual channels" ' which were apparently 'very greedy indeed'.[12] Similarly Scott-Moncrieff has recalled that '[King] Boris of Bulgaria told me in 1938 that when the "Royale" was announced he longed for one, but to buy one would have been a major political blunder for the head of such a wretchedly poor country.'[13] King Zog of Albania is also said to have coveted a Royale but, in the words of René Dreyfus, he 'was told he couldn't have one [sic] because, as ... Bugatti said, the man's "table manners are beyond belief".'[14]

After Ettore had attended the 1927 Spanish Grand Prix, he continued to use his Royale, with its Packard touring body, for at least a further year. On one occasion, he took his confidant and racing manager, Meo Costantini and two of his drivers, Jules Goux and Louis Charavel ('Sabipa'), for a run in the car. They set off for Mont Saint Odile, to the south-west of Molsheim, and Charavel has recounted that the car:

> held the road amazingly well, even at 145kph [90mph]. But when the Guv'nor reached 177kph [110mph], I began to feel a little uneasy, because he hadn't driven so fast for a long time and I knew that we would soon come to a level-crossing with a right-hand bend directly afterwards. Well, he hadn't lost his touch! He was doing over 160kph [100mph], but he took the bend with the sureness of a professional. We three passengers looked at each other, feeling a lot more confident. On the straight the 3,200kg [3 ton] Royale sped along at 193kph (120mph).[15]

As already mentioned, Bugatti used this prototype

---

[12] Letter to *Bugantics*, vol. 49, no. 1 (summer 1986).

[13] David Scott-Moncrieff, *The Thoroughbred Motor Car* (Batsford, 1963).

[14] Dreyfus with Kimes, *My Two Lives*.

[15] L'Ébé Bugatti, *The Story of Bugatti*.

The second body on the prototype Royale, 41100, was this curious two-seater coach which appeared outside the Grand Palais in 1928 where the Paris Motor Show was being staged.

to attend the German Grand Prix in July 1928. However, the prototype appears to have been fitted with a different type of wheel by this time because the rims were secured by sixteen set screws rather than the thirty-two used earlier. It seems likely that the American body was removed soon after this because Ettore rightly recognized that he needed to create purpose-designed coachwork to grace the massive chassis. A considerable amount of research has been undertaken by historians to ascertain that this first Royale chassis was the recipient of no less than four different bodies.

A Change of Coachwork

The Packard tourer was the first and, in 1928, came the second. The bonnet, wings and running boards still appear to be the same as those employed during its Packard-bodied days, but Ettore created what was described as a 'Coupé-Berline "Napoléon 1er" ', an extraordinary two-door closed coach in his *fiacre* style. The space between its back and the rear-mounted spare wheel was occupied by a large travelling trunk.

Bugatti kept to his pledge that the Type 41 would not be exhibited at the Salon, although the body was completed by October in time for it. The Royale was displayed outside Paris' Grand Palais where the show was staged and must have lingered, like some fantastic taxi, awaiting customers who never came.

This body can only have remained on its chassis for a relatively short time because, probably early in 1929, came yet another body. Once again Ettore adopted his *fiacre* lines though this was a proper four-door, six-seater commodious saloon with curious oval side windows. The registration number of 3293–J4 was, incidentally, carried over from its Packard-bodied days. Perhaps Ettore was dissatisfied with the results because this body was soon transferred to a Type 46 factory car that was affectionately known at Molsheim as the *Vieille Vache* (old cow)!

Probably because he now recognized his limitations as a stylist, Ettore dispatched the chassis to the Paris coachworks of C T Weymann, where it was fitted with a handsome two-door, five-seater saloon body. Despite the fact that Weymann had built its reputation on the invention and licensing of fabric coachwork, this lavish creation was, according to a contemporary account of June 1929,[16] 'panelled in sheet steel from the waist-line downwards, the upper portion being in real leather. This change from fabric leather to metal has been made in order to meet the present demand for a brightly

[16] *The Autocar*, 14 June 1929.

The third body on the first Royale chassis was this four-door saloon in Bugatti's favoured fiacre *style, built in 1929.*

Later in 1929 the chassis of the prototype Royale was fitted with this handsome two-door Weymann saloon body which was far more in keeping with its proportions. The body was panelled in metal from below the waistline and the centre section was finished in yellow. The upper portion of the body was lavishly covered in black leather . . .

. . . and was equally impressive from the rear. The Hermes trunk was covered with pigskin.

It is possible that this Double Berline de Voyage *body, on chassis 41150, was built at about the same time as the* fiacre *saloon and is effectively an open version of it. It is pictured here in America after restoration by the Harrah Auto Collection.*

finished car, which is cellulose painted in yellow and black.'

However, the wooden body frame followed the usual Weymann principles, in that the joints were secured with angle irons, so there was no wood to

Here the Berline de Voyage's *engine is being overhauled. The unit is lacking its cambox revealing its three-valve layout and the rear plate from the block, so exposing the individual cylinders. Similarly the massive crankshaft can be seen and, interestingly, the construction of the cylinder block which supports the nine main bearings.*

wood contact which made for a quieter car. A pig skin-covered trunk by Hermes completed an extremely handsome vehicle, which immediately won first prize in a Parisian *concours d'élégance*. This Royale would be the recipient of three such accolades, one of which was *L'Auto* magazine's *Grand Prix d'Honneur*. Ettore would use this car as his personal transport for the next two years.

There was, at this stage, probably only one Royale, and many writers consider that the next car to be completed was the lovely Jean Bugatti-styled open two-seater, delivered to Armand Esders in the spring of 1932. However, I think it far more likely that the numerically final chassis, number 41150, fitted with another of Ettore's eccentric *fiacre* creations, a four-door touring car called a *Double Berline de Voyage*, was the next car to be completed. It could also have conceivably existed by the time that the Weymann saloon was built.

It was Paul Kestler who first pointed out that this '*Double Berline* . . . is far closer stylistically to the rather horsey Coupe and saloon bodies . . . which successively replaced the Packard on the prototype.'[17] However, he considers it follows of the fourth rebuild of the original car, while I think it was an even earlier creation for reasons that will become apparent. The fact that this car was allotted the last of the Type 41 chassis numbers is unimpor-

[17] Paul Kestler, *Bugatti, Evolution of a Style* (Edita, 1977).

tant because most Bugattis were delivered out of numerical sequence.

Ettore may have created this body because, with the disappearance of the Packard one, he had no touring Royale to show potential customers. It is, effectively, an open version of the four-door saloon body briefly fitted to the first frame during 1929. This means that it could have been built at any time between about then and 1931. The latter date applies because, as will emerge, it was during that year that the prototype Royale was rebodied for a fourth time with magnificent coupe de Ville coachwork by Jean Bugatti.

It is distinguished by two features which appear on *all* subsequent Royales. These are two small flutes, introduced on either side of the radiator shell, which perpetuated the lines of the body mouldings, while there was a new design of bonnet fitted with movable ventilation flaps in place of the fixed louvres used earlier. The *Berline de Voyage* Royale has neither of these features which suggests that it must pre-date the 1931 rebodied prototype.

A Royale Destroyed

This brings us back to this Weymann bodied Royale and how it was, once again, dramatically transformed. It was in 1931 that Ettore had the misfortune to crash this car while driving on his familiar route between Paris and Molsheim. It is said that he dozed off at the wheel after a good lunch, an incident that was almost an exact repeat of an accident he experienced twenty-three years previously when he damaged the Deutz he was driving in the 1908 Prince Henry Trials! Fortunately, as then, Ettore was unhurt after the incident though his daughter L'Ébé, who was also travelling in the car, broke her arm. The other passengers were Mme Bugatti and M. Band from the firm's Paris depot.

It seems highly probable that, after the accident, the car's damaged frame was discarded though its 41100 chassis plate was allotted to a new one which shares the same 4,300mm (14ft 2in) wheelbase of the other Royales.

When it came to creating a new body for his own car, Ettore seems to have belatedly recognized his deficiencies in that respect and gave his son, Jean, the opportunity to lavish his proven skills on the mighty chassis. The younger Bugatti came up with an audacious design which was both archaic yet also adventurous in concept. This was a *coupé de ville* body in the spirit of Ettore's weird 1928 creation, though an infinitely superior and more assured design in which all the unwieldy elements are successfully united. It was similarly and

It was while driving this car that Bugatti crashed in 1931, on his way back from Paris to Molsheim, it is said after falling asleep at the wheel after a good lunch! By this time the car appears to have been resprayed black all over.

After the crash, the first Royale's chassis plate of 41100 was allotted to a new chassis frame and Jean Bugatti created this magnificent coupé Napoléon *body. Ettore's chauffeur, Toussaint, is at the wheel with Lidia, centre, and Mrs Grover-Williams on the right.*

immodestly enhanced with the *Coupé Napoléon* name.

Ettore's chauffeur, Toussaint – no more of Bugatti taking the wheel himself – was therefore exposed to the elements with the passengers travelling in a small coach-like compartment at the rear which, apart from the usual side and single front windows, was also novelly lit by four separate panes of glass let into the roof. The car's wing line was reminiscent of that adopted for the contemporary Type 50 and the edges of the wings were similarly moulded. The body sides featured Jean's familiar sweep panels and, instead of the bonnet mouldings stopping short of the radiator, they were daringly perpetuated on the shell itself, which was accordingly stepped at these points. The original surround had been mangled in the crash, a state of affairs which gave Jean the opportunity to subtly redesign it.

As far as the new bonnet was concerned, the louvres fitted to all the previous Royale bodies were dispensed with and replaced with a more sophisticated system with eleven movable flaps introduced to either bonnet side. When the occasion demanded it, such as hot weather or when the car was being driven in heavy traffic, the front five could be opened to trap incoming cooling air, with the interconnected remaining six angled in the opposing direction to let the heated air out. Ettore would use this Royale as his personal means of transport from the time of its completion until after World War II.

By 1931, Bugatti was beginning to feel the economic chill and, four years after the Royale's announcement, he had not sold one car. Despite his earlier statement to the contrary, he decided to exhibit an example on his usual stand at that year's Paris Motor Show.

As he was going to make an attempt actually to market his huge model, Ettore required a technical description of it. But in view of its special status, his daughter L'Ébé recalled that he wanted it 'written in a different style from an ordinary catalogue. So he organized a little competition among motoring correspondents, including the well known Charles Faroux [of *La Vie Automobile*], and then invited them to Molsheim to choose the best text submitted. His idea was to have a general description of the car without too many technical terms, as these would mean little to the wealthy clientele forming the small market for the Royale.'[18]

The assembled company was left to decide the version it liked the best but, prior to his departure, Ettore produced an unsigned text which he said he had not written himself. The majority of votes went to this account and then Ettore revealed that it had, in fact, been the work of his elder daughter, L'Ébé, and 'she had avoided using technical terms because she did not understand them . . .'

Having found a suitable text, L'Ébé says that copies were then 'placed inside elegant, soft leather folders in "Bugatti blue", hand-sewn by some of the most skilful of the factory's leather craftsmen. The two bottom corners were given a silver facing, and the right-hand one bore the signature of Ettore Bugatti.'

The third Royale to be completed was fitted with

---

[18] L'Ébé Bugatti, *The Bugatti Story*.

A superbly proportioned Kellner body on chassis 41141 was completed for exhibition at the 1931 Paris Show and it was also displayed at the London event in the following year. Note that it is not fitted with an elephant mascot.

a coupe body by Kellner, on chassis 41141, and finished in time for the 1931 Paris Motor Show. Probably at this stage, it carried the prototype's 3293 J4 registration number, the car itself then being reconstructed following its crash.

This Kellner-bodied car is often considered to have been built in 1932 because commentators have overlooked the fact that it was displayed at the 1931 Paris Salon as well as the following year's London event. At the French show, the Royale dwarfed everything else on the Bugatti stand, most noticeably Ettore's curious Type 56 personal electric vehicle, incongruously parked alongside it. Fitted with a magnificent two-door, five-seater coupe body, number 113136, the respected Carrossier Kellner of Paris was well known for its costly though beautifully proportioned, understated bodies on Hispano-Suiza and Duesenberg chassis. Described by the company as a 'Coach' it is arguably the finest example of surviving Royale coachwork.

The First Sale

Although this car did not sell at the Grand Palais, it was possibly as the result of its display there that Bugatti, at last, made his first Type 41 sale. The customer was the Parisian clothing manufacturer, Armand Esders. This wealthy, motoring-minded couturier, had, in 1927, gone to the trouble of having two bespoke miniature electrically powered Hispano-Suizas, each capable of around 40kph (25mph), created for his children. No run-of-the-mill production Type 52 electric Bugatti for his offspring!

Esders daringly specified a two-door open body on this Royale chassis and, in a surviving letter to Jean Bugatti, now in the Hucke archives, Esders says he chose the Royale because he was seeking the ultimate in flexibility, which would mean no gear changing and, in any event, he never exceeded 88kph (55mph). In addition, he specified that he did not require any headlamps. 'They

ETTORE BUGATTI
FABRIQUE D'AUTOMOBILES
MOLSHEIM (BAS-RHIN)

SUCCURSALE DE PARIS
SERVICE RÉPARATIONS
75, RUE CARNOT, 75
LEVALLOIS-PERRET
(SEINE)

TÉL. : PÉREIRE 42.40 / 42.41

R. C. Seine 423-206

FACTURE N° 2 51

Monsieur BRIGGS S. CUNNINGHAM
Greens Farms
Connecticut (U.S.A.)

N° DE DOSSIER : Rép. 1496
TYPE DE VOITURE : 41.141
Mot. 41.141

LEVALLOIS-PERRET, LE **12 JANV 1951**

Cl. L. - 17.309

	RÉFÉRENCES	DÉSIGNATION DES TRAVAUX ET FOURNITURES		
	1	Voiture d'occasion BUGATTI type 41 (Royale) N° 41.141 Carrosserie Conduite Intérieure Kellner		650.000.--
		Pièces détachées et fournitures pour la réparation		
		Fournitures :		
		4 Joints MP. de 14 X 20	6.00	24 00
		1 Boulon 6 pans de 7X40		85 00
		0m40 de durite 45X52		259 00
		1 bouchon vidange radiateur		70 00
		2 joints MP. de 12X20	6 00	12 00
		2 joints fibre de 7 m/m	2 00	4 00
		2 gicleurs de 14X25	163 00	326 00
		2 raccords orientables	266 00	532 00
		1 écrou de 14 m/m		61 00
		10 colliers d'eau (grand)	31 00	310 00
		2 joints MP. de 9X14	11 00	22 00
		1 joint liège		10 00
		2 vis tête carrée de 5X13	34 00	68 00
		2 " " " 5 X 21	38 00	76 00
		2 " " " 6X22	42 00	84 00
		2 écrous 6 pans de 7 m/m	30 00	60 00
		2 " " " 5 m/m	28 00	56 00
		5 graisseurs Técalémit	46 00	230 00
		5 mètres 50 lanière caoutchouc 3X4	27 00	148 00
		6 rondelles de 7 m/m	6 00	36 00
		6 " Grower de 4 m/m	0 50	3 00
		1 mètre lanière caoutchouc de 3x3		45 00
		2 ressorts prise d'air	165 00	330 00
		2 arrêts écrou de Bendix	6 00	12 00
		1 vis arrêt de Bendix		60 00
		à reporter :		652.923 00

N. B. - Passé un délai de huit jours, les pièces ne seront ni reprises ni échangées.

The first page of Briggs Cunningham's seven-page receipt for the Kellner Royale. The total bill for this car was 1,029,049fr. (£1050).

would be of no use,' he told Jean, 'as I never drive at night.'[19]

With the *Coupé Napoléon* already to his credit, Jean Bugatti came up with his second, impeccably proportioned Royale body, the wing line of which was essentially that of his equally magnificent Type 55 roadster of the previous year. Although this car is often referred to as a two-seater, it could, in fact, carry four people as there was also a dickey seat with its own separate windscreen.

This body was built at Molsheim and Noël Domboy, Bugatti's technical director, has subseqently recalled inspecting it with Jean Bugatti at the factory over the Easter of 1932, which fell between 25 and 27 March. The car was delivered soon afterwards, on 4 April, and was literally looked upon as a publicity vehicle by its new owner.

It was probably later in 1931 that Bugatti made his second Royale sale. The buyer was a wealthy German gynaecologist, Dr Josef Fuchs, from Nuremberg who ordered it at the height of the economic depression then engulfing his country. He liked fast cars, and indeed raced them, and was already a Bugatti owner, having in 1931 bought a Type 50 which was then fitted with a two-seater convertible body by Karosseriefabrik Ludwig Weinberger of Munich, established in 1898. It was the work of young Ludwig, the son of the firm's founder and Fuchs asked him to scale up the lines of the 50's body for the Royale. Its chassis cost him RM75,000 (£5,263). Prior to delivery, the engine underwent 50 hours of running in at the factory at the same time as the unit from the Esders car.

In January 1932 a Bugatti mechanic delivered the chassis to Weinberger's premises at 41 Zepplein Strasse, having driven it from Molsheim. Although he ostensibly remained in Munich for six weeks to 'supervise' the work, he was mostly absent in the town perhaps spending his time sampling the local brew and, according to Weinberger, 'never touched a hammer'.[20]

[19] Griffith Borgeson, 'Migrating Monuments' in *Automobile Quarterly*, vol. 24, no. 4, 1986.

[20] Erik Eckermann, 'Coachbuilder Weinberger, Munich' in *Veteran and Vintage Magazine*, vol. 16, no. 3 (November 1971).

The ultimate indulgence? The first Royale to be sold was 41111, completed in 1932, which was fitted with this magnificent Jean Bugatti-styled two door body. There was room for two further passengers in the dickey seat which possessed its own folding windscreen.

The Royale body was built in Weinberger's small first floor workshop behind his showroom where the chassis was displayed. Not only did the Royale resemble the Type 50, but the colour scheme was also similar to it. Both cars were painted black with yellow mouldings on the doors, which extended to the rear of the body, while the edges of the wings were also finished in yellow. The running boards were covered in white rubber.

The body was upholstered in Hungarian pig skin 'because skin and pores were bigger than domestic ones'. Weinberger was assisted in this work by Josef Reitmejer and his staff and, on completion,

A side view of the Esders Royale with a wing line similar to that used on the Type 55.

the body was lowered by lift and then fitted to the chassis. The finished car 'stood in the yard in all its splendour, witnessed by a crowd so big as to render photography impossible'.

After a brief spell on display by Weinberger, where it shared his showroom with a diminutive Austin Seven-based BMW Dixi fitted with the firm's roadster body, the Royale was officially

In about 1938 the Esders Royale was rebodied by Henry Binder for French politician Raymond Patenotre. After the war it came to Britain and was owned for some years by Freddie Henry. Here a policeman is taking an interest in 41111 as it still carries its French registration number. Henry eventually had to pay the long-avoided customs duty on the car.

A close up of one of 41111's wheels which were fitted with these nave plates when the car was rebodied by Binder.

handed over to Dr Fuchs on 26 May 1932. His final bill was RM82,000 (£5,754) which meant that the body cost RM7,000 (£490). The new owner clearly did not care for Ettore's elephant mascot because it was delivered with an unadorned radiator cap. The doctor probably took the opportunity of dispensing of his Type 50 on delivery of the Royale because the latter also bore the same registration number of 11N 15349. For his part, Weinberger bodied a further fourteen or so Bugattis though concentrated mostly on BMW, having acquired a sub-agency for the marque in 1931, and approximately 300 examples were so endowed between 1933 and 1939.

A London Purchaser

Compared with the bleak years between 1927 and 1931, it seemed as though 1932 would be a good year for Royale sales. In March of that year, W F Bradley reported[21] on both the Esders and Fuchs purchases, though the latter was unnamed and referred to as 'a famous German surgeon'. It was also stated that 'among those said to be awaiting delivery are the ex-King of Spain [again!] and M. André Citroën.' Neither of these two sales ever materialized. In the latter instance, the extravagent, mercurial Citroën would undoubtedly have responded emotionally to the Royale though, at this time, he was deeply engrossed in the development of his celebrated *Traction Avant* model which appeared in 1934.

Despite this, Bugatti decided later in 1932 to take the unsold Kellner coupe to Britain in the autumn of that year to exhibit at that year's London Motor Show, held at the Olympia. There, it was exhibited at a price of £6,500 and as such was the most expensive car to be offered for sale in Britain during the inter-war years. The chassis price was £5,250, which was well over two and a half times the price of the contemporary Rolls-Royce Phantom II. Once again publicly displaying the model may well have borne fruit because, at about this time, Captain Cuthbert Foster, a wealthy Gloucestershire land owner, ordered a Royale.

He was, apparently, unconcerned by the car's projected running costs or by the fact that, in addition to its purchase price, it was rated at 77.4hp, so would command an annual £78 road fund licence, which was allotted at the rate of £1 per RAC horsepower.

The chassis, number 41131, was sent over 'on a lorry, accompanying by M. Band, one of the engineers from Molsheim'[22] and delivered to Foster's chosen coachbuilders, Park Ward, at its works in 473 High Road, in the north-west London suburbs of Willesden. Perversely, the Captain had specified that the body follow the lines of his 1922 Daimler so the outcome was a four-door seven-seater limousine body, with interior glass partition, that looked positively dated when it was completed in the spring of 1933.

[21] *The Autocar*, 25 March 1932.

[22] Bradley, *Ettore Bugatti*.

Park Ward experienced a major problem when their work was completed because they found it impossible to start the car's engine. So Jean Bugatti came over to London to coax the recalcitrant Royale into life. His system was unorthodox, to say the least, for he first had the big Bugatti moved to the front forecourt of the works and he then 'proceeded to start quite a man's size fire beneath the engine with the aid of a gallon of petrol, and stood back while the flames licked upwards towards the inside of the bonnet'.[23] The heat no doubt had the effect of thinning out the oil, making it easier to turn the engine over, and the 12.8-litre straight-eight burst into life.

On 30 June 1933, Jean took the opportunity of handing this car, the last Royale to be sold, to its new owner. Yet like Dr Fuchs in Germany, Foster also declined Rembrandt Bugatti's elephant and, instead, his car was adorned with a Rolls-Royce Spirit of Ecstasy mascot. Despite this outward conflict of interests, the Captain was clearly delighted with his purchase, so much so that W F Bradley says that every Christmas, 'he sent a message to the engineer who initiated him, to assure him that it was still going strong. The chauffeur-driven Royale was registered ALB 2 with London County Council though it soon became known as "The Chauffeur's Nightmare" for there were few main roads in London where the car could be turned from the road into a side street without reversing.'[24]

The Kellner Royale was not the only example of the breed in London during the latter part of 1932. On 16 December, Ettore and Jean attended the annual dinner of the Bugatti Owners' Club, established just three years previously, and held at London's Grosvenor House Hotel. They arrived in the newly minted prototype Royale bearing its extraordinary *coupé de ville* body and *Bugantics*, the club's magazine, commented that 'the whole of Park Lane was stirred by its majestic presence outside Grosvenor House.' At the event, Ettore

---

[23] 'The coachbuilder and his art', letter from F W Gilbert in *Bugantics*, vo. 39, no. 1 (spring 1976).

[24] *Ibid.*

THE BUGATTI 'ROYAL'.

SPECIFICATION.

TREASURY RATING 77.4 H.P.   **TAX £78.**

ENGINE 8 cylinder in line monobloc. Bore 125mm. Stroke 130mm. Cylinder capacity, 12,760 c.c. The length of the cylinder block is 4'7", weight approximately 2cwt. 15lbs. The 9-bearing crankshaft, weight just under 2 cwt, is machined out of the solid and has circular webs. It is carried direct on the cylinder block and the water jacket is brought down between the cylinders to the main bearings. 3-valves per cylinder, placed vertically in the head, 2-inlet and 1-exhaust. Nine bearing camshaft. The cylinder block is attached to the frame members by three tubes going through it from side to side, giving an unusual degree of rigidity. The connecting rods are I-section forgings with bronze backed white metal bearings, and split cap with 4-bolts. The pistons are of aluminium alloy with split skirts. Pressure lubrication, dry sump. 1-pressure pump, 2-suction pumps. Dual ignition, 2-plugs per cylinder.

CARBURETTOR Bugatti Patent, specially designed to ensure a correct mixture at all engine speeds. Independent air and petrol supply control facilitates best results being obtained under varying climatic conditions. Petrol supply by electric pump.

CLUTCH This is mounted in the chassis separately in a casing in a position approximately under the front seat.

GEAR BOX & BACK AXLE These form a single unit. Three gears are provided: the first as emergency gear and, if wished, for starting; the second is direct drive, utilisable from a speed of approximately 2 m.p.h. upwards to approximately 90 m.p.h.; the 3rd. speed is geared up, principally for use on very long straight roads, when maximum speed is required.

WHEELBASE 14'2"   **TRACK** 5'3"

WHEELS Cast aluminium—Bugatti Patent.

TYRES 36 × 6.75

WEIGHT OF CHASSIS Approximately 32 cwt.

IT IS CLAIMED that the BUGATTI 'ROYAL' is the most outstanding motor car in the world, embodying unique constructional features, and giving a performance unapproached by any other motor car chassis in the world.

In spite of the quite exceptional size of the car it is so well proportioned that, excepting in comparison with others, it does not look inordinately large, and it can be handled with ease in congested Paris or London traffic, or on tortuous roads.

The performance of the car is charactised by extreme silence at all speeds, remarkable road-holding properties and unequalled ease in handling, these characteristics making it a car quite suitable for a lady to drive.

ETTORE BUGATTI,
1 - 3, BRIXTON ROAD, LONDON, S.W.9.
Telephone: Reliance 3165/6. Telegrams: Bugattimo, Claproad, London.

The nearest Bugatti's British company got to producing a sales leaflet for the Royale. This may be a proof copy of the finished product because it states that the car has a six-cylinder engine, the crankshaft weighs 7cwt and Weight is spelt Weicht!

responded to Earl Howe's speech of welcome in French, which was then translated by Colonel Sorel.

Although the Royale continued to be listed,

Captain Cuthbert Foster's Park Ward-bodied Royale, chassis 41131, pictured when in Jack Lemon Burton's ownership after the Second World War. It looks rather older than its year – it was completed in 1933 – and was finished in dark blue. The fittings were of nickel rather than the more fashionable chrome.

The rather stark rear compartment of the Foster Royale pictured on its completion in 1933. It was possible to accommodate three passengers on the rear seat and there were a further two rear-facing occasional seats. The cabinet on the division contained a Philco radio and the microphone for the passenger to communicate with the driver can be seen below the ashtray. There was another one on the opposite side of the car.

certainly by Bugatti's British sales outlet until 1935, the reality was that, from 1933 onwards, Molsheim made little attempt to market the model. Its appearance at Olympia in 1932 was the last occasion that a Royale was seen at an international motor show.

Not only were the model's mechanical specifications deeply rooted in the 1920s, and were therefore becoming progressively outdated as the following decade proceeded, but in 1933, Bugatti had delivered the first of ninety-one railcars, some of which employed no less than four Royale engines apiece for their motive power. This extraordinary transformation of a car which had been a chronic liability to the Bugatti company, to that of its saviour at a financially fraught time, will be more closely examined in the next chapter.

It should, nevertheless, be recorded that the railcar project absorbed no less than 216 Royale engines, which was a far greater figure than Ettore, even in his most optimistic moments, could have ever contemplated for the sales of a massive, costly and exclusive motor car.

Why so Few Buyers?

Why did the Royale fail to sell? From the very outset, little thought appears to have been given to marketing the car. Ettore naively seems to have imagined that wealthy and titled customers would have queued up for Royales in much the same way that their younger counterparts responded to his visually impeccable Type 35. But the truth was that at the time of the 'Golden Bug's' announcement in mid-1926, although Bugatti's Grand Prix cars were forging a redoubtable reputation on the motor racing circuits of Europe, he had not begun to establish himself as manufacturer of luxury cars in the manner of the Rolls-Royce, Hispano-Suiza or Isotta Fraschini. Prior to the arrival of the Type 41, his most expensive models were the eight-cylinder Type 30 and its unhappy Type 38 successor, and these were of a mere 2 litres capacity.

Ettore's quantum leap in producing a chassis costing an astronomical half million francs and with an engine of over six times the size of his largest existing model might have worked, had it been displayed from the outset with coachwork befitting such an expensive car. If Ettore had let Jean style the Royale's body earlier than he did, and had it been cloaked with coachwork in the spirit of the *Coupé Napoléon* in, say, 1927, its impact would have been considerably greater. Having said that, the younger Bugatti was only eighteen at this time and such a commission might have been beyond even his precocious talent.

The first glimpse that the public got of the Type 41 was, frankly, unimpressive, wearing as it did ageing and borrowed transatlantic coachwork. No potential purchaser could have possibly taken that weird 1928 coach seriously and it was not until the spring of 1929, a wasted two years after the car's announcement, that the prototype Royale finally received coachwork, by Weymann, worthy of its status as the world's largest and most expensive car. But it arrived just six months before the 1929 Wall Street financial crash which sharpened the climate for the sale of such costly vehicles.

Ettore was probably placing great faith in King Alfonso of Spain's purchase of a Royale but the monarch remained loyal to his native Hispano-Suiza though did opt for a Duesenberg. Similarly the other crowned heads of Europe seem to have been content with the products of Derby, Paris or Milan. And would any king have really bought a car as ostentaciously titled as this big Bugatti?

Also, the sheer size of the Royale must have been a disincentive to buyers. Its cubic capacity, of 12,763cc, has never since been exceeded by a production car. In addition, it is sobering to reflect that its huge 4,300mm (14ft 1in) wheelbase has only been surpassed in the 1960s by the eight-door, twelve-seater Checker Aerobus Limousine which measured 4,800mm (15ft 9in) between its wheel centres and was used by an American hotel to ferry customers to and from airports. It was, in truth, a 'stretched' version of a standard model but even the gargantuan American finned monsters of the 1950s were a good 450mm (1ft 6in) or so shorter than Bugatti's Type 41. The largest European car of recent years, the six-door Mercedes-Benz 600 intended for use by diplomats, had a 3,911mm (12ft 10in) wheelbase though this was still no less than 380mm (1ft 3in) shorter than the Bugatti's.

So much for the present. But what of the surviving Royales? As will have been apparent, only one car was destroyed and that was when Ettore himself crashed the prototype in 1931. However, all the remaining cars survive and their subsequent histories are as fascinating as the circumstances of their manufacture. These will now be charted, in order of completion, to the time of writing (1991).

Circa 1930 Royale, with *Double Berline de Voyage* body by Bugatti. Chassis no. 41150

Although this car bears the final Type 41 chassis number it is, in all probability, the oldest surviving car. Styled by Ettore and built at Molsheim (it bears the 'Carrosserie Bugatti' body plate) this *Berline de Voyage* is effectively a touring version of the *fiacre*-styled saloon fitted to the prototype chassis in 1929. Its landaulette leather roof folds down but the car's rather quaint appearance is accentuated by the fact that the bonnet does not quite line up with the scuttle! This car resembles a nineteenth-century double berline coach, and this is its strictly correct title though it is usually referred to in the abbreviated form shown above. This Royale's body is, appropriately, made of wood – it was built up from small pieces of the material, which were then covered with a wood veneer, and painted.

Not surprisingly, there were no takers when the car was new. This Royale remained in the hands of the Bugatti family though it was probably moved, along with the historic cars from Molsheim to Bordeaux in 1939. Subsequently, it was transferred to the Château at Ermenonville near Paris, which Ettore had bought just before the war. According to Briggs Cunningham (*see* chassis 41141), who in 1950 purchased this and the Kellner-bodied car, three years after Ettore's death, L'Ébé Bugatti told him that all three Royales, these two and the *Coupé Napoléon*, were hidden during the war in an old stable block at the Château. They were successfully concealed by the building of a 'brick wall across in front of them and they'd stayed hidden like that during the war and were never found. When they knocked down this wall after the war, there they were – untouched.'[25]

Cunningham's visit to Bugatti's eldest daughter had been prompted by his friend, D Cameron Peck from Chicago, owner of no less than 250 old cars. The millionaire racing driver was in France because he had entered a pair of Cadillacs at the 1950 Le Mans race (he was placed eleventh), and has recalled that Peck: 'had called me before the trip and asked: "If you're going over there, why don't you see if you can purchase the Royales? If you can I'll buy one from you. OK?" '

Fortunately for Cunningham, he had a friend in Paris named John Baus, who knew Roland Bugatti, and called him and asked whether 'it would be OK for Mr Cunningham to see the cars'. Roland agreed and Cunningham and the French-speaking Baus, who would act as translator and intermediary, had tea with L'Ébé at Ermenonville. Cunningham asked whether the Royales were for sale and he recalls that she said ' "I don't want to sell my father's car [the Coupe Napoleon] but I'll sell the other two." '

'So we went down to this big old brick . . . stable, I guess it was . . . and up against the back wall were three radiators sticking out, the three Royales all parked side-by-side, backs against the wall.'

They then talked about terms and Baus 'made the deal with her to buy the cars.'[26] Cunningham bought both Royales. He paid 1,006,221fr. for the *Berline de Voyage*, which was the equivalent of $2,875 or £1,027 at that time. He was, however, purchasing it on Peck's behalf because, it 'just left me cold as a car, silly-looking thing. I would have taken that one and had the [Esders] roadster body built on it. Boy I would have *loved* to have had that car!' Then L'Ébé asked Cunningham if it would be possible for him to obtain a couple of ice boxes, as 'The Germans have taken ours . . . and I'd very much like to get a couple of Frigidaires – General Motors Frigidaires.' Baus responded: 'I think we

---

[25] Doug Nye, ' A king's ransom' in *Autocar*, 12 August 1987.

[26] *Ibid.*

can manage that.' He was well placed to accede to the request as he was responsible for representing American industries in post-war Europe.

The sum that Cunningham had paid was conditional on the two Royales being ready for the road. Fortunately Bugatti's Paris depot in the rue de Debarcadère, Levallois, was still operative and the sales invoice that the American received for the cars stated, 'aux ateliers de Levallois, pour remise en ordre de marche du chassis, après plus de 10 années d'immobilisation' (to the Levallois factory to return the chassis to running order, after more than ten years of immobility).[27]

As it happened, it took some months for this work to be effected. It was not until the beginning of 1951 that the two Royales were ready for shipment and Cunningham says that 'John drove one car to Le Havre and Marco [from Bugatti's Paris depot] the other'. On 16 January 1951 the SS *de Grasse* set sail for New York, complete with its cargo of Bugattis. Once in America, Cunningham retained the Kellner coach but immediately disposed of the *Berline de Voyage* to Peck for $2,937.33 (£1,049). Writing to Cunningham on 11 April 1951, he declared that: 'The Bugatti is in excellent condition and I am very pleased with it. You were right in advising me that the body was very interesting [sic], and that it should be preserved as it is. The car runs beautifully and is amusing to drive in traffic because of its huge size, coupled with really good performance . . . enclosed is my cheque.'[28]

Peck kept the Royale for a relatively short time because his entire collection was sold by auction in 1952. The Bugatti was bought by Dr Benedict Skitarelic of Cumberland, Maryland and, by the 1960s, it was the property of Jack Nethercutt, a Los Angeles cosmetics tycoon. In 1964 he sold the car to the fast growing Harrah's Automobile Collection of Reno, Nevada for $45,000 (£16,129). Back in 1948, William Fisk Harrah, who owned the world's largest casino complex at Reno, had purchased a 1911 Maxwell. This was the starting point of a collection which would also create a world record in having, at its height, about 1,500 vehicles, of which 1,100 were on display.

At the time that Harrah bought the Bugatti, another Royale, the Binder *Coupé de Ville*, was under restoration by John Griffin of Montgomery, Alabama and Harrah's staff was so impressed with the standard of the work that the *Berline de Voyage* was also sent there. The outcome was that, later in the year, Harrah also bought that Royale and they were displayed alongside an example of Bugatti's miniature Grand Prix car.

The Auto Collection continued to grow but Harrah died suddenly in June 1978, aged only sixty-six. Within a year his casino and hotel complex, which included the car collection, had been sold to the Holidays Corporation, best known for its Holiday Inn hotel chain. It decided to slim down the Collection which, at that stage, involved selling one of the two Royales, and a series of three auctions were held to that end. The sale of the *Berline de Voyage* was set for the last of them, staged on 27 to 29 June 1986. Although there was an impressive line-up of cars for this three-day event, it was the Bugatti that attracted most attention because it was the first occasion on which an example of the model would be coming under the auctioneer's hammer. Commentators speculated that it might fetch a record $10 million (£7 million).

The Royale was offered for sale on the first day of the auction, on 27 June. Christie's Robert Brooks reported for *Classic and Sportscar* that it was sold at a special evening party: 'where everyone who had paid $75 to register as a general bidder was welcome, but only people with a letter of guarantee for $1 million were allowed to raise their hands!'[29] The Royale did not reach the anticipated $10 million but, nevertheless, went for $6.5 million (£4.6 million). Ironically the Bugatti model which was so disregarded when it was new and, indeed, an example which had attracted no customer interest at the time of its manufacture, had become the world's most expensive car.

---

[27] Christie's press release.

[28] *Ibid.*

[29] *Classic and Sportscar*, September 1986.

The buyer was real estate tycoon and car collector, Jerry J Moore of Houston, who had opened the bidding at $3 million, but he only retained the Royale for less than three months before he was approached by Domino's Pizza Company's Thomas Monaghan. He has said that he first fell in love with Bugatti's Royale when he first saw the ex-Dr Fuchs' car in the Henry Ford Museum. Monaghan paid Moore a staggering $1.6 million (£1.1 million) more than its Reno price and the Royale was his for ... $8.1 million (£5.7 million).

After its purchase, a spokesman for Monaghan said that his boss regretted that he had not bought the car at Harrah's in June. Already a dedicated car collector, Monaghan started delivering pizzas in 1960 and began his business with only $77 in the bank. By the time he bought the Royale in 1986, the company had an annual turnover of $2.5 billion. Monaghan says that he loves:

> ... the scale of the Royale, its nice proportions, its massiveness. I've always thought that the Royale is the best car in the world.
>
> I've always wanted to use a great car as a promotional tool for the company. And I felt that I could justify the cost of the car because it is *so* spectacular, whereas people just aren't interested in good Packards. This is now the best known collector's car in America.[30]

Today the *Berline de Voyage* is displayed at Domino's Ann Arbor headquarters.

In August 1990, the Bugatti returned to France for the first time in thirty-nine years to take part in an exhibition, held at the *Musée national de l'Automobile* at Mulhouse where all the Royales were reunited. This was not the first occasion on which the full complement of Type 41s was seen in one place. The previous occasion had been five years previously in America when, in 1985, all six cars were brought together for the first time.

Circa 1932 Royale, with *Coupé de Ville* by Bugatti. Chassis no. 41100

Unlike the other Royales, this car with its spectacular Jean Bugatti-designed *Coupé Napoléon* body, has hardly ever left France. Ettore was chauffeur-driven in it from its creation until the outbreak of war. As mentioned previously, it was interred, along with the two other unsold Royales, at the family's Château at Ermenonville during the war. With the ending of hostilities, it was exhumed and Ettore used it to attend the first motor race of the post-war years, held in Paris on 9 September 1945 at the Bois de Boulogne. There, he had the satisfaction of seeing Jean-Pierre Wimille, in the 4.7-litre single-seater Bugatti, win the third of the races staged, the *Coupe des Prisonniers*, for cars of over 3 litres capacity.

Ettore died in 1947 but the Royale was retained by L'Ébé Bugatti at Ermenonville. As mentioned in the history of the *Berline de Voyage* Royale, in the summer of 1950, she was approached by American millionaire, Briggs Cunningham and, as a result, sold him two of the Royales but not the *Coupé Napoléon*, as it was her father's car. Cunningham has recalled that she told him that 'she felt it ought to stay eventually in a museum in France.'[31] It was not until thirteen years later, in 1963, that the Royale eventually left the Château.

That year Fritz and Hans Schlumpf, who were creating a private museum at Mulhouse, bought all the historic Bugattis at Molsheim and those in the family's possession. At about the same time the *Coupé Napoléon* was joined at Mulhouse by the ex-Captain Foster Royale, which was the jewel in the crown of the Shakespeare collection of Bugattis from Illinois, USA, which the Schlumpfs had also purchased. However, in 1977 and faced with bankruptcy, the brothers fled across the Swiss border and their fabulous collection of Bugattis is today open to the public as the *Musée national de L'Automobile*. Although L'Ébé Bugatti died in 1980, in 1978 the French government had declared the collection a 'Historic Monument', prior to its achieving museum status. Today the *Coupé Napoléon* has pride of place at Mulhouse. L'Ébé could only have approved.

One of the very few occasions on which this

---

[30] *Sunday Times* colour magazine, 14 May 1989.

[31] Nye, 'A king's ransom'.

After the war the Kellner saloon was bought by Briggs Cunningham and exhibited at his museum at Costa Mesa, California until 1987 when it was sold for £5.5 million at a memorable Christie's auction, staged at London's Royal Albert Hall. It is now in Japan.

Royale left France was in the summer of 1985 when, on 25 August, it graced the lawns of the Pebble Beach Country Club, near Montery, California, USA, which was the first time that all six Royales had been assembled in the same place.

1931 Royale with two-door coupe body by Kellner of Paris. Chassis no. 41141

Like the *Coupé Napoléon* and *Berline de Voyage* Royales, this Kellner-bodied car, which was displayed at the 1931 Paris and 1932 London Motor Show, was unsold. It therefore remained at Molsheim where L'Ébé Bugatti is said to have used it for 'visiting locally and . . . for shopping!'[32] It was stored along with its sister cars at Ermenonville, when Briggs Cunningham bought the two Royales from L'Ébé Bugatti in 1950. At the time, Cunningham had already begun collecting historic cars. His first purchase had been made soon after World War II and was a sought-after 1912 Mercer Raceabout which was soon followed by a 1912 Alfonso XIII Hispano-Suiza, an outstanding European car which set the tone of what was to develop into the Briggs Cunningham Automotive Museum at Costa Mesa, California.

Briggs Swift Cunningham's family fortune came from the Proctor and Gamble soap powder company and railroad interests on his mother's side. Just pre-war, along with Charles Chayne (soon to be a Royale owner, but then chief engineer of Buick) he had built the Bu-Merc racer, a Buick/Mercedes hybrid. A long-time car enthusiast, Cunningham (born 1907) had been introduced to the Bugatti marque at school. 'The first time I'd seen anything about Bugatti was when I used to read *The Autocar* and *The Motor* at my last year at school in 1926.' But he'd never seen a Royale until his visit to Ermenonville in the summer of 1950. The Kellner Royale was the slightly more expensive of the two and cost Cunningham 1,029,049fr. ($2,940/£1,050), which made a total of 2,035,270fr. ($5,815/£2,076) for the two Royales.

On arrival in America in January 1951, both cars were driven to California and then the *Berline de Voyage* was collected by Cameron Peck. Cunningham had the car repainted in its original black

[32] Christie's press release.

but he changed the yellow of the window moulding for blue, which he preferred. The car also acquired an elephant mascot during Cunningham's ownership, one not being fitted to the car when he acquired it. The duplicate was made of sterling silver.

The Bugatti was thereafter displayed at Cunningham's California museum and, in 1987, he reflected that 'I doubt if we have put 1,000 miles [1,609km] on the car since it was overhauled in France . . . I never went to the factory, never went to Molsheim, I wasn't nearly as much of a Bugatti enthusiast as some other people and why did I keep the car as long as I did? Darned if I know . . .'[33]

In 1986, at the age of seventy-nine, Cunningham recognized that the time was right to dispose of his entire collection. All the cars were bought by the son of a one-time racing partner, but its recipient, Miles Collier, insisted that the Royale remain in Cunningham's name. But it had been decided to sell the Bugatti.

However, instead of auctioning the car in America, as the *Berline de Voyage* had been in 1986, the sale was handled by Christie's of London which, in May 1987, had staged a spectacular auction of Ferraris and Bugattis in Monaco. Robert Brooks, director of Christie's motor car department, who was to mastermind the Royale sale, commented on the announcement of the car's forthcoming auction: 'At one time the US was considered the *only* place to buy and sell collector's cars. Now certain blue chip cars have become international currency and our decision to bring the Royale to London was based on current confidence and buoyancy in the European market.'[34]

Auction at the Royal Albert Hall

The Royale required a special venue. The original idea was to stage the sale in a marquee in Berkeley Square, but it was a chance remark to Brooks from his historical consultant, Doug Nye, to the effect 'I expect you'll want the Royal Albert Hall next time' that set the wheels in motion for the Royale being auctioned there. Held on 19 November 1988 before a 3,000-strong audience, the Bugatti was the star of 'Ten Important Motor Cars' and it sold for a record-breaking £5.5 million ($9.8 million), so surpassing the Harrah Royale's price by $1.7 million. It was to stand as the world auction record until 1990.

The buyer was London dealer, Nicholas Harley, who was acting on behalf of Hans Thulin's Swedish AB Consolidator. Just eighteen months later, on 29 April 1989, an attempt was made to auction the Kellner once again with a reputed reserve price of $15 million. But the bidding stopped at $11 million and the Royale remained unsold. However, a year later, in April 1990, Harley was again the intermediary when the Bugatti changed hands, this time for a staggering £9 million.

Quoted in *The Sunday Times*, he said, 'I sold the car to a bank which asked me not to reveal its name.' However, the paper added that, 'it is understood that the Swiss bank was acting for a consortium of Japanese businessmen investing for the first time in a classic car market that is producing bigger returns for owners than ever.'[35] The Royale's present owner is Fusaro Seikiguch in Japan.

1932 Royale, with *Coupé de Ville* body by Henry Binder of Paris. Chassis no. 41111

This is the car that started life as the lovely open two-seater roadster for Armand Esders. It is not known how long he retained it but the Royale was displayed 'in various *concours d'élegance* and was used for sedate drives in the Bois de Boulogne'.[36] It may have passed through a number of hands before being bought, in about 1938, by the respected English Bugatti driver and specialist, Jack Lemon Burton. He bought the car in Paris for £600 but 'to have it imported would have cost me an extra £200 . . . So I disposed of that in France,

---

[33] Nye, 'A king's ransom'.

[34] Christie's press release.

[35] *The Sunday Times*, 29 April 1990.

[36] Clifford Penny, 'A Royale romance' in *Veteran and Vintage Magazine*, vol. 22, no. 1 (September 1977).

for £500, at a loss.'[37] In about 1938 the Royale is said to have been purchased by French politician Raymond Patenotre.

There is a characteristic twist to the story. An alternative version is that the car was destined for King Carol of Rumania. It came from David Scott-Moncrieff who, in 1947, was representing the Bugatti Owners' Club at Ettore Bugatti's funeral. There he met M. Band, who said that he had built all the Royales and had recorded maintenance details carried out on the various cars. He informed Scott-Moncrieff, 'that King Carol had placed an order for the *coupé de ville* but [it] was never completed'.[38] Quite independently, Frederick Henry, who bought the Royale in about 1948, claimed that 'it was the property of King Carol of Rumania'.[39] The truth of the matter has yet to emerge and the monarch abdicated, in favour of his son, Prince Michael, in September 1940, which was nineteen months after the Royale was completed. The former King spent the remainder of his life in exile in Mexico.

What is certain is that the ex-Esders Royale was taken to the Paris works of coachbuilders, Henry Binder, where the two-seater body was removed. What happened to the discarded coachwork is also a matter of conjecture. Pierre Dumont[40] says that 'It was preserved in pieces but all together in a depot belonging to Binder in Paris for a few months and was destroyed in June 1940 during a bombing raid on the Citroën works close by.'

In 1950, when Briggs Cunningham visited L'Ébé Bugatti, 'she showed me her scrapbook and she tore out a very nice picture of the Esders' roadster and gave it to me.'[41] Cunningham was so impressed with its lines that he contemplated scrapping the *Berline de Voyage* body on the second Royale and getting Binder to build a replica of the two-seater on it or try to obtain the original. Yet when he approached Binder, he was told that 'it had been cut up and taken away by the Germans for aluminium.'

In place of the roadster body, Binder created a *coupé de ville* body, clearly inspired by Bugatti's own Coupe Napoleon, even down to the moulded sweep panels. No doubt in an attempt to update the Bugatti's appearance the wheels were fitted with aluminium discs. But the overall outward impression was deceptive because Henry wrote in 1949 that, 'the rear compartment is bullet proof, with half inch [12.7mm] glass.'[42] Beneath the upholstery was amoured plate 4.7mm (3/16in) thick. Yet this protection did not extend to the chauffeur and, when asked why after the war, Binder countered with, 'it was to give the impression that none of it was.' Nevertheless, this formidable protection, which pushed the car's weight up to about 178kg (3.5 tons), suggests that the Royale was destined for a person of some wealth who feared assassination. The completed car was delivered, according to Binder's records, to its new though unknown owner, on 3 March 1939. Six months later to the day, World War II broke out. During hostilities, according to Henry, the Royale 'was lowered into the Paris sewers and jacked up to avoid capture by the Germans'.

From Binder to Britain

The Bugatti's immediate post-war history is equally shadowy. It may have belonged to a French businessman but when Captain Charles Lewis, a former Guards officer, brought the car to Britain, he did so on carnet. In view of its supposed temporary presence, the Royale retained its French registration number of 2250–RM7 and was kept at Lewis's London home at Belgrave Mews West.

At this time, and about five minutes drive away in Carlos Place, Mayfair, wealthy young man-about-town and Herefordshire antiques dealer,

---

[37] Jack Lemon Burton with Kevin Desmond, 'Bugatti Businessman' in *Classic and Sportscar*, March 1984.

[38] Penny, *op. cit.*

[39] 'Vintage Bugs' in *The Autocar*, 11 November 1949.

[40] Pierre Dumont, *Bugatti Thoroughbreds from Molsheim* (e.p.a., 1975).

[41] Nye, 'A king's ransom'.

[42] 'Vintage Bugs', *op. cit.*

Soon after 41111 had left Henry's ownership, it went to America and was subsequently restored by Mills B Lane of Atlanta, Georgia.

Frederick Henry, was lunching with a friend at the Connaught Hotel when he heard that a Bugatti Royale had arrived in London. He had developed 'a love of cars. Curiously, however, I had never owned a new one...'[43] His stable was to include 'a superb blown 16/90 AC, built for the 1940 New York World's Fair', a 3.5-litre Bentley and a Rolls-Royce Continental Phantom II drophead coupe. Henry's interest was fired by news of the Royale, 'and that it was believed to be for sale. Maybe I'd lunched a bit too well that day, for the upshot was that I made my way to the Mews, met ... Lewis and bought he car on the spot for £2,000.'

The trouble was that because the Royale was only supposed to be in Britain temporarily, hence the carnet, if it was to remain in the country for much longer, import duty was going to be payable on it. Henry therefore thought that a way around the problem was to obtain another carnet, which meant that he had to briefly become French-domiciled!

He and Charles Lewis therefore took the Bugatti back to Paris. There they took it out 'on the town', both wearing turbans inherited from a previous prank when Henry had won a bet by impersonating the Maharajah of Magador, the hero of a popular tune of the day, and had lunched, with a twelve-strong retinue, at the Connaught. In Paris, they 'even dined at Maxim's, the Royale attracting an appreciative crowd outside'.

On the return trip, they were joined by racing driver Lance Macklin, a friend of Lewis's, on what 'turned out to be a terrible night, pouring with rain, and the *pavé* was in its most slippery state ... and I had quite a job to hold the monster on the road.' When they arrived at Dover, Henry was duly regarded as a French tourist and was provided with coupons sufficient for him to buy 91 litres (20 gallons) of petrol, even though the Royale consumed fuel at around 1.8/2km/l (5/6mpg) on a long run. At best, it would have only taken the Bugatti about 160km (100 miles) though he was informed that he could apply for more.

The Royale was taken by its new owner to the Herefordshire town of Leominster, which was where Henry's antiques business was located. Needless to say, petrol became a major problem, it would remain rationed until 1950, and the Royale took 58 litres (36 gallons) at a time. Henry was able to get around that particular problem by obtaining 2,272 litres (500 gallons) of unrationed white spirit from 'a London spiv' and his local garage diluted it with the infamous 'Pool' petrol of the day.

Henry subsequently left Leominster and moved

---

[43] Frederick Henry, 'Remembering my Bugatti Royale' in *Veteran and Vintage Magazine*, vol. 22, no. 1 (A hilarious article and well worth reading in toto!)

to the village of Rudford, just outside Gloucester. As will have been apparent, by this time petrol was no longer a problem but tyres were. 'The original Dunlops on the car had been made in the company's French plant at Montlucon but during the war the moulds had either been stolen by the Germans or scrapped.' Welcome assistance came from Jack Lemon Burton, by then the owner of the Royale originally owned by Captain Foster and, 'he solved the problem by cutting away the rear wings slightly on his car so that gun carriage tyres could be fitted.' He supplied Henry with six similar covers for which he was most grateful though 'lifting the wheel off was a two-man job'.

With the tyre problem to some extent resolved, Henry found he could begin to extend the 12.8-litre Bugatti though, 'the brakes weren't at all impressive.' But 'the car being such a slow revver, the only suitable high-speed stretch I ever found on my way up to London was the Oxford–Witney by-pass, on which I could wind the Royale up to 100mph (160kph).' Also because of the slow turning engine, he considered the three-speed gearbox a limitation. This consisted of 'a very low first, or "garage" gear, direct drive and overdrive. The overdrive was in the normal "fourth" gear position (bottom right), but the lever came back a very long way, almost touching the bench front seat. It was an enormous lever with a large ivory control knob ... a fourth between direct and overdrive would have been much appreciated.'[44]

Running a Royale

Henry found that the really impressive aspect of the big Bugatti was the steering which was 'featherlight, and I'd never have believed that it was possible to turn those huge wheels so easily. The wheel itself was a beautiful wood rimmed affair, with horn buttons on the reverse side of each spoke.'

It was while the Royale was probably in Henry's ownership that it was encountered in about 1950 on the Braywick Straight, between Maidenhead and Windsor, by the respected motoring historian, the late Michael Sedgwick in his Fiat 500. Sedgwick would recall: 'We couldn't understand why a fast moving Bugatti – it must have been doing 75 – took so long to pass the Fiat fully extended at 50. It wasn't until [it] had passed that we realized that we'd come face to face with a Royale.'[45]

In the three or so years he owned the Bugatti, Henry took the opportunity to familiarize himself with the car, both outside and in. It was while examining the small travelling compartment that he made a discovery which perhaps gave a further indication of the importance of the car's one-time or prospective owner. 'We found a handsome cosh behind the rear seat, a real brute of a weapon consisting of a handle of thick coil spring with lumps of lead on top.' Also revealed were 'some strange brackets, a straight piece, then a right-angle to the right, and finally a right angle to the left with a hinge-piece on the end. We never figured out their purpose, though the newspapers got hold of the story and said that they were machine-gun mountings.'

Following Henry's move to Gloucestershire, in 1951 Customs and Excise officers finally caught up with him, though a letter he wrote to *The Autocar* in October 1949, complete with a photograph of the Royale still retaining its French plates, in response to an article on the ex-Foster Royale, can have hardly helped! The Bugatti was accordingly impounded at the garage of C Healey and Son in Westgate Street, Gloucester and its owner had to pay £1,000 duty to reclaim it. The car lost its French registration number at this time and, in June 1951, was registered KDG 456 with Gloucestershire County Council.

This meant that, by mid-1951, the Royale had absorbed some £3,000, while its tremendous running costs must have strained even Henry's comfortable circumstances. The world's largest production straight-eight was hardly ideal for everyday motoring. In short, in the latter half of 1951, Henry started seriously thinking of selling his Bugatti with an asking price of around £8,000.

[44] Henry, *op. cit.*

[45] Henry, *op. cit.*, postscript by Michael Sedgwick.

He had, in fact, been contemplating such a course of action since at least 1950, and he had corresponded with international collector, D Cameron Peck of Chicago.

On 15 November 1950, the American wrote to him, pointing out that 'he had just secured another Bugatti in Paris [the *Berline de Voyage*], adding, "I think you might find it rather difficult to secure the price you are asking, although, goodness knows, there is no fixed market for cars such as yours."[46]

Henry at last disposed of the car, exchanging it with Simmons of Mayfair, probably in 1951, for an equally unusual vehicle, a Rolls-Royce Phantom II, rebodied in 1947 by Figoni and Falashi for Nepalese royalty. From Simmons the Bugatti went, in 1952, to Connaught Engineering of Send, Surrey. This was because, as Charles Meisl who then worked for the company explained in *Classic Cars*,[47] 'Rodney Clarke, the managing director, was just passionate about Bugattis.' The trouble was: 'We had such a job to sell it when it was taking up space in the stock of second-hand cars at Connaught.'

The Royale passed to Len Potter and then Lieut. Carl Montgomery, then serving in the US Army. The latter exported it to America where it was bought, in 1954, by Dudley G Wilson of West Palm Beach, Florida. In 1961, Wilson sold the car to Mills B Lane of Atlanta, Georgia for his Stable of the Thoroughbreds Museum.

A Much Needed Restoration

Lane had first heard of the Bugatti from a Rolls-Royce public relations man who, 'in the course of the conversation said he had been in West Palm Beach and heard that there was a Bugatti Royale there with its body at one end of town, its engine in another.' Lane sought the car out and found that it was, by then, in one piece. Although the body was in poor condition, he was able to drive the car, and found that 'the engine ran and the smoothness of the steering . . . was absolutely amazing.'[48]

A deal was struck and the Bugatti was delivered by truck to Atlanta in June 1961 though it was not until March 1963 that work on the car's restoration began. This was entrusted to the one-man business of John Griffin of Montgomery, Alabama and it took precisely one year and twenty days to complete. This was a chassis-up rebuild and, when the engine was stripped down, it was found that four of the pistons were broken. New ones were made, but otherwise the big eight was only suffering from the usual wear experienced by a large, thirty-year-old car. When it came to restoring the body, Lane chose a colour scheme of midnight blue and silver which already featured on his 1959 James Young-bodied Rolls-Royce Silver Wraith. Ten coats of paint and twenty of lacquer were applied. Replacing the bullet-proof glass was a problem until a national glass company came to Lane's aid and made a special run of the material for the Bugatti.

In 1964 news of the restoration, which was nearing completion, reached the ears of staff at Harrah's Auto Collection, over 3,000 miles away (4,830km) in Reno, Nevada, as it had just purchased the *Berline de Voyage* Royale. The quality of Griffin's work so impressed Harrah's representatives that their newly acquired Bugattti was sent to Montgomery, and it was there subsequently photographed alongside the Binder-bodied car. As a result of this association, in August 1964, Mills parted with the Royale to the Reno collection for $50,000 (£17,921). Bill Harrah thus became the proud owner of two of these exclusive but massive Bugattis and, when asked what a Royale was like to drive would delight in replying, 'which one?' In 1966, this car was the overall winner in the prestigious Pebble Beech *concours d'élégance*.

Following Harrah's death in 1978 and the Holiday Corporation's subsequent purchase of his collection, in 1986 the Binder-bodied Royale was one of eighty-nine cars sold, in a $27 million deal, to CalAir's founder, General William Lyon, and he is the car's current owner.

---

[46] Henry, *op. cit.*

[47] 'Looking back' in *Classic Cars*, January 1990.

[48] Mills B Lane, 'Royale' in *Bugantics*, vol. 27, no. 4 (winter 1964).

The cabriolet body of Dr Josef Fuchs' Royale was by Weinberger of Munich and was a scaled-up version of one that the same coachbuilder had fitted to his Type 50 Bugatti. It was completed in 1932 but not fitted with an elephant mascot and today...

1932 Royale, cabriolet by Weinberger of Munich. Chassis no. 41121

Although all six Royales survive, this one, originally owned by Dr Josef Fuchs of Nuremberg, is the example of the breed that came closest to being scrapped. Fuchs only used the car for a relatively short time in Germany for, only two years after he had taken delivery of the Royale, he and his car left Europe, bound for the Far East. Maybe prompted by Hitler's rise to power in 1933, it was on 9 May 1934 that the Bugatti left the Italian port of Trieste to be shipped to Shanghai. We can only imagine what the Chinese made of this German doctor and his gargantuan car. It is not known how long Fuchs remained in China but he subsequently moved to America and settled at Long Island, New York.

In 1937, the Royale was noticed by Charles A Chayne, one of America's leading automobile engineers and then responsible for Buick car design. Educated at Harvard and a graduate, in 1919, of the prestigious Massachusetts Institute of Technology, after experience with Lycoming, Chayne joined Buick in 1930 and became its chief engineer in 1936. The occasions on which he first saw the Bugatti were the practice days for the Vanderbilt Cup race, held at the Roosevelt Speedway, Long Island.

Unfortunately, soon afterwards, during the winter of 1937/1938, the Royale suffered a cracked cylinder block through its water freezing and it was left for years in the backyard of Fuchs' house, covered with a sail. Its owner had, apparently, been unable to get his car repaired, though on at least one occasion it was taken to a New York repair shop which was no doubt overawed by the magnitude of the task and it remained untouched.

A Good Buy

It was not until six years after Chayne had first seen the Royale that, in June 1943, he 'received a telephone call from a friend in New York to the effect that the car had been sent to a junkyard in New York City and, if I wanted it, I had better move fast. I immediately phoned Charles Stitch, who operates a foreign car shop in New York, and had him buy the car and store it.'[49] Chayne did not say what he paid for the Royale but it is reputed to have been in the region of $350 (£87) which, in retrospect, would make the purchase one of the greatest automotive bargains of all time.

The Bugatti was subsequently transported to

[49] Eaglesfield and Hampton, *The Bugatti Book*.

. . . can be seen in the Henry Ford Museum at Dearbourn, Michigan, USA after restoration by Charles A Chayne in 1946 and 1947.

Chayne's home at Flint, Michigan but because of the war and business commitments, he was unable to embark on the formidable restoration process until late in 1946. It was completed in time for its owner to compete in the Veteran Motor Car Club of America's 1947 Glidden Tour.

The main problem was, of course, the engine. On stripping it down, it was found that the outer walls were badly cracked and bulged above and below the crankcase cover plates. In view of the massive size of the casting, it was decided not to attempt to rectify the damage by welding but instead to resort to patching. The loose panels were knocked out, the bulged sections of the block milled flush and plates of 3.9mm (5/32in) mild steel attached to either side. After screwing them into position, 'we filled the casting with salamoniac solution, flushed it thoroughly, allowed it to dry and then made sure of the seal by filling it with a resin-type sealer under pressure which was then set permanently by baking the case for several days at a moderate temperature.'[50] It was found that the freezing had distorted the cylinders but these were rebored and new 0.30in oversize pistons fitted. New valves and springs were required and it was found necessary to make a new camshaft as the original was badly corroded.

[50] Eaglesfield and Hampton, *op. cit.*

So far so good. But as Chayne continued to restore the car, the engineer rather than the historical perfectionist surfaced in him and he embarked on a series of modifications which, in retrospect, appear garish and out of sympathy with the specifications of a 1932 Bugatti. Conversely, we should be eternally grateful that Chayne rescued the Royale from almost certain destruction, for there would otherwise be five of these extraordinary cars left today, rather than the full complement of six. And many of the changes that he effected are reversible . . .

One of the more radical 'improvements' that Buick's chief engineer made to the engine was to replace the single Bugatti carburettor with four single Strombergs with a balance tube connecting the four elbows. The original Bugatti braking system, admittedly a weak point, was replaced with hydraulic actuation. Chayne also believed that the twin Strombos air horns mounted externally ahead of the radiator, were 'in keeping with majesty of the car'. The Royale's cast aluminium wheels remained intact though Chayne added 'trim discs . . . to improve [sic] their appearance'. Tyres inevitably presented a problem but he managed to locate a 7.50-24 tractor mould and had a set specially made for the car.

Inevitably, as the years passed, Charles Chayne found that he had less time in which to drive the

The instrument panel of the Park Ward Royale. There is a handsome Bréguet stop watch in the centre of the wood rimmed steering wheel with a horn button located under each spoke.

Royale. In 1951 he had been promoted to the post of General Motors' vice president of engineering and eight years later, in September 1959, Chayne presented the car to its new owner, the Henry Ford Museum at Dearborn, Michigan, established by Ford in 1933. It has resided there ever since and provided the inspiration to Thomas Monaghan to buy *his* Royale.

1933 Royale with limousine body by Park Ward of London. Chassis no. 41131

The Royale owned by Captain Foster was purchased in 1946 by Jack Lemon Burton who, prewar, had briefly owned the ex-Esders car. He bought it from Foster himself, who had gone to America, and was disposing of his car after it had only covered about 56,325km (35,000 miles). It had, according to Lemon Burton, 'spent the duration of the war in Finchley, raised up on jacks'.[51]

Three years later, in October 1949, journalist Dennis May wrote of an encounter with the Royale in *The Autocar* in which he took his readers behind the car's steering wheel, setting down in his characteristic prose, the distinctive controls and instruments a Royale owner would encounter. Initially there were eight dials:

(i) A superb Bréguet stop watch mounted in the hub of the typically Bugatti steering wheel, wood rimmed and steel spoked; this watch cost £85 and Le Patron threw it in as a *cadeau*, in common with the silver elephant mascot surmounting the radiator cap. (ii) Top left of a sixsome facia group, fuel contents gauge, the needle of which moved about as far on the down-beat when motoring as on the up-beat on refills, because La Royale gives 8mpg [2.8km/l] if driven economically, or if not, not. (iii) Top centre, ammeter. (iv) Top right, water temperature. (v) Bottom left ... oil pressure gauge ... (vi) Bottom middle, 120mph [190kph] speedometer, which on ALB 2, tyred as it is [with 7.50-24 Goodrich Silvertowns, courtesy of a US Army mobile service gun] ... tells six per cent less than the truth. (vii) Way out of the group on the extreme port side, accumulator indicator serving the six, six-volt batteries which are coupled to give 12-volt supply.

The rest of the facia furniture, which includes a total of thirteen switches, is far too numerous to detail *in toto* but *en passant* one may mention the twin switches controlling one bank of plug apiece ... a hand throttle lever matching an

---
[51] Lemon Burton, 'Bugatti businessman'.

advance and retard ditto; three massive ivory knobbed items for, severally, the throttle stop, a main-jet regulator and an extra air device on the Stromberg-type carburettor; and a warning light which when one of . . . the Passengers thumbs a button in the back, glows red to tip you, the chauffeur, that your sovereign prince is coming through on the house-phone.

May noted that the Royale's 'gear positions are unusual. Up right for bottom, up left second, down left high.' He revealed that Lemon Burton had 'never exceeded 90mph [145kph] for a variety of reasons. One, it uses too much petrol at high speeds. Two, the tyres now fitted . . . are oversize and thought to have a deleterious effect on the front wheel deportment under certain conditions. Three, the Bugatti brakes, cable operated, lacking any form of servo and soon due for new linings, fall quite noticeably short of modern hydraulic standards of retardation.'[52]

From America to France

Lemon Burton retained his Royale for ten years, selling it in June 1956 for $10,000 (£3,571), to John Shakespeare of Centralia, Illinois, USA. The latter was in the process of building up a collection that would eventually amount to thirty Bugattis, and, at the time, the largest number then assembled. But in March 1964 he sold all his cars to the Schlumpf brothers for a reputed $250,000 (£89,605), and the Royale joined the ex-Bugatti Royale at Mulhouse where it can today be seen.

This completes the inventory of Royales but the summer of 1990 saw the arrival of a seventh which, it should be stressed, is not an original car but an impeccable recreation of the lovely Esders open two-door Royale by the *Musée national de l'Automobile* at Mulhouse. When the secrets of the Schlumpfs' extraordinary hoard were revealed in 1977, one of the projects being undertaken was to reproduce this Royale, using many of the spare parts which the brothers had obtained from Molsheim, though their bankruptcy brought the work to a halt. In 1984 the idea was revived by Jean-Claude Delerm, the museum's director, and was to be backed by the equivalent of £100,000 in subsidies from national and regional bodies. The project, which restarted in earnest from late 1987, took three and a half years to complete and the car was secretly tested for the first time at a local airfield in June 1990.

This Royale, finished in two-tone green, is powered by a railcar engine though the clutch, transmission and rear axle, with three-speed gearbox, are original, along with the suspension, steering brakes and wheels. It was completed in time for the gathering of all six Royales, held over the weekend of 8/9 September 1990. It was wholly appropriate that this should take place at the Parc de Bagatelle in Paris, the city which, over the past sixty or so years, had played host to most examples of this car, that most extraordinary of Ettore Bugatti's creations.

[52] Dennis May, *The Autocar*, October 1949. (Good, vintage May!)

10
Twin Cam Twilight

'Let us consider the fabulous Bugatti, prince of motors. Imagine a string-straight, poplar-lined Route Nationale in France on a summer's day. That growing dot in the middle distance is a sky-blue Bugatti coupe, rasping down from Paris to Nice at 110 miles per hour . . .'

Ken W Purdy[1]

If the 1920s were golden years for Bugatti, then the following decade was a time to rationalize and diversify. If Bugatti had had to solely rely on car sales in the 1930s, he would have undoubtedly followed many of his contemporaries into bankruptcy. A fashionable alternative was a merger with another company. But it is impossible to contemplate Ettore accepting an association of the type that Louis Delage, his great former rival on the race track, was forced to do.

The great irony was that the Royale, that whitest of white elephants, should lead to the railcar project that would just ensure Molsheim's survival until the outbreak of World War II. With no more such orders from 1939 onwards, Automobile Bugatti's days would have been numbered and the firm would have probably ceased car manufacture soon afterwards, had not the war intervened.

Inevitably, as the *automotrices* moved centre stage in the six years between 1933 and 1938, car production was downgraded and output radically reduced. To give this statement some perspective, it should be remembered that in his close on thirty years at Molsheim, Bugatti built approximately 7,800 cars. Of these around 7,000, or about 90 per cent of output, were produced prior to 1934 the 800 or so between then and 1939 represented a mere 10 per cent of the total.

The fact is that Ettore Bugatti, fifty-one years old in 1932, was beginning to lose interest in his cars and, from thereon, he spent an increasing amount of time away from Molsheim in Paris. This allowed his eldest son, Jean, to take over greater management responsibility which was extended from his original brief of coachwork design to overseeing the creation of the Type 57, effectively the sole Bugatti road car produced from 1934 onwards.

Yet ultimate responsibility was still vested in Ettore and he vetoed Jean's initiative that the 57 should become the first Bugatti road car to feature independent front suspension, while the model continued to employ mechanical brakes at a time when many of its contemporaries had switched to hydraulic operation. Maybe his pioneering experiences with the system had cast a long shadow.

This approach was also extended to the Type 59, the Grand Prix car created for the 750 Kilogramme Formula of 1934, which unleashed the sophisticated might of the German Auto Union and Mercedes-Benz racers on the motor racing circuits of Europe. The Bugatti was a visually triumphant though archaic design, as it was the last car from a major constructor which retained two-seater rather than *monoposto* bodywork. It also lacked the reliability of its Type 35 predecessor and, once again, Ettore remained faithful to non-independent suspension and mechanical brakes.

The team of factory cars was only run for one full

[1] Ken W Purdy, *The Kings of the Road* (Arrow Books, 1957).

season and then sold off to private owners. Thereafter Bugatti had a minimal presence in such frontline competition though the marque achieved unexpected success in the field of sports car racing with two victories in the Le Mans 24-hour race.

The repercussions of the Wall Street economic crash of September 1929 had left its mark on the decade, as far as the manufacture of low quantity luxury cars was concerned. European mass car production, by contrast, boomed. But the specialist car maker, unless it had diversified into other markets, either merged, ceased car production to concentrate on other fields or, quite simply, went out of business.

It was aero engine manufacture which underpinned the activities of such prestigious companies as Hispano-Suiza in France and Spain and Rolls-Royce in Britain, but this was by no means a universal panacea. A similar diversification was unable to assist Isotta Fraschini in Italy which effectively ceased car manufacture in 1935. Earlier in the decade, Alfa Romeo, the country's most famous racing and sporting marque, was heading for bankruptcy. It would have gone out of business in 1933, had it not been for Mussolini's government which came to the rescue to provide a bastion against the growing power of Fiat and to uphold the glories of his fascist regime on the race track.

In Britain, the proud Bentley Motors went to the wall in 1931, only to be taken over by Rolls-Royce, which also attempted to absorb a bankrupted Lagonda company in 1935 though that firm was rejuvenated under new management with fresh funding. Across the Atlantic in America, Duesenberg ceased production in 1937 as the Auburn Cord Duesenberg Corporation, of which it was part, collapsed. Cadillac, by contrast, was safe within the General Motors financial corral.

It was much the same story on mainland Europe. In Belgium the stately Minerva came to the end of the road in 1935, while in France the respected but ailing Delage company was taken over by Delahaye in the same year. Although Voisin continued in his idiosyncratic way, output was down to a trickle by the time that production ceased in 1938.

Origins of the Railcars

Bugatti might well have joined these companies, had it not been for a chain of events which began in 1929. The French railways had been partly nationalized under the ÉTAT name in 1923. The remaining private companies had been granted fifteen-year *concédés* due to expire in 1938 when they would become state property. In 1929, the French railways decided to initiate a programme based on the development of the diesel-powered railcar though other types of fuel were contemplated.

The self-propelled railcar was not, of course, new. Indeed the first, a steam-powered device, had run in Britain in 1847 and, in 1930, there were about forty similarly driven examples in France, along with a few petrol-fuelled railcars though all were relatively small and limited in their range. Nevertheless, it was believed that the introduction of more railcars would be less labour intensive than steam engines and, above all, would not use coal, France possessing few mines, apart from those in Alsace Lorraine.

A further ingredient was supplied that year when André Michelin, one of the partners in the famous tyre company which bore his name, slept badly on the night train from Paris to Cannes. The following morning he is reputed to have said to his brother, Édouard, 'Why don't they run trains on rubber so we can get a good night's rest?'[2] In the face of not unexpected hostility within the company, Édouard took up André's idea and, supported by Jacques Hauvette and Pierre-Marcel Bourdon, work proceeded, with experiments being undertaken later in 1929, on a 40hp six-cylinder Renault car modified to run on rails.

Next came a 20hp Panhard Levassor truck chassis which was to result in Michelin beginning its researches into steel braced pneumatic rubber tyres, work which would culminate in the firm producing the world's first successful radial road tyre, the Michelin X, in 1953. These rail tyre were limited to a weight of only 680kg (1,500lb) but this,

[2] Jan P Norby, *The Michelin Magic* (TAB Books, 1982).

in turn, led to the firm undertaking pioneering work into the principles of lightweight railcar construction.

On 23 July 1931, Michelin held a press conference and demonstrated no less than six Panhard-based railcars, which had been given the appropriate *Micheline* name, and run on the Issoudun–Florent route. Trials began with daily runs between Coltainville and Saint-Arnould, went on through the summer of 1931 and, in September of that year, came a new Hispano-Suiza based unit, with a light ten-seater body, operating the Paris and Deauville run at an average speed of 107.01kph (66.5mph). As a result, a regular service was introduced in the spring of 1932. This railcar could only carry twenty-four passengers though it led to a larger thirty-six seat railcar and, ultimately, in 1936 to the sleek 100-seater *Micheline*.

By 1939, the company would have a total of 140 of its rubber-tyred railcars in service. But Michelin had by no means a monopoly of their manufacture. By September 1933, no less than 173 had also been ordered from a variety of firms, including Renault, De Dietrich, Franco-Belge, Decauville and . . . Bugatti.

Bugatti Steps In

Ettore was fascinated by Michelin's experiments when they were publicized in 1931. Indeed, his old friend the Duc de Gramont, who had made his Paris laboratory available to him during World War I, has recalled: 'I remember reading him an article in *L'Illustration* about the first *Micheline* railcars, and he at once expressed his admiration for this form of transport, followed by a running commentary on the improvements and changes that could be made to them. And it was not long before he carried them out, applied to his own railcars.'[3]

It was in 1932 that Raoul Dautry, director general of the French state railway, ÉTAT, approached Ettore with a view to him undertaking railcar manufacture. According to Stella Vendromme, who was secretary of Ettore's Paris office between 1928 and 1940,[4] the introduction to Dautry came via two faithful Bugatti car owners, Bapifaut and Fabre. She has also recalled that the work was paid for in stages, with Bugatti billing the railway company as the construction proceeded. In 1935, when the programme was well underway, Ettore would write to Dautry, recounting that, 'I was pushed, thanks to you, into this new branch of industry.'[5]

The push could not have come at a better time, with the P3 Alfa Romeo giving Bugatti a tough time on the race tracks and Europe in the grip of a recession which damaged the market for expensive road cars of the type that Bugatti produced. There is little doubt that Molsheim was hard hit by the depression. In March 1930, Ettore's English friend, Llewellyn Scholte, had written to him following a visit he had made to Colonel Sorel, head of the firm's English branch. In the letter, Scholte writes that Sorel, 'told me that business was very bad and that since October, there was no business at all and that he didn't believe it was worth making announcements or competing'.[6]

Trade did pick up though never reached the levels attained in the 1920s, so a diversification could not have been better timed. But, above all, the railcar contract gave Ettore a heaven-sent opportunity to put the 12.8-litre eight-cylinder Royale engine to an alternative use, for although the majority of French railcars used diesel engines, the pioneering *Michelines* employed, despite the fire risk, petrol-fuelled units and Bugatti would follow suit.

The only trouble was that there was no building at Molsheim large enough to take a railcar and, as Bugatti team driver, René Dreyfus, records: 'A big shed was hastily built . . . and typically too Bugatti pulled men out of the racing car shops to work on it. Most of them, as a matter of fact. The project really piqued his interest.'[7]

---

[3] L'Ébé Bugatti, *The Bugatti Story*.

[4] Hugh Conway, 'Another little bit of history' in *Bugantics*, vol. 45, no. 4 (winter 1982).

[5] Conway and Sauzay, *Bugatti Magnum*.

[6] Sholte/Bugatti correspondence.

[7] Dreyfus with Kimes, *My Two Lives*.

The first Bugatti railcar being manhandled in May 1933 from the factory through the streets of Molsheim to the railway station. Later, lines were laid to the works. The entire operation was meticulously planned by Ettore Bugatti.

When T P Cholmondeley Tapper, who was later to campaign his white-painted Type 37A Bugatti in Europe, visited Molsheim in 1934, he was told that Ettore had 'first thought of the design while walking in Strasburg. It is said that he sketched his idea on the pavement and then sent for a factory draughtsman to copy his rough drawings.'[8] Bugatti's eldest daughter, L'Ébé, says that the work on the railcar's layout began in earnest late in 1932 and continued well into 1933, when Ettore was 'ill with influenza for much of that winter and he worked in the bedroom of his Paris home at 20 rue Boissière, using the mantlepiece as a table'.

Noël Domboy, technical director at Molsheim, has recounted that, when Raoul Dautry visited the factory in the spring of 1933, 'he was shown a partly complete chassis with two of its four engines in place, and was able to make a short run of a few metres along a piece of track laid down in the shop.'[9]

Dautry proposed a near-impossible date of May though, incredibly, the work was finished in time. Then Bugatti was faced with moving the 1,626kg (32-ton) structure from the factory to the nearest railway station which was fortunately, in the town of Molsheim. At this time, however, there were no rails joining the two, though a link was subsequently laid.

Moving the First Railcar

Ettore meticulously planned the entire operation from his Paris home though he was present to supervise the delicate removal of this 22,300mm (74ft) long car. Dreyfus says of this memorable occasion, that 'Ettore . . . suggested that everyone – and he meant everyone – join him to push the *automotrice* the 1.6 kilometres [1 mile] down to the depot.'[10] L'Ébé Bugatti, who was also there, recalls:

> There were a number of hazards on the way, not least being two sharp bends in the road. There was also the usual traffic, augmented by a long line of people all the way to the station who had come on foot and by bicycle to wait to see this 'train' go by, giving encouragement but forming something of a hindrance.
>
> The railcar was brought out of the factory gates at an angle, after part of the wall had been knocked down, so that it would be in line with the road, later a turntable was installed at the exit.[11]

The most difficult part of the entire exercise, which was effected by moving the 'car on two short lengths of track, was at the approach to the square in front of the railway station. The removal was finally completed at about two o'clock in the afternoon of the following day, having taken over 24

---

[8] P Cholmondeley Tapper, *Amateur Racing Driver* (Foulis, 1954).

[9] Conway and Sauzay, *Bugatti Magnum*.

[10] Dreyfus with Kimes, *My Two Lives*.

[11] L'Ébé Bugatti, *The Bugatti Story*.

hours to complete. Ettore himself had been present throughout and, even when he snatched a few hours' respite, he slept in one of the railcar's seats. As for the rest of the workforce, L'Ébé remembers that glasses of beer and sandwiches in great numbers were brought from the Hôtel de la Gare to keep up morale.'[12] Once the car had reached Molsheim station, there came a further tricky operation when it was removed from its two bogies. The body was then loaded on to a flat truck and the bogies stored in another wagon. The dismantled railcar was then transported by train to Montrouge, near Paris where it was reassembled, so that testing could begin.

This took place on a disused section of line at Gallardon between Paris and Chartres, which had previously been used for evaluating rolling stock. Speed was therefore restricted to 90kph (56mph), due to the poor state of the track. With Jean Bugatti in the high, centrally positioned cabin, he immediately attained a speed of 125kph (78mph) and Domboy says that the 'railway engineer, who only appreciated the achieved speeds after examination of the Flaman recorder, suffered retrospective apprehension but did authorize Jean to continue up to 135kph [84mph]'.[13]

After these trials, the railcar was returned to Molsheim, where the only major modification required was the introduction of a two-speed gearbox, Ettore presumably having considered one unnecessary. During the test, Jean had simply slipped the protesting clutches to move off and accelerate. He subsequently visited Daimler in Britain and, as a result, a fluid flywheel was introduced into the design which resulted in a smooth, if slow, take-off. It should be stressed that although the overall conception of railcar design was Ettore's, much of the subsequent work was undertaken by Jean.

The first series of railcars were diplomatically called Le Présidentiel in honour of the French president, M. Albert Lebrun, who made the first official journey in the prototype on 30 July 1933. This took place between Paris and Cherbourg, when the car averaged 115kph (71mph) for the 371km (230-mile) run which impressively reduced the previous travelling time by an hour. Later, the first Bugatti railcars were introduced on the nearby Paris to Deauville route.

Inspired by the Tanks

This is not the place to take a detailed look at the entire railcar programme. However, the overall design of the *Présidentiel* is of interest as it illustrates how Ettore ingeniously adapted the Royale engine for railcar use. Also, the *Présidentiel*'s profile was closely related to the bodywork of one of his racing cars and accordingly sprang directly from aerodynamic experiments he had undertaken ten years previously.

Bugatti was clearly concerned to reduce drag when he came to design his railcars and their clean, angular lines, devoid as they are of all decoration, shows that his sense of proportion and visual balance remained as impeccable as ever. In 1935 he informed a visiting English locomotive engineer that, 'he had carried out a large number of wind tunnel tests to determine the best form for the railcars but considered the results unreliable. He then started road experiments and arrived at a wedge-shaped nose and tail.'[14]

What Ettore did not say was when he had undertaken this work and, in view of the speed in which the first railcar was conceived, it seems likely that he was referring to earlier experiments for, we have it on his own authority that the *automotrice*'s lines were inspired by those of the controversial 'Tanks' that he had run at the 1923 French Grand Prix.

As far as the 48-seater *Présidentiel*'s mechanical layout[15] was concerned, the four Royale engines, making a total displacement of 51.2 litres and

---

[12] L'Ébé Bugatti, *op. cit.*

[13] Conway, *op. cit.*

[14] John Bolton, 'Les automotrices, a railman's view' in *Bugantics*, vol. 29, no. 3 (autumn 1966).

[15] GOP Eaton, 'Les automotrices', in *Bugantics*, vol. 24, no. 1 (spring 1961).

The heart of the Bugatti railcars: the massive eight-cylinder 12.8-litre Royale engine distinguished by its twin down-draught Zenith carburettors and associated manifolding.

800hp, were positioned at right angles in the centre of the car though offset to one side to permit passengers to pass to and fro. The minimum of modifications were made to the 12.8-litre straight-eights, apart from the carburation. The single Bugatti-designed unit was dispensed with and replaced by two downdraught Zeniths and their associated seprarate manifolding. Fuel was a petrol-benzol-alcohol cocktail and was consumed at the alarming rate of 1.1 litres per kilometre which is the equivalent of 10 gallons every 12 miles.

The Mechanical Layout

Each pair of engines was coupled together and drove the nearest pair of wheels on the front and rear bogies, via fore and aft propeller shafts. There were four axles per bogey and eight wheels and, although Ettore did not attempt to emulate Michelin's rubber tyres, these were fabricated in three parts with rubber inserts sandwiched in between each section, which were intended to reduce noise and vibration. The drum brakes were pneumatically actuated though were cable operated in view of Bugatti's traditional distrust of hydraulics! As already mentioned, the cab was located directly above the engines and, while visibility was excellent from this central lofty perch, it did prove difficult for drivers to see the front of the car, which presented problems during shunting.

There was little doubt that, like any Bugatti, the *Présidentiel* was fast. Soon after its completion the prototype gained a world record, the first of many that Bugatti would delight in publicizing, when it achieved a speed of 175kph (108mph) and later raised this to 196kph (122mph). It should be said, however, that the original version was later joined by more popular two-engined types which, although slower, consumed less fuel.

In addition to their supply to the French state railways, Bugatti railcars were also delivered to two of the other private rail companies, namely the Alsace-Lorraine Cie and the Paris–Lyon Midi. Their impressive performance was not in doubt and this attribute was particularly noticeable on the latter service for they enabled Paris businessmen to go to Lyon and back in the day for the first time.

How did these railcars compare with the behaviour of their contemporaries? There can be little doubt that they were fast and although Bugatti claimed that they were able to stop from 150kph (93mph) in 800 metres (about half a mile), in practice this did not prove to be a consistent figure. *La Vie du Rail* commented that 'The drum brakes were the weak point on the Bugattis. If they allowed a stop from 150kph in 800 metres, it was not possible to repeat the performance a second time. It was very difficult to distribute the braking effort equally between all the brake drums, which resulted in the overheating of particular linings, accelerating their wear and reducing brake performance.'[16] As a result, the drivers only used the

[16] Conway, *op. cit.*

brakes when it was absolutely essential, first reducing speed and then applying them at about 68kph (40mph). Like Ettore's cars, the rail variety was intended to go, not to stop.

A British Viewpoint

The railcars' ride also left something to be desired. In 1938, M. Dumas addressed Britain's Institution of Locomotive Engineers, when he presented a paper entitled, *The Development of Rail Motor Services in France*. In the ensuing discussion, one member complained that:

> He had the pleasure, and it was a real pleasure, in travelling in one of the diesel railcars that were employed on the fast service between Lyon and Paris. He was told that, on that particular day, he would enjoy a comfortable trip because it was the Renault twin diesel railcar but that if it had been a Bugatti he would not be so comfortable, because the latter was supposed to jolt the customers around a great deal.

Another engineer expressed similar sentiments and declared that '... he had to confess, as one who had travelled a good deal in railcars in France at one time or another, that he had experienced as rough riding in certain cars there, particularly Bugattis, as he had ever experienced in this country.'[17]

A further contemporary impression of a journey in a Bugatti railcar has come from Oliver Bulleid, assistant to Sir Nigel Gresley, chief mechanical engineer of the London and North Eastern Railway (LNER). In the early 1930s the LNER was studying the effect of drag on trains, with a view to making its locomotives more aerodynamically efficient. The first to reflect these disciplines were the legendary A4 Class Pacifics which appeared in 1935, of which *Mallard* is the most famous example.

Bulleid had kept Gresley closely informed on the performance of the Bugatti railcars and both men travelled to France and made a number of journeys

The light and comfortable interior of a Bugatti railcar.

between Paris and Deauville in examples though Bulleid found it 'a hair raising ride, for the cars travelled at 80*mph* [his italics] into the centre of Paris and they rode very badly'. However, the LNER engineers appear to have admired the railcar's aerodynamic efficiency and were 'greatly impressed by the wedge shape as they stood at the back of the car and watched the flow of wind hitting the track 50 yards behind them. The look of Gresley's streamlined locomotives was already forming in their minds.'[18]

In June of 1935, the year in which the first of the streamlined Pacifics appeared, Bulleid and Gresley accepted an invitation from Ettore himself to travel in a railcar. On meeting Bugatti, these sober-suited engineers were somewhat taken aback by his appearance and Bulleid's biographer has set down

[17] Bolton, 'Les automotrices'.

[18] Sean Day-Lewis, *Oliver Bulleid* (George, Allen and Unwin, 1967).

his subject's impressions, who said of Bugatti 'If you met him at Newmarket you might well think he was a jockey. He was small and short and invariably dressed in riding breeches with a bright yellow sweater.'[19] During the journey they 'were sitting right on top of the petrol tank. When Gresley asked what would happen if there was a leak and whether it might not lead to a fire, Bugatti seemed quite unconcerned and replied that there could easily be a fire but it would be 90 metres (100 yards) behind the vehicle before it broke out.'

There were, in all ninety-one Bugatti railcars produced between 1933 and 1938 and, in the peak year of 1935, thirty examples were delivered. A further two were built after the war. The Bugattis continued to be used well into the post-war years, when they ran on three routes; between Paris and St Etienne, Le Havre and Trouville. The last railcars, the two-engined *Surallonge* (extra long), which had entered service in 1937 were finally withdrawn in 1958. All have now been scrapped, except one, number XB1008, which was commissioned by the European Office for Research and Testing (ORE) and spent its last years as a travelling laboratory, crammed with scientific equipment, running not only in France but Belgium, Germany, Denmark and Sweden. Today it resides at the *Musée des Chemins de fer* (Railway Museum) at Mulhouse, only a few kilometres away from the fabulous collection of Bugatti cars at the *Musée national de l'Automobile*.

Labour Troubles

The switch in emphasis from cars to railcars at the Bugatti works was a factor in Molsheim being a victim of the industrial unrest that began sweeping through France in 1936, which coincided with the election of a Popular Front government there. As a result, Ettore who, in any event, was spending an increasing amount of time away from the factory in Paris, left Alsace completely.

These momentous happenings are best described by L'Ébé Bugatti, who says that 'The men stopped work, there were protest marches, the Red Flag was waved and the factory was occupied for a sit-down strike. It was a terrible shock, a thunderclap without a storm. Bugatti was embittered by what he took to be so much ingratitude and an attack on him personally. He left Molsheim altogether; and so it was then that Jean, then aged twenty-six, took over entirely.'[20]

W F Bradley was probably right when he described Bugatti as a 'benevolent dictator'[21] as far as the running of his factory in the 1920s was concerned. Like many of his contemporaries, Ettore worked long hours and expected his workforce to do the same. Fortunately, we have an insight to his attitude to labour from his own hand. It appears in the surviving correspondence, from the years between 1924 and 1927, that he conducted with British industrialist and Type 35 and 30 owner, Sir Robert Bird, Bt., chairman of the Birmingham-based custard manufacturer, Alfred Bird and Sons. He was also, for twenty-three years between 1922 and 1929 and 1931 to 1945, Conservative Member of Parliament for Wolverhampton West.

The letter that Bird sent to Bugatti, dated 12 May 1926,[22] is of particular interest because it was written on the last day of the General Strike, which had begun on 4 May. As might be expected from a loyal member of Stanley Baldwin's second administration, he paints a gloomy picture of the conflict, and hints of communist manipulation of the crisis.

Bird confirms A J P Taylor's[23] opinion that there was a 'strange Conservative delusion that the unions were influenced or even inspired by Communists.' Bird wrote to Bugatti:

> It is an assault on the State, that is to say a revolt, but a revolt which is intended in the first place to be effected by means of the rigid and menacing discipline of the Unions. Those who run the

[19] Day-Lewis, *op. cit.*

[20] L'Ébé Bugatti, *The Bugatti Story.*

[21] Bradley, *Ettore Bugatti.*

[22] HGC, 'Sir Robert Bird' in *Bugantics*, vol. 38, no. 1 (spring 1975).

[23] AJP Taylor, *English History 1914–1945* (Oxford, 1965).

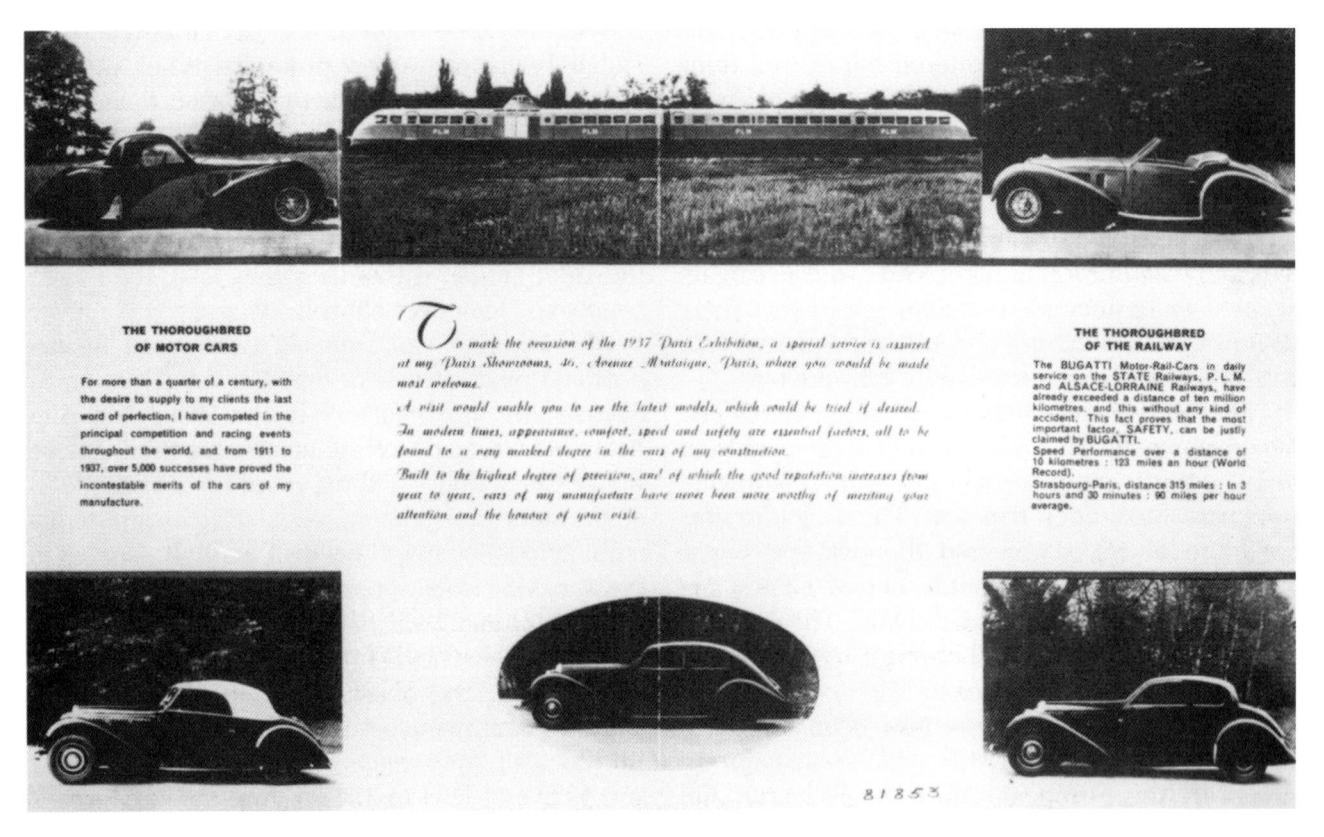

Bugatti produced this leaflet for English customers at the 1937 Paris Exhibition which linked the railcars' impressive safety record with the Type 57 car range.

Unions at the moment are almost without exception extremists. They have worked for a long time to organize an abominable grouping ... The extremist leaders have deployed considerable resources ... One notes that the 4,000,000 strikers are dithering a little. The immense majority of these men are good citizens and not at all revolutionaries.

On 17 May, Bugatti responded:

I admire the firmness and solidarity which reigns in England against disorder and anarchy. I think that the proof has been demonstrated decisively that nothing can be put above the will of self-respecting people set against a minority seeking by menace to put into slavery the best in the society of a country. The Russian system is not applicable there where the people have good sense and are civilized.

I was very interested to receive the explication which you so kindly gave me of the unfolding of this strike which was, above all, an outrage against the State which is certainly the elite of your nation.

A Response to Trade Unions

Despite his views of the growth of organized labour, L'Ébé Bugatti says that her father 'was aware of the progress of the trade union movement in France, but was somewhat perplexed by its development and demands ...' In 1929–1930 he therefore introduced 'a system of social benefits among his employees'. Molsheim employees were, he considered, well paid and therefore thought they should save and also improve their living conditions. L'Ébé says that he 'decided to encourage thrift by increasing the wages of workmen who could prove they were living within their means and either putting money by or by acquiring possessions'.

This was clearly a difficult policy to implement because it meant scrutinizing workers' spending habits and smacks of Henry Ford's ill-fated Sociological Department, a relatively short-lived exercise in paternalism which followed the introduction, in 1914, of the $5 day. Ettore therefore gave up about one day a week to his employees 'who wished to take advantage of his offer and were prepared to reveal their financial situation . . . that is to say the property or possessions they had acquired . . . *since* starting to work at the Bugatti factory'.[24] If they had provided evidence of thrift, Ettore would sanction a cash payment to the individual in question. But few workers approached Bugatti and, as L'Ébé says, 'the Guv'nor ended by smiling at it himself . . .'

Yet in 1936, Molsheim was overwhelmed by national politics. This was the era of the Popular Front, which arose in response to the Fascist dictatorship in Italy and Germany. The Popular Front consisted of a union of left wing and centre parties, pledged to preserve democracy by means of social reforms. In June 1936 a Popular Front government came to power in France, headed by mild-mannered intellectual, Léon Blum. However, French industrialists regarded him with the deepest suspicion, fearing the shadow of Russia. These sentiments were fortified by stories of communist atrocities in the nearby Spanish civil war which had broken out only two months prior to the government's arrival. For electoral purposes, the Front had been supported by radicals, socialists and communists, in addition to the trade unions. Despite the government having been elected by a large majority, the communists refused to participate and, instead, reserved the right to undermine it by means of strikes.

The automobile industry was a prime target for such disruption, which had begun in May 1936, with factories occupied in Le Havre, Amiens and Toulouse. These upheavals reached Paris when the Hotchkiss factory was paralysed though the campaign only began to receive national attention when the capital's 25,000 workers at Renault, the country's largest car maker, downed tools and occupied the Billancourt plant.

L'Ébé Bugatti says that her father believed that his factory would be unaffected by these events. He looked upon himself as *Papa Bugatti*, which is what Ernest Friderich had called him when he was awarded the *Légion d'Honneur* by the French government. He was, in any event, *Le Patron*, the Master of Molsheim. ' "I've nothing to worry about" he kept saying, "My workforce know me, they're part of the family." '

As industrial anarchy swept through the country, it 'embraced a wide variety of industries, from the big *metallos*, the textile and building trades, the stores and insurance companies, down to cafés and restaurants'.[25] Inevitably, the Bugatti works also succumbed though W F Bradley considers that the railcar programme was a factor. Because of the railcar's construction, Bugatti 'had been obliged to take in certain men who were quite unknown to him. Among them were extremists who cared nothing for craftsmanship, who scoffed at the ideas of a "family" in any industrial organization.' By this time there were about 1,200 workers on the Molsheim payroll.

Ettore Leaves Molsheim

As a result of the strike, 'the order was issued to prevent *Le Patron* entering his own factory. The blow was severe. Touched to the quick, refusing to argue with the men or rally them around him by persuasion, Bugatti saw only the cruel injustice of the mass which allowed itself to be led away by a few fanatics. Impulsively, he announced that he would leave Molsheim.'[26]

Bugatti had, in any event, been spending in-

---

[24] L'Ébé Bugatti, *The Bugatti Story*.

[25] Anthony Rhodes, *Louis Renault* (Cassell, 1969).

[26] Bradley, *Ettore Bugatti*.

creasing amounts of time in Paris but Bradley says that, after 1936, he 'installed his office at his showrooms in the Avenue Montaigne. Perhaps it was a coincidence, but next door there was a riding school managed by an Englishman.' So Ettore continued to work 534km (331 miles) away from his factory and although he did return there occasionally these instances were rare. Says his biographer, 'Molsheim, once the pride of his life, had become an open sore.'

Far from living in self-imposed exile in Paris, Ettore's younger son, Roland, considered that his father 'was even more productive after leaving Molsheim than before because he had handed over all the problems of administration to Jean. He didn't abdicate; he just said, "Since the workers don't consider me a good boss, I'll put my son in charge and they will be governed only by the law, leaving the heart out of it." '[27] Jean continued to liaise closely with his father by making perhaps two journeys a week to Paris to discuss the problems arising from the increasingly precarious state of Automobiles Bugatti's finances.

Roland Bugatti has said that, once installed at his office in the French capital, his father 'worked most of each night and then all day with an equipe of draughtsmen, which ranged from three to ten men'. These assignments embraced a range of projects in which Ettore expanded his horizons to embrace, not only familiar arenas of land and air, but also sea. These accordingly consisted of cars, aircraft and boats but, it has to be said that, although most of these inventions reached the experimental stage, none ever attained production status.

At Molsheim, despite the all-out strike soon being over, the manufacturing constraints it produced lingered on until well into 1937 and things at the Alsace factory were never quite the same again. Although Jean was then in sole charge, he had, in any event, been responsible for the overall concept of the Type 57 car, announced in 1934, and destined to be the last series production Bugatti of the inter-war years.

[27] Barker, *Bugatti*.

Enter the Type 57

Prior to the arrival of this model, the firm, in common with many of its contemporaries, had produced a jumble of touring, sports and racing cars of varying chassis lengths and engine capacities and configurations. It will be recalled that the last four-cylinder Bugatti, the Type 40A, had ceased production in 1933 and thereafter only eights were built. The slow selling, luxurious Type 50 lingered on until 1934, and the last of the lovely Type 55 sports cars was delivered in mid-1935. But with the appearance of the 57, the Bugatti range was drastically rationalized.

This 3.3-litre car was effectively the sole model built between 1934 and the outbreak of war and this single design was ingeniously created for the tripartite role of high speed tourer in the *Grands Routiers* manner, while it also formed the basis of the 57 sports car, and the engine was also used to power the last significant Grand Prix Bugatti; the visually glorious though flawed Type 59.

The 57 was also intended to open a new era of refinement at Molsheim and was completely new throughout, though its 72 x 100mm bore and stroke and 3,257cc were inherited from the single cam Type 49. The engine had originally been of 2.8 litres capacity and that was increased by upping the stroke from its original 88mm.

The established Molsheim formula of a fixed head block, a dated feature by this time, mounted on an aluminium crankcase was perpetuated. The six main bearing crankshaft was a single piece component and this and the connecting rods employed plain bearings while the pistons were of the Bonalite invar-strut variety. Thankfully, a pressurized lubrication system was established Bugatti procedure by this time and there was wet sump lubrication.

On all previous Bugatti engines, the drive for the overhead camshaft, or shafts, had been at the front of the unit but, on the 57, it was at the rear and took the form of a train of helical gears driven from the crankshaft. There were two valves per cylinder, inclined at 94 degrees, though they were operated via the fingers of the earlier single cams rather than inverted cups used on the Miller-derived twin cams.

At this stage the engine was unsupercharged and relied on a single Stromberg UUR-2 updraught carburettor mounted on the right, which was the same side as the inlet camshaft driven AC mechanical petrol pump. The Scintilla distributor for the twin coil ignition was similarly actuated. On the opposite side of the engine, the water pump, mounted alongside the crankcase, was driven from the rear gear train and actuated the dynamo. Gone were the twin 'bunch of bananas' exhaust manifolds and replaced by a single ribbed component with a forward-located flange. In this form the Type 57 developed 135bhp at 5,000rpm.

For the first time on any Bugatti, the gearbox was mounted in unit with the engine. Therefore out went Ettore's distinctive multiplate clutch, to be replaced by a single dry plate unit. The gearbox itself was new though incorporated features intended to simplify and quieten the process of gear changing. This consisted of introducing helical constant mesh gears on second, third and top gears which were engaged by dog clutches. The gear lever was a centrally-mounted ball change unit. Drive was by a propeller shaft, accompanied by the traditional Bugatti torque arm, to the rear axle which was originally of the sort used on the Type 44 and 49 though its internals were soon replaced by stronger Type 46 gearing. The axle was sprung by the customary reversed quarter-elliptic springs.

The model was only offered with one chassis length, with a 3,300mm (10ft 10.5in) wheelbase. Cable-operated brakes conservatively perpetuated previous practice and had 350mm (13.7in) diameter drums. Out went the costly cast aluminium wheels with integral brakes of the sort used on its Type 46 and 49 predecessors, to be replaced by cheaper Rudge Whitworth spoked wheels.

An Articulated Axle

Despite the fact that the Type 57 retained the established Bugatti beam axle, it was articulated to provide a modicum of independence. The beam itself was split in two pieces though joined internally and secured by a screwed collar. In truth, the system does not seem to have worked very well. It was soon replaced by a one-piece beam. This device smacks of compromise because, as originally conceived, the Type 57 was to have had independent front suspension. The work of Italian engineer, Antonio Pichetto, who was responsible for the abortive Type 53 four-wheel drive car, the prototype Type 57 shared a similar layout, employing twin transverse springs.

Jean Bugatti also created a refinement of the Bugatti horseshoe-shaped radiator, retaining the essentials of its lines though introducing a more pronounced V to a slatted grille with a central rib. Both this and the independent front suspension system are shown on a Molsheim coachwork drawing, dated 15 July 1932, and would appear to have been introduced into the design during Ettore's absence in Paris.

By all accounts, Ettore was unaware of these refinements and, on discovering them, ordered Jean to revert to the conventional and soon to be outdated beam with perhaps the split axle agreed as a compromise. Consequently no Molsheim-built Bugatti was subsequently fitted with an independent front suspension system and that included the eccentric Type 251 racing car of 1955, which appeared eight years after Ettore's death.

This imposition is significant and, sadly, is a pointer to the decline of the Bugatti marque with Jean, on the one hand, wanting the Type 57 to incorporate the latest technology of the day and Ettore displaying all the characteristics of the self-taught engineer.

It was in 1933, just before the 57's introduction, that Mercedes-Benz announced the 380 model with its innovative coil and wishbone system which was, incidentally, extended to its W25 Grand Prix car of the following year. Also in 1933 the British Alvis company introduced a cruder transverse leaf system on its Speed 20 model, a layout derived from that used on the firm's pioneering front-wheel drive model, a sequence which also had parallels with Jean Bugatti's thinking. Admittedly these cars were exceptional and, in 1933, most of Bugatti's competitors still retained the conventional beam axle. But, unlike the cars from Molsheim, they soon adopted ifs. Delahaye and Delage did so in 1934, Alfa Romeo and Rolls-Royce in the following year and Lagonda in 1937. The systems adopted by the mass producers were, in some instances, of questionable worth. However, a sophisticated and impressive torsion bar and wishbone layout, complete with hydraulic brakes, were just two of the outstanding features of Citroën's celebrated *Traction Avant* model of 1934, introduced in the year of the Type 57's announcement.

Non! Independent Suspension

Bugatti's defenders may well have argued that he was perhaps waiting to satisfy himself of the worth of the various systems and, in any event, the Type 57 was a fine car, sprung as it was. But this was the negative aspect of Ettore's single-minded, idiosyncratic approach, which had produced such sensational results in the past, and there was real tragedy in the fact that Jean was not permitted to incorporate independently sprung front wheels on the 57's Type 64 successor, intended for the 1940 season. Sadly, it perpetuated the once magnificent but by then obsolete front beam axle.

If Jean lost the battle, as far as the 57's suspension was concerned, he was more fortunate with his design for a new Bugatti radiator, despite initially having to adopt a more conventionally shaped one. This, ironically, did reflect current manufacturing trends in that the lovely German silver radiator of the earlier cars was discontinued and replaced by a new one which consisted of a separate core and chromium-plated surround with painted, thermostatically operated slats though these were subsequently chromed. Jean did introduce a characteristic touch to the radiator, in that the lines of the bonnet mouldings were perpetuated by small flanges in the shell itself, a feature that he had introduced on the *Coupé Napoléon* Royale. But the original V-shaped design would emerge, in a magnificently refined form, on the model's potent Type 57S derivative which appeared two years after the original car's appearance, in 1936.

Although the Type 57 chassis could be bodied by individual coachbuilders, the vast majority of cars were sold complete with a range of Jean Bugatti-designed, mostly Molsheim-built coachwork, the names of which were inspired by Alpine passes. In the first instance, these consisted of four bodies; the two-door Ventoux, which was available in two and four light forms, and the Galibier four-door saloon. There was also a two-door drophead coupe called the Stelvio and a short-lived sports roadster named the Grand Raid.

Apart from the last-named body, these names would endure for the next four years though their

The Galibier pillarless four-door saloon was available from the type 57's outset in 1934. This is an early example.

There was also the Type 57 two-door Ventoux, with echoes of the Profilée *bodies of the Type 50. This is a Series 1 car.*

lines would progressively develop and more than keep pace with contemporary thinking. Their appearances would eventually outpace the model's mechanicals which remained essentially the same. As this coachwork evolved, they would underline Jean Bugatti's growing stature as an international stylist of the highest order.

Molsheim-Built Bodies

In its original form the Ventoux can be seen as a perpetuation of the Profilée bodies that Jean had introduced on the Type 50, with their distinctive sloping windscreen, moulded sweep panels and wing line incorporating running boards. The Stelvio drophead coupe was essentially similar though was built for Bugatti by Gangloff of Colmar. The Galibier was a four-door pillarless saloon with sweep side panels. Curiously, however, the rear doors were distinguished by their absence of external handles though they were fitted with internal ones. The Grand Raid, only listed for 1934, was the most expensive of the 57s and cost 88,000fr. (£1,060). The Galibier, on the other hand, was the cheapest. It was listed at 79,800fr. (£961) though,

The Stelvio drophead coupe was available from the model's arrival and this is an early example. Although a factory offering, it was the work of Gangloff of Colmar.

from 1935, its price was increased to 83,000fr. (£1,000), making it more expensive than the Ventoux. The model was substantially cheaper than the 4.9-litre supercharged Type 50 that preceded it.

This made the Bugatti excellent value, when compared with the straight-eight 4-litre D8 Delage which sold complete for £1,350. Similarly the 3.5-litre Bentley, which appeared the year before the 57, was also more expensive and cost £1,460 in saloon form while the 30hp 4.5-litre Hispano-Suiza was listed at £1,050 in chassis form, no less than £175 more than the 57. These two last-named cars were powered by relatively inexpensive pushrod six-cylinder engines. Both cars were well tailored for the harsher economic climate of the 1930s, being also cheaper to run and manufacture than the straight-eight unit which had come to prominence in the previous decade.

The Ageing Eight

The eight was well in decline in Europe by the mid-1930s. Such respected names as Isotta Fraschini and Austro-Daimer ceased production at this time, taking their respective eights with them. Delage listed no less than eight variations on the theme in 1935, a dilution of resources which was a contributory factor in that year's take-over by Delahaye. From thereon the once proud Delages gradually became badge-engineered Delahayes. Darracq also dallied with the straight-eight but, after its 1935 demise and purchase by Anthony Lago, it concentrated exclusively on a new range of sixes. In Germany, Horch continued to produce its stately eights until 1939 but, from 1933 and with depression and rationalization in the air, it was allied with the Audi, DWK and Wanderer to form Auto Union. If such firms as Alfa Romeo, Daimler and Mercedes continued to produce eights until the outbreak of war, they all had more popular six-cylinder lines to supplement declining sales.

Bugatti had a peculiar problem because his eight also had to do double duty in a Grand Prix car and accordingly had to employ costly twin overhead camshafts. But in view of the Type 59's poor showing, there was no publicity to provide a spur to sales of road cars. Also Ettore, with his single model and long standing commitment to the configuration, was unable to contemplate a six-cylinder Bugatti. He therefore had to rely solely on what was becoming an increasingly outdated concept for road cars, though the twin cam eight was a strong sales plus.

However, the model was destined to be the third best-selling Bugatti model after the earlier 16-valve cars and the Type 44, with a total of 685 were produced during its six years of manufacture.

The 57 proved to be a thoroughbred in the best traditions of its predecessors with a top speed of around 153kph (95mph). It had a smooth, torquey engine and its steering was excellent and became progressively lighter as the car's speed increased.

Roadholding was, as might be expected, good with the 57 really coming into its own on long runs. It was an ideal car in which to convey its owners, speedily and reliably from, say, Paris to Monte Carlo or Cannes. It was, in short, the archetypal *Grand Routier*, happiest when eating up the kilometres during all-day runs. Because of its rather heavy controls, it was less happy in traffic and for town driving. The model's weakest point was the brakes which were poor, requiring considerable pedal pressure and with a tendency to grab on the pre-1937 cars.

Arrival of the Atalante

In 1935 the Galibier, Ventoux and Stelvio were joined by a fourth factory body, a new two-door fixed head coupe, called the Atalante. It was named after Atalanta, the Arcadian heroine of Greek mythology renowned for her fleetness of foot and, like *La Royale*, the model name was applied to the car in the feminine gender. The first 57 Atalante was delivered in April 1935 and the model remained available until 1939. It was produced in addition to the two light Ventoux and became the most expensive model in the Type 57 line, selling for 90,000fr. (£857).

This body immediately looked more modern

The Type 57 two-door Atalante coupe arrived in 1935 and was listed until 1938. This example dates from the latter year.

than its other Molsheim contemporaries because of its absence of running boards. In addition, Jean adopted a new type of body moulding in the form of an ellipse which began at the forward end of the rear wheel arch, bisected the door and tapered along the bonnet sides to the radiator. There was also a drophead version.

For the 1937 season, the Type 57 concept was expanded with the new cars unveiled at the 1936 Paris and London shows. Both mechanical and body modifications were made to the original model and a supercharger was introduced for a Type 57C (for compressor) derivative. In addition there was the splendid and radically reworked 57S sports version.

First the 57 proper. For these Series 2 cars, the coachwork was updated and the running boards discontinued in the manner of the Atalante and the boot became more integrated with the body. Modifications were made to the engine's cams and timing, while the exhaust outlet moved from the front to the rear of its manifold. As these changes coincided with the arrival of the 57C, the crankcase was altered so that the supercharger could be driven from the rear right-hand side of the engine, though this was blanked off on the mainstream model.

In an effort to reduce the noise produced by the train of timing gears, the engine was flexibly mounted on 'Silentbloc' rubber bushes. As it had hitherto reinforced the chassis, a half cruciform bracing was introduced to the frame.

The rear axle was strengthened by the fitment of a stronger Type 46 differential. All-important

Introduced for 1937, the 57C was available until the war interrupted production in 1939. The addition of a Roots supercharger pushed engine power up from about 135 to 160bhp.

The Series 2 Type 57 arrived for 1937 and one of the principal under-bonnet changes was the introduction of rubber engine mounts and here such a unit is undergoing testing. At around the same time, the crankcase was redesigned to facilitate the provision of a supercharger on the 57C though, as can be seen, this was blanked off on the basic model.

changes were also made to the braking system. In an effort to prevent brake grab, sprocket-guided chains were let into the front cables to improve their angle of operation.

As the accent was on refinement, changes were also made to the shock absorbers. Originally the Type 57 had been fitted with all round Hartford Telecontrol friction units. While these were retained for the rear, for the year 1937 they were discontinued at the front and, perhaps to compensate for the lack of independent suspension, were replaced by a pair of de Ram shock absorbers, marketed at the formidable cost of the equivalent of £170. They were essentially a combination of friction and hydraulic units and each contained thirty-six plates apiece.

Unlike the relatively crude Hartfords, in which the driver manually adjusted their setting according to the road surface, the de Rams automatically compensated for such changes. They were therefore relatively free in response to slight axle movements but stiffened up considerably when subjected to severe road shocks.

A Series 2 Type 57 Stelvio, as catalogued by the factory.

A Supercharged Version

In 1936 the Type 57's chassis cost 62,000fr. (£590) while the supercharged 57C version, was 14,000fr. (£133) more, at 76,000fr. (£723). As will be apparent, these pound sterling equivalents, which date from September 1936, are less than when the model was first introduced because the franc was devalued that month. It therefore made the Type 57 extreme good value for money for British customers and, indeed, in other export markets.

The Roots-type supercharger which the 57C employed was similar to that previously used by Bugatti on his Grand Prix cars. Driven off the rear timing gear train and located on the right-hand side of the engine with the Stromberg carburettor mounted underneath, it ran at 1.17 times engine speed and blew at about 5 to 6psi. Such cars carried the C suffix to their engine numbers.

The introduction of a supercharger increased the model's top speed, which edged towards the 193kph (120mph) mark, an impressive figure in those days before the XK120 Jaguar of the post-war years made such speeds commonplace.

The third of the trio of Type 57 derivatives was the handsome and potent 57S, though a sports racing version of the original model had briefly appeared in the previous year. On 7 September 1935, two works-entered 57s ran in the Tourist Trophy race, held at the Ards circuit near Belfast in Northern Ireland and driven by Earl Howe and the Hon. Brian Lewis. Both cars were fitted with lightweight, open two-seater bodies, Howe's having been made of Elektron and Lewis's of blue painted Duralumin.

A further 57 was present in the form of a private Irish entry from P M Dwyer, and driven by Belfast driver, Hugh McFerran, who had already owned two Type 35s. In the event itself, despite Lewis attaining the third fastest lap, he did not finish, while McFerran crashed at Newtownards. By contrast, Howe drove an excellent race and finished third.

A 57 resembling the Howe/Lewis cars and marketed as the Tourist Trophy model was shown on the Bugatti stand at the 1935 London Motor Show. Fitted with a 'streamline' sports torpedo body, and priced at £1,150, it was £145 less than the Type 57 that the James Young Airflow saloon also displayed. The TT car was also designated the 57T, but the concept was not perpetuated perhaps because, on the same stand, was a sensational coupe which represented the starting point of what was to become the Type 57S in the following year.

The Fabulous Atlantic

The coupe was simply listed, at this stage, as the Competition model though in its ultimately refined form was titled the Atlantic coupe. This is perhaps the most extraordinary of Jean Bugatti's bodies and, in my opinion, the supreme expression of his genius. It is a highly idiosyncratic yet totally controlled and superbly proportioned body of a truly sculptural quality, a synthesis of Jean's talents in which the vitality of his grandfather Carlo's furniture and Ettore's impeccable eye for line and form are all apparent.

Automobiles Bugatti was left without a sports car when the last Type 55 roadster was delivered in July 1935 and it was not until the summer of 1936 that it was replaced by the 57S. As already mentioned, its origins are rooted in the prototype Atlantic coupe which first appeared in 1935. Allotted Molsheim body number 1076, it was originally given the Aérolithe name. As a sports coupe, it reflected a growing European interest in the science of aerodynamics and the realization that a closed body designed to these disciplines had a lower drag coefficient than an open one and was therefore faster. However, it is unlikely that the coupe's lines were subjected to wind tunnel testing prior to its creation.

Yet unlike the Atalante which, elegant as it was, still resembled an open car with an added roof, on the Aérolithe coupe the cabin became an integral part of the whole, with the roof line extended right to the back of the car. In the interests of lightness, the prototype was probably made of Elektron, a manganese alloy, and the British press referred to the body as having been made of the material.

However, there are contemporary French references to it also having been made in the aluminium alloy of Duralumin. The fact that the 1935 Type 57 TT cars' bodies were made of the respective materials suggests that, at this time, Molsheim was evaluating the respective merits of the two alloys.

The production versions of what became the Atlantic were made of aluminium, though the method of construction demanded by the original material was perpetuated. At this time both Elektron and Duralumin, particularly the former, were proving difficult to weld satisfactorily and it was not until after the war that Elektron sheet was successfully joined by using helium and argon arc techniques. In view of this problem, the two halves of the coupe's body were riveted together in a manner which echoed that used on the tail of the Type 59 racing car.

But this joint was dramatically extended to form a central rib which began at the radiator cap. Projecting about 12.2mm (0.5in) above the centre of the bonnet, it ran upwards between the divided windscreen and over the roof to the extreme rear of the car, where it was elongated and extended to form a projecting fin around 50mm (2in) proud of the body. The latter feature was a theme that echoed a distinctive aspect of that aerodynamic *tour de force*, the Czechoslovakian Tatra 77, which had appeared in the spring of 1934.

Daringly on the Bugatti the button head rivets were exposed, aircraft-like and the concept was successfully extended to the peaks of the front and rear wings. There were no running boards while the doors, with their semi-ellipsoidal windows, were extended into the roof and opened at a rakish angle which also reflected contemporary aviation practice.

First the Aérolithe

The prototype Aérolithe was displayed on the Bugatti stand at the 1935 London Motor Show though it differed from the production versions in a number of important respects. It was, in the first instance, approximately 150mm (6in) higher, because it employed a shortened Type 57 chassis with a 2,980mm (9ft 9.5in) wheelbase and was

The prototype 57S Atlantic, displayed at the 1935 Paris and London Motor Shows, used a shortened Type 57 chassis and therefore its radiator, and consequently was higher than the production cars. These had lower lines, the handsome V-shaped radiator and free-standing headlamps in place of the rather unhappily located integral units of this car.

accordingly fitted with its standard, thickly slatted radiator which meant that the front of the car required refinement. Because of the height of the radiator and, therefore the bonnet, the adjoining wings did not visually relate to them and consequently all appeared as a number of unrelated concepts rather than a fully harmonized entity. The headlamps were, in the interests of aerodynamic efficiency, mounted integrally with the wings.

By the time that the first production coupe was delivered in September 1936, it had been superbly refined and the model renamed the Atlantic. It should be made clear, however, that this car was not the first production 57S. That had been delivered in the previous month and will be described later in this chapter. All examples of the 57S differed from the 57 and 57C because of its redesigned chassis frame and a new radiator which combined to lower the entire car.

The 1932 Type 57 prototype represented the starting point for the new radiator and it remained faithful to Carlo Bugatti's influence, in that its outline resembled that of an inverted egg. But instead of adopting the flat plane which had featured on all previous Bugatti radiators, Jean extended the centre line to produce a pronounced V which was accentuated by the presence of thin, chromium-plated slats.

On the Atlantic this shape perfectly complemented the rib that extended the length of the car. In addition, the radiator shell was scalloped on either side of the cap with the bonnet appropriately contoured to fit. Another departure was the low-mounted headlamps, which were integral with the wings on the prototype and similarly positioned, though free-standing on the production Atlantic.

The 57S chassis retained the 2,980mm (9ft 9.5in) wheelbase of the Aérolithe coupe but was otherwise completely new. It echoed the profile of the Type 35 chassis in that it tapered to its point of greatest stress which was just behind the rear engine bearers. Also, rather than slinging the chassis below the axle, as was the customary method of lowering a car's lines, the side members were

The Atalante body was also extended to the 57S and its radiator was a worthy successor to the traditional horseshoe. Note how much lower than the 57 the model appears, on account of a new chassis in which the rear axle passed through apertures in the side members.

expanded at the rear so that apertures could be introduced into them through which the axle could pass.

Suspension inevitably followed previous Bugatti practice of half-elliptic springs at the front, and rear reversed quarter-elliptics though, on the 57S, and unlike the 1936 model 57s, de Ram shock absorbers were also fitted to the rear as well as to the front axle. On the S, the latter was split and articulated in the manner, if not the detail, of the early 57's though torque reaction arms were introduced.

The 57S engine also differed from the 57 one in that it was a dry-sump unit in the manner of the Type 59 Grand Prix car and was supplied from a 16-litre (3.5-gallon) oil tank located under the bonnet outside the left-hand chassis members. The twin ignition coils were dispensed with and replaced a dash-mounted Scintilla Vertex magneto driven, Type 51 wise, from the rear of the exhaust camshaft. Not only was it more efficient but it meant that the projecting inlet camshaft-driven distributor of the standard 57 could be deleted, so that the bonnet line could be lowered. Compression ratio was also upped, to 7.5:1, and all these modifications contributed to raising output from 135 to 170bhp. In view of this extra power, the single plate clutch was replaced by a twin plate one. The gearbox and rear axle were essentially carried over from the 57.

The Potent 57SC

It will be recalled that a supercharged version of the 57, the 170bhp 57C, was announced at the 1936 Paris Salon and the S was to benefit similarly, the resulting supercharged 57SC developing a formidable 200bhp. It is therefore the fastest and most expensive of the Type 57s. Its 1936 chassis price of 95,000fr. (£905) was 15,000fr. (£143) more than the 57S's though, like the 57C, the option did not become available until late in 1937.

On the supercharged 57S, the Bugatti-built Roots-type blower was driven at 1.17 times engine speed and its manifold was liberally ribbed. It ran at around 3–4lb psi, though the engine's compression ratio was reduced to around 6.7 to 1. Just one example of the 57SC was produced, as owners preferred to buy a 57S and then have it so converted by the factory. The Type 57, it will be recalled, gained rubber engine mountings in 1936 but the S's engine was initially bolted directly to the frame. Later cars were equipped with rubber-mounted engines in which the front de Rams were incorporated with the flexible mounts. These later engines were fitted with 14 rather than 18mm sparking plugs.

As introduced at the 1936 Paris Salon, the chassis price of the 57S was listed at 80,000fr. (£761), which was 18,000fr. (£171) more than the 57. The topline Atlantic coupe cost 117,000fr. (£1,114). As already mentioned, the first example (57374) was delivered in September 1936. Its recipient was British domiciled and only twenty-six years old, the Hon. Victor Rothschild who, in 1937, became the third Lord Rothschild. This Atlantic, registered DGJ 758, was supercharged by the works in 1939 and thus became a 57SC.

In October 1936, a month after the dispatch of the first Atlantic, came the second (57473). That car went to a Mr Holtschub, of which nothing else is known, although he may have been dissatisfied

Rear view of the Atlantic with its distinctive and memorable central fin.

The first owner of this 57S Atlantic coupe, 57374, delivered in September 1936, was the Hon. Lionel Rothschild. It later passed to Robert Arbuthnot's High Speed Motors. In 1939 it had been converted to 57SC specifications by Bugatti. Exported to California in 1947, it has remained in America ever since. The faired headlamps are a post-war refinement.

with his Bugatti because he does not seem to have kept it for a very long time. The coupe was soon being driven by the Bugatti works driver, Williams, when it carried the 5800 NV3 registration number.

It was to be a further eighteen months before Molsheim sold its third, and last, Atlantic (57591). This also went to another British customer, Richard Pope, chairman of the Pope and Pearson Colliery, based in the West Riding of Yorkshire. It was his tenth new Bugatti. A tall man, Pope dealt directly with Jean Bugatti and, at his request, the car's roof line was raised by about 12.7mm (1in). Pope took delivery of the car in May 1938 and the Atlantic was sold after being registered EXK 6. In the following year it was returned to Molsheim and fitted with a supercharger, so bringing the coupe up to 57SC specifications.

The 57SC Atlantic, 57591, was owned from new in 1938 for around thirty years by Richard Pope and then by Barrie Price.

The Atlantic Evaluated

There were thus only four examples of the Atlantic, this most celebrated of Bugattis: one prototype and three production cars. In view of the rarity of the breed, few people have ridden in one, let alone driven an example. But in 1937 Williams brought a works Atlantic to Britain and gave *Bugattiste*, C W P Hampton, a demonstration run on Hertfordshire's Barnet by-pass. Hampton was to recall:

> It was simply terrific: 112mph [180kph] still accelerating over the cross roads past the Barn – and the road cluttered with the usual Friday evening traffic. Along the next stretch we did 122mph [196kph] and I thought, under the circumstances, that was enough and said so in no uncertain fashion. Thereafter we 'cruised' along at a mere 90–95mph, [145–153kph] and doing just over 100mph [160kph] in third gear. Except, possibly, for the run I had with Jean Bugatti in France, it was the most alarming experience ever; yet Williams drove superbly, absolutely at ease and complete master of every situation . . .[28]

In view of the Atlantic's undoubted qualities, why then did Molsheim sell so few examples of what was one of the world's fastest pre-war sports cars? Obviously the market for such models was limited and this Bugatti was expensive; it sold for £1,700 in Britain. Perhaps the best insight to the Atlantic's virtues and vices has come from Barrie Price, dedicated Bugatti owner, one-time chairman of the Bugatti Owners' Club and today chairman of the Bugatti Trust. In the 1960s he bought the last Atlantic coupe to be built from Richard Pope, its original owner. It should be noted, in the light of the foregoing comments, that Pope had had the car's exhaust system modified to quieten it, so we can only guess at what it must have sounded like originally . . . For Price's chief recollection of the car was:

> . . . the tremendous amount of *noise* inside – conversation at more than 40mph [64kph] was impossible.

> Nor was the handling, in my opinion, up to that of the standard open 57S. I think that the additional weight of the body over the rear wheels spoilt the balance or something: it would oversteer at the drop of the hat.
>
> To me, the best features were undoubtedly its looks, its quality and its performance which was incredible for its day and even now is outstanding. There was loads of torque and it was incredibly tractable because of the blower.
>
> It was beautifully made, very rigid and stiff and there were never any rattles. On top of that it was very reliable.
>
> But oh! That engine *noise*.'[29]

The Surviving Cars

Before saying farewell to this most visually impressive of Bugattis, it seems appropriate to chart the histories of the three production Atlantics, all of which survive: two in America and one in France. Lord Rothschild's car, the first to be delivered, became, in 1945, one of the delectable Bugattis to pass through the hands of Continental Cars of Send, Surrey. It was bought that year by an American enthusiast, Robert Oliver of Los Angles, and exported to the USA in 1947.

In 1953 the car was returned to Molsheim for a mechanical overhaul when it was fitted with a Cotal gearbox. It was also taken to Motto of Turin to be painted blue, and modest modifications were made to the body, of which the most noticeable was fairing the headlamps into the front wings. The 57SC resided in California until Oliver's death and it then passed to the President of the American Bugatti club, Dr P Williamson.

The Holtschub Atlantic has had a more chequered history. According to M Mortarini,[30] a former owner, after the war it went to a dealer named Panibon, then to another named Baraquet and yet a third, Bodel of Cannes. He sold the car to Bugatti specialists, Lamberjack in Paris, from where it was bought by Mortarini who, in turn, passed 57473 to

[28] Purdy, *The Kings of the Road*.

[29] *Old Motor*, February 1982.

[30] Hugh Conway, 'The Atlantic Type 57S' in *Bugantics*, vol. 38, no. 3 (autumn 1975).

M. Chatard. In August 1955 he was killed in it when the car was hit by a train on a level crossing. It was later rebuilt by P Berson of Montargis and fitted with engine 576445–473 from a touring Type 57. The car looks good but there are minor departures from the original specification, namely a louvred bonnet top and the door profiles are not exactly as they should be. Nevertheless we should be grateful for M. Berson for ensuring this car's survival.

As already mentioned, Richard Pope's car remained in the hands of its original owner until the 1960s. A keen tennis player, he had used the Atlantic on runs to participate in tournaments in the south of France though it was otherwise little used. Barrie Price took it over and, by 1974, it still had less than 50,000 miles (80,465km) on the clock. From Price the car went to Anthony Bamford and then across the Atlantic to Thomas J Perkins of San Francisco to join his stable of supercharged sports cars which included a Type 55 and a 57S Atalante. Perkins embarked on an extensive restoration and corresponded with Pope in Britain to ensure that all important details were correctly perpetuated. Perkins subsequently sold his cars though the Atlantic is still in America.

Whatever became of the prototype? In February 1989 *Classic and Sportscar* magazine published a letter from Robert MacKenzie of London, Ontario, Canada who enclosed a photograph of a Bugatti he had encountered during the war. It was clearly an Atlantic and was said to have been owned by an American pilot who ran the car on the 100 octane fuel reserved for his aircraft! The car was later dispatched by sea to New York and it seems to have disappeared. So is there an Atlantic awaiting discovery somewhere in North America?

However, this extraordinary coupe only accounted for seven per cent of 57S production which amounted to a mere forty-one cars delivered between August 1936 and May 1938. The Atalante coupe, introduced in the Type 57 and also perpetuated on that model, was the only factory body fitted to the 57S chassis: there were, in all, nineteen examples built. The model also attracted the attention of the such specialist coachbuilders as Gangloff and Vanvooren in France and Vanden Plas and Corsica in Britain.

In the Tourist Trophy

Corsica were responsible for bodying the first 57S (57375) which was delivered to its owner, Nicholas S Embiricos, in August 1936. Fitted with an open grey, two-seater body, it was entered in that year's Tourist Trophy race, held at the Ards circuit on 5 September. Regrettably, after only completing one lap, and just before the Moate curve, he had the misfortune to tangle hubs with Mongin's Delahaye which had been the last car to leave the start and the Bugatti was damaged in the ensuing crash.

The 57S was then taken over by Guilio Ramponi and prepared for Richard Seaman to drive in 1937.

In Britain, Corsica bodied no less than seven of the fifteen 57Ss imported and these are four of them. All are different because, unlike most coachbuilding companies, Corsica did not employ its own stylists. Instead the lines were the work of the car's owners. Corsica had been established in 1920 by coppersmith Charles Stammers and his two brothers-in-law. The firm took its name from its original premises in London's Corsica Street near King's Cross but later moved to nearby Grimaldi Street. It finally established itself in the north of the capital at Cricklewood Broadway NW2. The first 57S, chassis no 57375, was fitted with a Corsica body and raced by Nicky Embiricos in the 1936 Tourist Trophy race though he crashed there.

Early in 1938, the car was bought by Ronnie Symondson, who took it back to Corsica and his suggestions were incorporated in the body it still retains. He is seen here in one of his many appearances at a Bugatti Owners' Club Prescott meeting. Symondson retained the car until 1984 when it was bought by Neil Corner.

However, when he joined the Mercedes team, the car was put up for sale. It was bought, early in 1938, by engineer and Type 57 owner, Ronnie Symondson. He took the car back to Corsica and the resulting car was a synthesis of both talents. Symondson and his 57S would continue to enliven Bugatti Owners' Club events for around forty years. He retained the 57S until 1984, when ill-health prevented him from using it.

Two 57Ss were displayed on the Bugatti stand at the 1936 London Motor Show and *Motor Sport* echoed the esteem in which the marque was held by British enthusiasts when it commented:

> Such is the reputation that Ettore Bugatti has built up, since his sixteen valvers that did 63mph [101kph] intrigued British enthusiasts at very early Olympias, that each year Col. Sorel merely has to take two or three of the latest Molsheim motors and put them on a simple stand, and throughout the period of the Show, people gather round and worship.[31]

The 57S was destined to last for another eighteen months and it was discontinued in the spring of 1938. Demand for the model had declined during the latter half of 1937 and would probably have ceased production that year, had it not been for the requirements of the British market which greatly benefited from the falling value of the franc. Following the devaluation of the French currency in September 1936, it had been left to find its own level though its worth continued to be eroded. This meant that, over a period of two years, the price of the 57S chassis fell by a substantial £440, from £1,300 in 1936 to a mere £860 in 1938 which was when five 57Ss were delivered to Britain.

The price of the Type 57 was to also decline

[31] *Motor Sport*, October/November 1936.

The original owner of this 1937 Type 57S was Sir Robert Ropner Bt. Chassis 57503 was delivered to Jack Barclay in November 1936 and Corsica built this lovely open four-seater touring body. Originally registered DUL 351, like many of the Bugattis featured in this book, it was owned by Continental Cars' Rodney Clarke who had the misfortune to crash it badly in 1946. It was then rebuilt by H H Coghlin and remains in Britain.

The body of this 1937 two-seater 57S, chassis 57531, was designed by its first owner, Sir Malcolm Campbell. Called Bluebird in recognition of his record-breaking achievements, it was catalogued by Bugatti in Britain as a sports model but only the one was built. It is appropriately pictured at Brooklands soon after its completion. Campbell only ran the car for about 800 miles (1287km). It was then returned to Brixton Road and, early in 1938, was sold as a new car to R E Gardner of Warninglid, Sussex. He retained it for about forty years when it went to Neil Corner. It passed to the marvellous collection of supercharged sports cars owned by Tom Perkins of San Francisco. By 1980 it had only covered around 15,000 miles (24,239km).

Eric and his brother, Colonel Godfrey (Goff) Giles were great Bugatti enthusiasts and, between 1929 and 1947, no less than nineteen passed through their hands. This particular car, a 1938 Type 57S, chassis number 57593, was owned by Godfrey Giles and the coachwork was designed by his brother. Eric Giles was a distinguished interior designer and the outcome was one of the finest bodies to have graced a sports Bugatti. Its memorable registration number had appeared on a number of the brothers' earlier Bugattis, including a Bertelli-bodied Type 49 tourer and a 1934 Type 57 from the same coachbuilder which Eric Giles also styled. GU 7 was converted to 57SC specifications by Bugatti in 1939 and was swapped by Giles for William Harges' James Young-bodied 57C fixed head coupe. After the war it was exported to America where it remains.

This 1938 car, which belonged to Peter Hampton, was the last 57S to be imported into Britain. It is chassis 57602 and was bodied by Corsica to its owner's design, which clearly owed something to the Atalante coupe. It can be seen today at Mulhouse.

proportionally. But in view of continuing downward pressure on the franc, a new lower limit of 179 to the £ was set in May 1938, the month in which the last three 57Ss crossed the English Channel, one of which was Richard Pope's Atlantic coupe.

Hydraulic Brakes at Last

Soon after the 57S's demise, the Types 57 and 57C entered their final phase of manufacture. At long last improvements were made to the brakes, modi-

Vanden Plas only bodied two 57Ss. One had a grey and red open body, costing £257, on chassis 57541, delivered in August 1937 for a Mr Rand of New York and this was the other, 57572. London's Green Park Garage took delivery of the chassis in August 1937. The coachbuilder described the grey painted body, number 3613, as a Concealed Hood Coupe and design 1488 which cost £395. Because of the long bonnet, the steering column had to be extended by 63mm (2.5in). It was completed in November 1937 and was later owned by Lord Beaverbrook's son, Max Aitkin. After the war it was owned by Lord O'Neill in Norther Ireland. Today it can be seen at Mulhouse.

The Series 3 Type 57 appeared at the 1938 Paris Show. Modifications included the fitment of vertically mounted telescopic shock absorbers and the front wing line was raised and modified to accommodate them. Headlamps were usually integral with the wings. Hydraulic brakes were, at long last, introduced and the range slimmed down to the Galibier saloon and Stelvio coupe.

fications which were displayed at the 1938 Paris Salon. The established mechanical actuation was replaced by a Lockheed hydraulic system.

La Vie Automobile was impressed and considered them: 'magnificent; having had to brake in Paris on a wet wood surface on two or three occasions, the car always stopped in the minimum distance without showing the slightest sign of lateral skidding; bravo for the Lockheed hydraulic controls . . .'

Modifications were also made to the cruciform chassis frame, which was strengthened, and to the 57's suspension. The combination of de Ram and Hartford and telecontrol shock absorbers was discontinued and they were replaced with all round telescopic Allinquant units. However, because the latter were vertically mounted, unlike the de Rams which were horizontally located, the line of the inner front wings was raised to accommodate them.

Changes were also made to a number of factory bodies available. The two-door Ventoux saloon and the Atalante coupe were continued for 1939 though there was the new top line Aravis drophead coupe which sold for 125,000fr. (£757), and was built for Bugatti by Gangloff. This left the existing but cheaper Stevio at 115,000fr. (£697) and the Galibier four-door saloon which was listed at 116,000fr. (£703).

Jean Bugatti took the opportunity to update the Galibier's lines. In Paul Kestler's opinion this ultimate Type 57 Molsheim-built saloon had a 'flowing roof line in the Chrysler Airflow manner . . . also adopted by Peugeot on the 402'.[32] The front wings (fitted with either free-standing or integral headlamps) were enlarged, the spare was removed from the rear of the car to the left-hand bonnet side where it was concealed by a metal cover. This meant a larger boot.

Distinctive features were the windows of the doors which were fitted with swivelling ventilators while the model could be specified with four separate glass panels let into the roof which echoed an idea that Jean had been first used on the *Coupé Napoléon* Royale. He had already used transparent roof panels on the discontinued Ventoux.

The Type 57 was only destined to endure for another ten months in production as manufacture was halted in August 1939 because of the impending outbreak of World War II. Possibly only one car was delivered that month, a Galibier saloon (57837), though the final Type 57 chassis number is 57841. As the Molsheim records do not show a delivery date, this Letourneur and Marchand 57C could have been delivered prior to or after that date.

This brought the total number of Type 57s built to 687 which consisted of 552 examples of the standard model, while the 57C accounted for a further 94 and the 57S a more modest 41. In any event, a Type 64 replacement was destined for the aborted 1939 Paris Salon. But before examining that particular car which was, alas, stillborn, we should retrace our steps to consider the last Bugatti Grand Prix car, the Type 59, which first appeared in 1933. Jean Bugatti had taken over design responsibility at Molsheim at the time of the 59's creation though the hand of Ettore is clearly apparent in the car's overall conception, more so than in the Type 57 which was emerging in tandem with it.

[32] Kestler, *Bugatti, Evolution of a Style*.

This 1938 Type 57 Atalante was owned by that great Bugatti enthusiast, Lord Cholmondeley, while his wife ran . . .

. . . this 1939 Type 57 with body by Figoni and Falaschi. This chassis, 57739, was imported in February 1939. Today it survives in Switzerland.

A 1939 Type 57C in an appropriately English setting. This is chassis 57787, supplied by Jack Barclay to its original owner, William P Harges, an American, on 31 March. It was seen by Colonel Godfrey Giles and Harges admired the latter's 1938 Corsica-bodied 57SC, so an exchange took place. Giles christened the car 'Charmaine' and one of its post-war owners, from 1949 until 1951, was Miss Dorothy Padget, who had financed the 'Blower' Bentley team of the late 1920s. It still survives in Britain. The body is by James Young.

The Racing Years

It will be recalled that, in 1928, the so-called *Formule Libre* (free formula), which took no account of a car's engine capacity, was introduced to the motor racing circuits of Europe and was, in effect, to endure for the next six years. However, on 1 January 1934, a new formula was introduced which had been 'intended to tidy up International motor racing so that it would once again take some tangible form'.[33] Details had been unveiled by the AIARC organizing body in October 1932 and it was intended to run for three years, from 1934 until the end of 1936.

Although the rules effectively perpetuated the *Formule Libre* practice of not specifying a limit to a competing car's engine capacity, their principal requirement was that the cars should not exceed a dry weight of 750kg (14.73cwt), though less driver, fuel, oil, water and tyres.

This limit was chosen in the hope that it would favour such small and efficient racers as the Type 51 Bugatti, of 2.3 litres and the 2.6-litre *monoposto* Alfa Romeo, which weighed about 750 to 800kg (14.7 to 15.7cwt), and penalize less desirable manifestations as the 4.9-litre Type 54 and the 12-cylinder Type A from the same respective stables.

In addition to the all important proviso of weight there were two further requirements. One was that competing cars should present a minimum cross-sectional area of 85 x 25cm (33 x 9.8in) at the driving seat. It also specified that races should be run for a minimum distance of 500km (310 miles). Bugatti's Type 59 was therefore created to conform to these parameters.

When work on the new car began in 1932, Molsheim's fortunes were at a low ebb. Sales of road cars had been badly hit by the world depression, the Type 51 racing car was not proving to be anything like as successful as its predecessor and the life-giving railcar contract was still a year away. Also with the withdrawal from the race track of such stalwarts over the previous decade as Delage, Talbot and Ballot, it fell to Bugatti alone to uphold France's Grand Prix prestige, for the only other Gallic representative in the shape of the newly created and complex SEFAC, rarely reached the starting line.

When cars for the new formula appeared in 1934, the main challenges came from Germany, with its Mercedes-Benz and Auto Union racers, and the Italian Alfa Romeo concern, all of which were subsidized by their respective fascist governments while Molsheim benefited from no such direct state support.

As ideas first began to gel at Molsheim for a new Bugatti Grand Prix car, there were thoughts about creating a racer based on the Type 51 chassis though powered by what was to become the Type 57 engine. This was not proceeded with though at least one drawing does exist showing such a change.

The Type 59 Described

Instead it was decided to use, in essence, another existing frame. Like that of the Type 54 Grand Prix car, the 59's was based on the chassis built for Ettore's curious 16-cylinder cars of 1929. But while the 54 used the long Type 47 frame, the 59's was based on the 2,600mm (8ft 6in) Type 45 one. Its rear springs therefore projected beyond the bodywork as the side members were not contoured in the manner of the Type 35/51 one. Although the 1933 Type 59s used plain side members, all subsequent examples were drilled for lightness.[34] These holes were cut uniformly on the first six or seven chassis though the pattern varied thereafter.

Suspension followed Bugatti's established half elliptics at the front and reverse quarter rear springs. But as *Motor Sport* put it, 'how then had [the layout] been adapted to deal with speeds in the neighbourhood of 170mph [273kph]? Simply by changing over to de Ram shock absorbers... Monsieur de Ram has been for many years

---

[33] 'Denis Jenkinson' in *The Racing Car Development and Design*.

[34] Martin Dean, 'The Type 59 and its development, parts I and II' in *Bugantics*, vol. 47, no. 4. (Two first-rate and detailed articles.)

engaged in research in connection with springing ... a state of affairs made easier through his personal friendship with Monsieur Bugatti, came to the conclusion that the only satisfactory type of shock absorber, especially on a car intended for high speeds, was the friction type.'[35]

Details of the hydraulically loaded units which resulted have already been chronicled, for it will be recalled that these shock absorbers were also used on the Type 57 and 57S of 1936–1938. Yet Ettore's apparent answer to his suspension problems was dearly bought because the de Rams sold for the equivalent of £170 a set, plus a £30 fitting charge, which was about the price of a new MG Midget in Britain, although Molsheim obviously did not pay this full retail price.

At the front of the car, the de Rams were mounted in the engine compartment just behind the radiator. The pivots projected through apertures in the body sides and damped the front axle via arms which did double duty as radius rods that ran parallel with the springs. Their respective mountings were the reverse of Type 35 practice as they were shackled at their inner ends and located by sliding trunnions at the front. As ever, they passed through apertures in the hollow front axle. This was split and the two halves were secured by a sleeve with left- and right-hand threads which resulted in a modest degree of articulation, a feature that was inherited, in essence, from the Type 54 and was subsequently adopted on early Type 57 road cars.

Lower, Longer and Wider

Ettore was clearly determined to reduce the frontal area of the Type 59 which, although longer and wider than the Type 51, was significantly lower. The engine was therefore low mounted in the chassis and was essentially that used on the Type 57 though the 59 had its own drawings and therefore had its specially made components. It was thus a twin overhead camshaft eight-cylinder unit with a fixed head cast iron cylinder block mounted on an alloy crankcase, though with spur rather than helical teeth used for the timing gears. The single piece crankshaft similarly ran in six plain bearings. This latter feature marked a European move away from roller to plain bearings, a trend which had already been set by Alfa Romeo and Maserati. The Type 59's capacity was 2,821cc, with a 72 x 88mm bore and stroke, which was initially shared with the 57.

The Bugatti-built Roots type supercharger was mounted on the right-hand side of the engine and driven, like the 57C, from the rear timing gear train which incorporated three idlers. But unlike the road car, which had an additional cog introduced between the blower and the lower of the idlers, on the 59 the supercharger gears meshed directly with it and the unit therefore ran in the reverse direction and blew downwards rather than up. This was because of the low slung engine: it was not possible to mount the carburettors underneath the blower, as hitherto had been the case, because there was insufficient room. The twin downdraught Zenith units were therefore mounted, for the first time on a Bugatti, above the supercharger.

The same space constraint resulted in the employment of a dry-sump lubrication system. The oil tank was located alongside the driver, because the bottom of the crankcase was a mere 128mm (5in) from the ground. There were two pumps, the smaller of the two circulating lubricant around the engine while the larger scavenged any oil that had dropped into the crankcase. This was returned to the cockpit-located reservoir via a number of large diameter copper pipes exposed on the left-hand chassis members which served as a crude oil cooler.

Nearby was the aperture for the side-mounted starting handle which was also located there, Miller-like, because the low position of the engine precluded the customary frontal arrangement. It engaged with the rear timing gear train. The car's low build was also the reason for the exhaust pipe running alongside the body rather than out of sight beneath it.

As on the Type 51 the exhaust camshaft drove a

[35] 'A modern Grand Prix car' in *Motor Sport*, April 1935.

The Type 59's engine. It externally resembled the Type 57 though, as a Grand Prix car, its components were peculiar to it. The supercharger is out of sight though the twin Zenith carburettors, covered by a mesh grille, are apparent. This is the engine of a 59 being run as a road car by Rodney Clarke after the war; hence the horns and front wings.

dashboard-mounted Scintilla Vertex magneto, though the engine had a different firing order to that of the Type 57: 1,5,3,7,4,8,2,6 rather than 1,6,2,5,8,3,7,4.

Gearbox and Rear Axle

Unlike the Type 57 which featured unit engine/gearbox construction, Ettore's hand is apparent in the use of his multiplate clutch. The drive thereafter went, via a short shaft, to the gearbox. This was essentially the four-speed Grand Prix unit though strengthened to cope with the more powerful engine, its gears having wider teeth while the square section shafts were thicker than hitherto. Instead of being mounted on twin tubular cross members, as in the case of the Type 35, the 59 followed the Type 54 and even earlier Brescia practice, in that the casing was flanged, drilled in the case of the 59, and extended either side to reinforce the frame. The wooden-handled gear lever was mounted outside the body and followed past precedent in that it moved forward to engage second and top gears.

An open propeller shaft thereafter took the drive to the rear axle and was accompanied by the usual torque arm. The external radius rods, which had been a distinctive feature of the Type 35, were deleted. The drive was taken through the springs. The rear de Rams were mounted inboard, ahead of the axle, with their arms exposed to view and lightened by drilling.

No such considerations appear to have constrained Bugatti in the design of the 59's double reduction rear axle, which *The Autocar* described as 'massive compared with previous Bugatti practice.'[36] This was the price paid, both in weight and cost, for the low-mounted drive train because, after the stubby prop shaft, 'comes the bevel pinion shaft, which drives a crown wheel to which is bolted a massive spur gear which drives up to a very much bigger spur gear machined from solid and containing the four spindles of the differential bevel pinions'. The 'enormous' robust drive shafts were fully floating for the first time on a Bugatti and incorporated the rear hubs so that, if the shaft broke, it would not result in the loss of a wheel.

A New Wheel

It may well have been that, because of the weight of the rear axle unit, Bugatti decided to design a lighter type of wheel for the 59, as the lovely 'piano

[36] *The Autocar*, 19 April 1935.

wire' ones, which are forever associated with the car, appeared at a fairly late stage in its development. Factory drawings, dated 1932 and 1933,[37] show the 59 fitted with eight-spoked aluminium wheels of the sort used on the Type 54. The four new wheels, in total, saved a further 18 precious kilogrammes (40lb).

The design was, if anything, of even greater beauty than the original aluminium one. It was distinguished by two rows of thin gauge spokes but, unlike conventional wire wheels, they did not transmit drive and braking reactions. That was the role of a large serrated aluminium flange, to which the brake drum was bolted, riveted to a standard Rudge Whitworth splined hub. The teeth meshed with their opposite numbers on the inner edge of the rim. Like their predecessors, when the wheel was removed, the aluminium shoes of the mechanically actuated brakes were exposed to view for, if necessary, instantaneous replacement.

The 59's bodywork followed, in essence, that of the Type 35 of eight years previously though, in the interest of weight saving, it was made of Elektron. It came complete with the famous horseshoe radiator, while the elegant tail was riveted along its centre line because of the material used, a theme which would recur on the 57S Atlantic coupe. As with the Type 51, there were twin fuel fillers for the 136-litre (30-gallon) tank. The car was therefore outwardly a two-seater though the space for the non-existent mechanic was occupied by the oil tank and half of the cockpit was covered by a detachable cowl of the sort introduced on the 35.

All these features combined to produce a car of great subtlety of proportion to delight the eye and uplift the spirit, but was the Type 59 good enough to challenge the formidable opposition which was taking shape to the east of the Rhine?

Looks Are Not Enough

The truth was, alas, no. Ettore had produced a car of great beauty, whose lines were, if anything,

[37] Dean, 'The Type 59 and its development'.

superior to those of the Type 35, but which fell badly short of expectations. Its specifications echoed the previous decade rather than reflected the technological advances of the 1930s. The bodywork, in particular, was outdated. Every other manufacturer which had designed cars for the 750kg formula dispensed with two-seater bodywork and opted instead for the single-seater bodywork of the type which had arrived with the P3 Alfa Romeo in 1932. It was not until the new formula was two years old, in 1936, that the Type 59 tentatively appeared in single-seater guise.

Similarly the cart-sprung suspension, although aided by the costly de Rams, had appeared in essence on every Bugatti since 1913 and the mechanically actuated cable brakes also differed little from those fitted to the Type 35 of 1924. In fairness to Bugatti, mechanical brakes were still the norm on racing cars at the time of the 59's creation but hydraulics were revived for Grand Prix racing by Maserati in 1933.

The rival Alfa Romeo concern employed cart-sprung suspension and mechanical brakes on its celebrated P3 in 1934 but quickly switched to the Dubonnet independent front suspension and hydraulic brakes in the following year. This was to match such actuation and the all-independent suspension of the German Mercedes-Benz and Auto Union of 1934, which heralded a new era of motor racing technology, and closed the chapter on cart springing for Grand Prix cars. But Ettore remained firmly wedded to the beam axle, even if his son Jean would have probably wanted ifs, and hydraulic brakes did not appear on Bugatti racing cars until 1939.

In addition to the 59's dated specifications, it soon shattered Bugatti's hitherto almost inviolate reputation of reliability. The car's Achilles' heel was its gearbox, which developed a tendency to jump out of gear, while bottom could fail altogether through the gear cracking between the root closest to the square shafting. Even the magnificent piano wire wheels were unpopular with the works drivers. The trouble was caused by the flange and its serrations which meshed with the inner rim of the wheels.

Still essentially a two-seater in the monoposto *era. René Dreyfus pictured in 1933 at Molsheim in a type 59 on which the glorious 'piano wire' wheels appeared for the first time. The chassis of these early cars were not drilled.*

Rear view of an early Type 59. As it used the type 45 chassis, the rear springs projected beyond the bodywork, unlike the 35/51s which were faired into the tail. This car has been registered for the road.

René Dreyfus, who drove the 59 in its first two seasons, has recalled that 'when the clutch was let out, a little bit of slack . . . would create a knocking noise. It sounded like a loose rear end, but it wasn't – only the wheel. We complained about this on occasion to M. Bugatti. "Well it's too bad" he would laugh "You'll have to get used to it." '[38] This annoyance, coupled with the more serious gearbox trouble meant, says Dreyfus, that none of the works team 'cared for the car much in the beginning. We did love the look of it – the 59 was gorgeous . . . a breathtaking machine to behold. But it felt strange to drive.'

Destined to be enlarged to 3.3 litres in 1934, the Type 59 was no sluggard. Brivio was timed at 256kph (159mph) at the 1934 Coppa Acerbo and claims of nearly 290kph (180mph) were subsequently made.[39] By the time of the 1934 season, the 59 developed 240bhp at 5,500rpm, or 73hp per litre, which compared adversely with 88 for the

[38] Dreyfus with Kimes, *My Two Lives*.

[39] Denis Jenkinson, *Motor Sport Racing Car Review 1947* (Grenville).

2.9-litre P3 Alfa Romeo and 107 for the Mercedes W25, which was also powered by a 3.3-litre twin overhead camshaft eight-cylinder engine. The other pillar of German hopes rested on the mid-V16-engined 4.8-litre A type Auto Union, which developed 295bhp, but proportionately less than the Bugatti at 68hp.

A Debut in 1933

With the 750kg formula due to come into force in 1934, Molsheim was anxious to have some cars ready for 1933, so as to rectify any shortcomings that might manifest themselves. Whatever his own thoughts, René Dreyfus says 'that the Type 59 would return Bugatti to pre-eminence in racing was the singular thought at the factory'.[40] Approximately twelve chassis would be laid down though, of these, only ten would be used. As will emerge, the cars' mechanical and body lines were to progressively evolve over the following six years.

The 59 was destined to make a competition debut at the 1933 French Grand Prix, held at Montlhéry on 11 June. Molsheim entered four cars but two of the practice days passed without any appearing. Albert Divo and Williams were at the circuit but were evasive when questioned by the French press and countered enquiries with: '*Nous attendons les instructions du patron.*' ('We are waiting for the boss's instructions.')

On the Friday evening before the Sunday race, Jean Bugatti reported that one car had been completed and, after a brief road test, would be sent to the circuit for Varzi to drive. Divo assured the ACF that all the cars were on their way and asked that the scrutineering deadline be extended. For its part, the Club wanted a Bugatti presence so that there would not be a repeat of the Alfa Romeo victory of the previous year.

But Jean eventually had to admit defeat. The car in question was given a 100km (60-mile) test but was not considered to be *au point* and had to be withdrawn. Unfortunately the only other available car, the Type 51 which had just won the Monaco Grand Prix, was awaiting overhaul. Therefore only privately entered Bugattis ran in the Grand Prix which was a triumph for the Italians. Their cars took the first six places. Campari gave Maserati its first French Grand Prix victory and the remaining five positions were taken by Alfa Romeo.

In the following month, Molsheim entered a team of three cars for the Belgian Grand Prix, to be held on 9 July. A Type 59 participated in practice but team manager, Meo Costantini and Varzi, who would have driven it, were unhappy about its performance and the latter drove a Type 51 instead.

The Type 59 made its first competitive appearance in the Spanish Grand Prix at San Sabastian on 24 September with a team of three cars, driven by Benoist, Dreyfus and Varzi. Despite, says Dreyfus, the carburettors working loose, he finished sixth. Varzi was fourth though with Benoist unplaced. The 59s were still running in 2.8-litre form at this stage.

Turning the Scales

The arrival of the 750kg formula in 1934 prompts a fascinating question. How much did the Type 59 weigh? It should not, of course, have exceeded the prescribed limit but, after the war, English *Bugattiste* Bob Roberts owned the ex-works Type 59s (59122) which had been sold by English racing driver, Lindsay Eccles, in 1935. By this time it had been equipped for road use and, in May 1955, weighed, all up, 2,370lb which is the equivalent of 1,075kg.

The combined weight of road equipment, preselector gearbox fitted pre-war in place of the original troublesome unit, fuel and tyres amounted to 560lb (254kg). This meant that, in its race-prepared form, this particular car turned the scales at 822kg (1,812lb), which amounted to 72kg (159lb) over the 750kg threshhold, and could not have begun to have approached it, even if some allowance is made for error.

In the 59s' heyday, it is said[41] that the Bugatti

---

[40] Dreyfus with Kimes, *My Two Lives*.

[41] Dean, 'The Type 59 and its development'.

First time out. Although Type 59s participated in practice in the Belgian Grand Prix in July 1933, they did not compete in the event proper. It was not until the Spanish Grand Prix of 24 September that they appeared, running in 2.8-litre form in the race. Here Bugatti team manager Meo Costantini confers with Varzi. He was placed fourth and Dreyfus was sixth.

Still in a 2.8-litre and less its cockpit cowl, Dreyfus pictured practising for the 1934 Monaco Grand Prix. The 59's chassis is now drilled and he did well to be placed third behind a pair of Alfa Romeos in the race itself.

factory went as far as removing the brake *shoes*, as well as the tyres, when the cars were weighed and that sympathetic French officials turned the Gallic equivalent of a blind eye while Molsheim mechanics placed tyres levers under the weighbridge and stood on them . . . Yet the weight of Nuvolari's car, which is believed to have been the one sold to Eccles, at the French Grand Prix of 1934 when the 59s were the heaviest entrants, is given as 747kg while Benoist's was 747 and Dreyfus's 749.5.[42] Intriguing.

The first major race of 1934 was the Monaco Grand Prix, held on 2 April, in which three 59s still

---

[42] Hodges, *The French Grand Prix*. (The weights of the other cars were: Alfa Romeo: Chiron, 720.5kg; Varzi, 730kg; Trossi, 721.5kg. Auto Union: Stuck, 740.5kg; Momberger, 738.5kg. Maserati: Zehender, 735kg; Etancelin, 748.5kg. Mercedes-Benz: Caracciola, 739.5kg; Von Bauchitsch, 737kg; Fagioli, 737kg.)

Three Type 59s were run in the 1934 French Grand Prix at Montlhéry held on 1 July. The event was notable for the first appearance of the new W25 Mercedes-Benz and the mid-engined P-Wagen Auto Unions though the German cars suffered from teething troubles. Here Nuvolari in a Type 59 is ahead of Carlo Trossi in a P3 Alfa who was placed third. The Bugatti dropped out on the seventeenth lap with gearbox problems.

running in 2.8-litre form, were entered. Varzi had departed for Scuderia Ferrari's Alfa Romeos by this time and his place was taken by the wealthy Italian marquis, 'Tonino' Brivio though he did not run at Monaco. Another new recruit who did was Jean-Pierre Wimille, who was joining the team from Alfa Romeo as junior driver.

Dreyfus and Benoist remained and drove 59s in the Grand Prix while the great Tazio Nuvolari forsook his Maserati to take the wheel of a Type 59 as an independent. Dreyfus did well to come in third behind two Alfa Romeos, while Nuvolari was fifth and Wimille retired after eighteen laps when a rear brake lining came adrift.

A team of 59s was entered for the Tripoli Grand Prix on 6 May though, during practice, Dreyfus's car was plagued with gearbox disorders. 'It is not a good thing to be approaching a curve at 241kph [150mph] in fourth, shift down and suddenly find yourself in neutral.'[43] Repairs were made to the car but he could only manage sixth place and he was the only one of the Bugatti works drivers to finish.

The French Grand Prix, held at Montlhéry, marked the first appearance in France of the new Mercedes-Benz and Auto Unions. They had already appeared on home ground at the Eifelrennen of the previous month and, as it happened, suffered from teething troubles, so Alfa Romeos took the first three places.

The three Type 59s, now enlarged to 3.3 litres, were all plagued by a variety of maladies. Dreyfus withdrew with supercharger problems on lap 16, and the Nuvolari car, which was also driven by Wimille, dropped out on the following one with gearbox trouble. Only Benoist, who had at one

[43] Dreyfus with Kimes, *My Two Lives*.

The dead park at the 1934 French Grand Prix. In the foreground is Dreyfus's Type 59 which withdrew on the sixteenth lap with supercharger trouble and, on the extreme right, is Nuvolari's car. Behind them is Fagioli's Mercedes-Benz and Prince zu Leiningen's Auto Union.

stage managed to get to fourth position, kept going though he suffered from misfiring and his engine proved temperamental after a pit stop. His was the only Bugatti still to be running when he was flagged off before he could complete the distance.

A Win at Spa

The 59s had better luck at the Belgian Grand Prix in the following month. It was won by Dreyfus who says that 'both my Type 59 and I behaved impeccably' after the two leading Alfa Romeos crashed. Brivio was second and Benoist fourth.

The Bugattis returned to the Spanish Grand Prix, on 23 September, for the last big race of the season. The new Mercedes were starting to find their form and came in first and second, Nuvolari, still driving as an independent, was third in his car which had been painted red during the season. Wimille was sixth and Dreyfus seventh. Molsheim also ran its 59s in second stream events, namely the Swiss Grand Prix. There, Dreyfus was third, while Brivio was similarly placed in the Coppa Acerbo at Pescara. Wimille finished the season off by winning the Algiers Grand Prix on 28 October.

It had been a very chequered first season and Ettore, having no doubt viewed the opposition and with railcar production moving centre stage at Molsheim, decided to sell the works cars. Thereafter, he maintained a reduced and diminishing presence in Grand Prix racing over the five seasons between 1935 and the outbreak of war. Four works Type 59s all went to English drivers. In March Lindsay Eccles took delivery of 59122. In April Earl Howe got 59123, also numbered 54208, C E C Martin 59121, and Noel Rees bought 59124, also allotted 54213, for the Hon. Brian Lewis to drive. They were thereafter campaigned with some success in British and overseas events.

Howe attained a speed of 228.3kph (138.34mph) on the Outer Circuit at Brooklands at the 1936 BARC Whitsun meeting, making it the fourth fastest car around the concrete bowl. The histories of these four works cars are set down separately and all, thankfully, survive. Today a Type 59 is one of the most sought-after Bugattis in the world and, because of their scarcity, a number of replica cars have been constructed.

A Slimmed-Down Team

René Dreyfus and 'Tonino' Brivio left the Bugatti works team and joined Scuderia Ferrari, and thus Alfa Romeo, in 1935. Along with Nuvolari, they

In 1935 the four Type 59 team cars came to Britain. This is Charles Martin in 59121 at the 1935 International Trophy race at Brooklands when no less than three Type 59s participated. The Noel Rees/Earl Howe entry (59123), driven by Brian Lewis, threw a connecting rod. Lindsay Eccles in 59122, broke its torque tube, locking the rear wheels and Martin retired with transmission trouble. He later recounted that, as a result, 'Jean Bugatti came over from the Works with some ... mechanics, and finally the cars were put in running order again.'

After Martin, his car passed to the Duke of Grafton but, unhappily, he was killed in it at Limerick in August 1936. It was bought by Arthur Baron, shown here, who rebuilt it with a new body and, unlike the other 59s, its tail is not riveted along its centre. He also replaced the troublesome gearbox with a pre-selector unit and used the car for sprints and hill climbs. It was later owned by George Abecassis and then Kenneth Bear who fatally crashed it at Jersey in 1949. It was bought in 1950 by E A Stafford-East, who has rebuilt the car and still owns it.

This Type 59, chassis 59124, was used by the Hon. Brian Lewis in 1935 though his only significant win that year was the Manin Moar race in the Isle of Man. The car then spent some time in America but, on its return to Britain, was owned by Ian Craig who repainted it in this rather striking ivory and white livery. Like Baron's car, it was used for sprints and hill climbs. After the war it passed to Rodney Clarke, who converted it for road use and it was registered LPG 211 in 1947. In 1952 it was bought by New York-based F R Ludington and has been in America ever since.

Prescott had only been open a year. The date is 11 June 1939 and the occasion a Bugatti Owners' Club meeting. On the left is Jack Lemon Burton with the Type 59 previously used by Lindsay Eccles, having bought it in 1938 from H W 'Bill' Papworth, who had dispensed with the cable brakes and replaced them with hydraulics. He enlarged the engine to 3,798cc by having a new long stroke crankshaft made. After the war, the car went to Bob Roberts, D Z de Ferranti in Ireland and it is now in America. On the right is Arthur Baron's Type 59 which made fastest time of the day with a climb of 48.71 seconds. Lemon Burton was second at 49.86.

may have been responsible for that year's enlarged 3.1-litre P3s being fitted with reversed quarter-elliptic rear springs, *à la* Molsheim, in place of the half elliptics previously employed. This left Benoist and Wimille as Bugatti works drivers though the latter invariably drove a 59 he had purchased from the factory. He was placed second in a 59 at the 1935 Tunis Grand Prix of 5 May.

Only one car was entered for the 1935 French Grand Prix held on 23 June. It was clearly a development of the original theme because, apart from being fitted with a cowled radiator, under the bonnet (which unfortunately blew off during the race) was a 3.8-litre Type 50 related engine. If the 3.3-litre car had exceeded the 750kg limit, goodness knows what this unit, with its iron block, must have weighed! As far as the power unit was concerned, 'the top half of the engine [was] more like the 4.9 with forward drive to the camshafts, inclined plugs, larger and centrally placed supercharger with much modified manifold.'[44]

It required the fitment of a new 'elephant's ear' oil cooler which sprouted from the bulkhead just ahead of the cockpit.

This engine represented an attempt to make the Type 59 more competitive. After its increase to 3.3 litres, the existing twin cam eight appears to have reached the limit of its development, although in Britain at least one car was extended to 3.8 litres by the fitment of a new Laystall crankshaft. But by drawing on the existing Type 50 unit, the so-called 50S Molsheim paved the way for a complete redesign of this unit. The gearbox may have been simultaneously modified for three-speed operation.

The race itself was marred by an incident that occurred on the sixth lap. A build-up of air pressure blew the bonnet off though, fortunately, did Benoist no harm. Miraculously he caught it and

Wimille in a Type 59 in which he won the 1936 Comminges Grand Prix. It was running in unsupercharged form.

later recounted that: 'my right eye was blocked, I could only see with the left one, and my goggles were covered with oil'[45] and drew up by the first chicane to regain his composure. He returned to his pits to throw the bonnet away and continued to race without it though retired on lap 16.

German Supremacy

The victor of the 1935 French Grand Prix was Caracciola in a Mercedes-Benz, and the silver single-seaters from Stuttgart also came in second and fourth. German cars were to win every other major race during that year, with the notable exception of the German Grand Prix, when Nuvolari in an ageing P3 scored the greatest victory of his career. But from thereafter until the outbreak of the war, Mercedes-Benz and Auto Unions were victorious in practically every other front-line race.

---

[44] Eaglesfield and Hampton, *The Bugatti Book*. There is some divergence of opinion about the capacity of this car. Hodges says of Benoist's Bugatti: 'This apparently had a 3.8-litre engine – a "4.9" blower probably led to contemporary reports that the engine was of that capacity . . .' William Court, in *The Power and the Glory*, also quoted 3.8. But Martin Dean's authoritative article 'The Type 59 and its development, parts I and II' in *Bugantics*, vol. 4, no. 2, says that Benoist's car had a 4.9-litre Type 50 engine. Conway does not quote any capacity.

[45] H. G. Conway, 'Benoist and the 50B' in *Bugantics*, vol. 47, no. 2 (summer 1984).

Four Type 59s were entered for the Belgian Grand Prix, to be driven by Benoist, Louis Chiron, Taruffi and Wimille but none, apart from Benoist, who was fifth, completed the race, held on 14 July. Their presence is, nevertheless, revealing because it indicates that there were no less than eight Type 59s in existence at this date, the quartet of 59s at Spa, plus the four works cars in Britain. A week later at the Dieppe Grand Prix, on 21 July, there were no less than six 59s present but only four started with just Wimille, who came in third, placed. He was less fortunate in the Italian Grand Prix held at Monza on 8 September. His car blew up but Taruffi in the 3.8 kept going to finish fourteen laps behind the winning Auto Union.

At the Spanish Grand Prix on 22 September, Wimille was fourth behind a trio of Mercedes. He was gradually emerging as the most successful Bugatti driver of the period. He had, earlier in the year, been placed second at the Tunis Grand Prix in May and followed this, in June, with a similar position at the Lorraine Grand Prix held in Nancy.

As the 1935 season was drawing to a close, it was becoming clear that if Bugatti was to maintain some semblance of a presence in Grand Prix competition, a new engine was desperately needed. The Type 50-based eight was five years old in 1935 and was, in any event, too heavy. As there was no upper engine limit to the 750kg formula, it was decided to re-design the Type 50 engine and the cast iron block was duly replaced by an alloy one. This unit was to do double duty as an aero engine. Having said that, it is unclear what came first: the demands of the race track or an aircraft application?

Also an Aero Engine

In October 1935 the French press carried a news item to the effect that Bugatti was intending to re-enter the aviation market with a new engine of 'between 3 and 6 litres, of very high power, capable of 4,000–5,000 revolutions per minute.'[46] In 1936 Ettore met the freelance aircraft designer, Luis de Monge, and asked him whether he could break the world's speed record with a pair of 500hp engines? De Monge believed that it was possible and there were also hopes of entering it in the prestigious *Coupe Deutsch de la Meurthe* air race.

Work on the project was under way by the autumn of 1937 but it took a little time to obtain French Air Ministry backing, though a contract was signed for the construction of two prototypes in August 1938. The resulting small wooden monoplane powered by two engines, which carried the 50B designation, with contra-rotating propellers was nearing completion by the outbreak of World War II. It survives in America.

The 50B engine was available in three forms. Initially there was the supercharged BI, with a capacity of 4,739cc. The dimensions of the BI came closest to the Type 50 car engine, though its bore was reduced from 86 to 84mm, but the 107mm stroke was retained. Next came the short 100mm stroke 4,433cc unsupercharged BII. The foregoing description applies to these two engines. The 3-litre BIII differed from them and will be described later as it was created for the subsequent 1938–1940 racing formula.

The Type 50 Re-Design

The engine's overall layout essentially followed that of the Type 50, but otherwise the later eight greatly differed from it in detail. The twin overhead camshafts were therefore driven from the front, unlike the Type 57 unit which had rear-located timing gears. Miller-type cam followers were used, though they were larger than the usual Type 50 ones and drilled for lightness. The exhaust valves were sodium cooled. As on the Type 50, the supercharger was driven from the front and mounted on the right of the engine. It aspirated through two C2 Bugatti twin choke carburettors located underneath. Similarly its opposite number was the water pump situated on the exhaust side.

The concept of a crankshaft with its nine plain bearings attached directly to the crankcase was

[46] Conway and Sauzay, *Bugatti Magnum*.

perpetuated though, once again, the design of the shafts and its attendant rods, which were beautifully machined all over, were peculiar to the 50B. A further Type 50 inheritance was the employment of a dry-sump lubrication system. The double oil pumps (and all others) used skew gears so that they would run in both directions as the engine was designed, from the outset, to so operate for aviation use. In 4.7-litre form, the 50B developed 485bhp at 5,500rpm, which was the equivalent of 103bhp per litre, and therefore a proportional improvement of the 59's 73. There was no question of the fragile Type 59 gearbox being able to cope with this extra power, so a new four-speed unit was specifically designed for the 50B. Not only was it stronger with wider teeth to cope with the close on 500bhp developed by the big eight, it was fitted with synchromesh on every gear and so was superior to the 57 unit. It was designed so that it could be fitted with a bell housing to be mounted in unit with the engine for sports racing applications or separately when used in a Grand Prix car.

A New Single-Seater

A 4.7-litre engine was installed in the Type 59 chassis which Benoist had run at Montlhéry in 1935 when he drove it in practice of the 1936 Monaco Grand Prix, held on 13 April. It was fitted with a single-seater body and was therefore the first *monoposto* Bugatti of the 750kg formula. Gone was the lovely horseshoe-shaped radiator. It was

After two years of the 750 kilogramme formula, Bugatti produced a monoposto. The occasion is the 1936 Monaco Grand Prix and this Type 59-based chassis has been fitted with a 4.7-litre version of the type 50B engine, the oil cooler of which can be seen below the radiator. Robert Benoist tried the car in practice though it did not participate in the race. Note that the radiator grille and bonnet are riveted to line up with the tail in the manner of the Type 57S Atlantic road car.

Rear view of the same car with well-louvred bonnet. An electric starter can be seen just beyond the car's tail.

A mechanic working on the single seater, though with its oil cooler repositioned, at the 1936 Swiss Grand Prix. Wimille only lasted for three laps until forced out with the inevitable gearbox trouble.

replaced by a cowl resembling that of the contemporary Mercedes-Benz W25. But there was a spark of Bugatti individuality there, because the shell was flanged and riveted together at the top and this feature was perpetuated along the bonnet in the manner of the Atlantic coupe of that year, and continued on the essentially standard Type 59 tail.

The oil cooler was moved from its curious position near the driver to a more orthodox location, this time beneath the radiator. The external exhaust pipe was perpetuated. But there were no radical changes made to the suspension or cable brakes. Perhaps Benoist was dissatisfied with the car's performance because it did not run in the race proper.

Two more outwardly conventional 59s did, their engines apparently enlarged to 3.8 litres,[47] maybe in the manner of Benoist's car of the previous year. Wimille was placed sixth and Williams was ninth

and last. It is sobering to reflect that only seven years previously, in 1929, he had won the inaugural Monaco race in a Type 35B. The 1936 event was the marque's final appearance on the Monaco circuit.

There was, however, a change in Bugatti's fortunes in the 1936 French Grand Prix, held on 28 June. The French public was aghast at the German triumph of the previous year and, it should be recalled, the last victory by a French car in the race had been five years previously when, in 1931, a Type 51 Bugatti had taken the chequered flag.

Since then foreign cars, first Italian and then German, had dominated the event. After the 1935 race, 'the French press was up in arms . . . regarding it as a national humiliation and insisting loudly that the possibility of a repetition could not be countenanced.'[48] They adopted the view that if a French victory could not be guaranteed, then the Grand Prix should not be held, or the rules be altered.

It was for these demonstrably chauvinistic reasons that the ACF decided to run that year's race as a sports car only event, which would effectively prevent a German victory. It would cover eighty laps of Montlhéry's Circuit Routier and there was a 60,000fr. (£800) purse and three classes: Group I, 750 to 2,000cc; II, 2,000 to 4,000cc and III over 4,000cc. Bugatti accordingly prepared a team of three 3.3-litre cars for the race.

As it happened, Molsheim was preparing a new car, designated the 57S40, with a Type 59 chassis and a 4-litre version of the Type 50B engine to run at Le Mans in 1936. However, the event was cancelled because of labour unrest in France which, it will be recalled, had culminated in Ettore leaving his factory and settling in Paris.

The new car was to have been fitted with the full-width, all-enveloping bodywork associated with the cars that first appeared at the 1936 French Grand Prix. That car's mechanicals differed radically from those of the projected Le Mans entry because they were based on those of a production model. They were therefore more closely related to the 57S

[47] Hodges, *The Monaco Grand Prix*.

[48] Hodges, *The French Grand Prix*.

Mechanics working on a Tank for a successful record attempt held at Montlhéry on the 19 and 20 November 1936.

rather than the 59 and were consequently un-supercharged. Designated the 57G, these cars were immediately known as Tanks in memory of the 1923 French Grand Prix predecessors. Their distinctive streamlined alloy bodywork featured large alligator bonnets with horseshoe-shaped apertures though they concealed the presence of conventional radiators. Although the event was being run in daylight, low-mounted, wing-located headlights were fitted. The cars' engines were essentially those of the soon to be announced 57S, with dry-sump lubrication, and lightened crankshafts and rods. The new synchromesh gearbox was mounted in unit with the engine.

Despite the fact that the 57G had the 2,980mm (9ft 9.5in) wheelbase of the 57S, the frame followed that of the standard 57 and the rear axle therefore did not pass through the side members. De Ram

Williams pictured during the 1936 record breaking.

shock absorbers were fitted all round and the springs were stiffer than usual. Type 59 wheels took the place of the usual wires so that the brake shoes could be changed quickly during the race.

A Change of Fortune

Three Tanks were entered, driven by Wimille and Sommer, Veyron and Williams and Benoist and de Rothschild. All the cars initially performed splendidly and, at 300km (186 miles), were running first, second and third, pursued by no less than eight Delahayes. But all the Bugattis' rear brakes suffered from overheating and changing the shoes, despite the fitment of the 'piano wire' wheels, proved to be a time-consuming business. A Delahaye moved into the lead and that make held it until lap 62, when Wimille managed to get ahead and stayed there to win at 125.28kph (77.85mph). The two remaining Tanks also kept going to the end. The Veyron/Williams 57G was sixth with Benoist and Rothschild thirteenth.

A week later Wimille in a 57G also won the Marne Grand Prix at Rheims, held that year for sports cars. Benoist was second and Veyron fourth, both in Tanks.

In the autumn of 1936, Wimille crossed the Atlantic to participate in the Vanderbilt Cup race, held at the dirt-track-like circuit of Roosevelt Raceway, Long Island, New York. He drove the single-seater that had been seen at Monaco earlier in the year, though it was minus its oil cooler and fitted with a large fuel tank and fairing above. The works also entered another Type 59 which was driven by local driver, Lou Meyer. Wimille did well to be placed second behind Nuvolari's Auto Union, though Meyer was less lucky and crashed in practice. That 59 remained in America but was dismantled. Its chassis, which is numbered eight, survives.

These were the highpoints of the year that was otherwise confined to entering the cars in secondary events. Wimille was third in the Tunis Grand Prix and he also ran in the South African Grand Prix where he took second place. He won at Deau-

What the other drivers saw: Bugatti fielded a pair of 57G Tanks at Le Mans in 1937; one, driven by Labric and Veyron, retired but this is Wimille and Benoist's car.

A splendid photograph of the winning Wimille/Benoist Tank at Le Mans in 1937. The car averaged a speed of 136.98kph (85.12mph), which was only broken by another Bugatti victory in 1939.

A celebratory postcard, illustrating the winning Le Mans car of 1937 and highlighting the record success of the previous year.

ville running in unsupercharged form and also won the Comminges Grand Prix with a car of similar specifications.

In 1937 the persuasive powers of his son Jean and of Robert Benoist were employed to convince Ettore of the need to enter the Le Mans 24-hour race, despite his distaste for the event following the débâcle of 1931. He agreed and two Tanks were prepared.

A Special Engine

What is thought to have been the spare 57G engine used in this race has, fortunately, survived[49] and it differs quite significantly from the standard 57S unit though these modifications could date from the previous year. The cups containing the sparking

[49] *See* letter by Bob Roberts in *Bugantics*, vol. 48, no. 1 (spring 1985).

Three 57G Tanks participated in the Marne Grand Prix at Rheims on 5 July 1936. The outcome was a win by Wimille, with Benoist second and Veyron fourth.

Rear view of a Tank at the 1936 Marne GP, reflecting the aerodynamic thinking of the day.

plugs were screwed, instead of being pressed, into the head. Rather than the magneto being driven from the exhaust camshaft it was geared off the idler directly below the twin camshafts and therefore mounted to the rear of the centre of the block. Electron pistons took the place of the aluminium ones usually used and the engine had a higher compression ratio than the norm, 8.95 rather than 8.5. The inlet and exhaust valves were 2mm (0.07in) greater in diameter at the seats than standard. The valves' caps also differed, in that they were mushroomed at the top, perhaps to accommodate stronger valve springs. Lower down the engine, the connecting rods were 24gm (0.84oz) lighter than standard and nicely radiused.

But 'when the month of March was reached, it was realized that the cars could not be ready for the race in June unless the men agreed to work overtime.'[50] As Molsheim was experiencing its share of labour troubles at this time, Jean and a colleague drove to Paris to mediate with M. Jouhaux, general secretary of the Labour Federation, and 'pointed out that they wanted to win a race which, for nine years, had been captured by British and Italian firms; the national reputation was at stake and unless permission was given for the men at Molsheim to work overtime, an Englishman or Italian might again win'.[51] The official was impressed by their arguments and the overtime ban on individuals working on the car was lifted.

Two tanks were entered, driven by Wimille and Benoist and Veyron and Roger Labric. Sommer in a Delahaye initially led but, on lap four, Wimille set a new lap record of 148.98kph (92.58mph) and moved into second place. On his next circuit he took the lead and the car stayed in front from thereafter. At 7.30 in the evening he handed over to Benoist who, on his first lap, once again broke the record at 155.16kph (96.42mph). At midnight he was still in the lead with the second Tank lying third though it was to later drop out on the 130th lap when in seventh position.

On the 23rd hour the Bugatti broke the 1933 distance record and Wimille went on to win to thunderous applause, to be followed by the playing of the *Marseillaise*. It was the first occasion that a French car had won the race since 1926.

Withdrawal at Montlhéry

The French Grand Prix held on 4 July was, once again, a sports car event. Bugatti entered two cars which were a derivative of the one conceived for the aborted Le Mans race of the previous year. These retained the Type 59 chassis, were powered by the unsupercharged 4.5-litre 50BII engine and were accordingly designated the 57S45. The full-width bodywork resembled that of the Tanks though the wings were more pronounced.

One of the cars arrived for practice and Wimille put in a respectable 5min 33sec lap and the organizers were requested to extend the scrutineering deadline so that the second car could participate.

[50] Bradley, *Ettore Bugatti*.

[51] Bradley, *op. cit.*

But then Robert Benoist, who combined his successful racing career with that of Bugatti's Paris sales manager, and in the face of official warning, used the Montlhéry circuit to demonstrate a private car to a prospective customer. As a result he was disqualified from competing. Bugatti responded by withdrawing both cars on the grounds that the factory had not the time to prepare them!

Wimille otherwise had a reasonable season with the Type 59 though mostly in sports car events. He won at Pau in unsupercharged form, and the Bone and Marne Grand Prix. In April Bugatti had won a 400,000fr. (£3,125) prize offered by the French government for the car which was fastest over 200km (124 miles) of Montlhéry's Circuit Routier. Wimille in a Type 59 completed the sixteen laps at an average of 146.65kph (91.12mph). However, he was unable to win the top 1,000,000fr. (£7,812) award in August when the car failed and it went instead to Dreyfus in a 4.5-litre Delahaye.

Despite the fact that the 750kg formula was intended to run for three years, it was extended for a further year and cars were run to this specification in 1937. The new formula for 3-litre supercharged cars and 4.5-litre unsupercharged ones, was therefore in force from the beginning of 1938 until the outbreak of World War II. The reappearance of limits to engine size was intended to keep speeds in check though there was a sliding scale of minimum weights and a 850kg (1,874lb) ceiling for the largest capacity cars.

The Disastrous 3-Litre

A 3-litre version of the Type 50B engine was developed for the new formula. The lower part of the wet liner unit was essentially that of the BI though differed from it in that the block was fabricated from dural plates screwed and bolted around a forged bronze cylinder head. It was originally designed with two superchargers, in the manner of the P3 Alfa Romeo, rather than the single unit employed hitherto, though it was not run in this form. The 50BIII can be regarded as an unhappy engine which was never fully developed and conceived at a time of increasing financial stringency at Molsheim. One of the two built blew up while being tested at the factory.

A new chassis was constructed especially for this engine with the right-hand rail bulged to accommodate the twin superchargers. But as it was run with a single blower, the 50BIII was instead fitted in that doughty workhorse; the frame which Benoist had first run in the 1935 French Grand Prix. New bodywork, at the front at least, was conceived for it.

The car made its competition debut in Ireland at the first Cork Grand Prix, held on 23 April. Wimille drove it over three unhappy practice laps. In the race proper the 3-litre was timed, on its seventeenth circuit, at 236.97kph (147.25mph) down the 2.6km (2.25-mile) Carrigrohane Straight which was the fastest speed recorded that day. But on the

A 3-litre version of the 50B engine was developed for the new formula which began in 1938. The chassis had evolved from the Type 59 unit and the suspension is still non-independent. Note the large and separate all-synchromesh gearbox developed for the 50B engine. The brakes are still mechanically operated.

Work underway on the 3-litre car at Cork in April 1938. This photograph was taken by the famous motoring cartoonist, Russell Brockbank.

Jean-Pierre Wimille, second right, pictured with the fragile 3-litre Bugatti at Cork. The Motor reported that: 'Wimille's Bugatti had been in trouble during its brief practice the day before and he started off handling it as if it were made of glass.' He withdrew on lap 21 with piston trouble.

following lap, the car's engine blew up (thought to have been caused by a broken piston) when Wimille was running third. The race was won by Dreyfus in a 4.5-litre Delahaye.

That year's French Grand Prix was staged at Rheims, rather than Montlhéry, on 3 July and Bugatti made a characteristically last minute entry in the 3-litre. It weighed in at 888kg (1,957lb) which was less than the lightest D type Auto Union at 890kg (1,962lb) and considerably more than the W163 Mercedes which was 978kg (2,156lb),[52] all of which, it will be noted, exceeded the 850kg ceiling. Wimille was placed at the rear of the field but the blue car only lasted for one lap when it withdrew with a damaged oil pipe. This was the last occasion on which a works Bugatti ran in this pre-war event. An exasperated Jean-Pierre left Bugatti and returned to Alfa Romeo.

There can be little doubt that the financial position of Automobiles Bugatti was deteriorating badly by 1938. The railcar contract came to an end that year, car sales were far from buoyant and there were redundancies at Molsheim. Stella Vendromme, who became Mme Tayssèdre in 1935 and a secretary at Bugatti's Paris office, has confirmed[53] the chaotic state of the firm's affairs during this period. Henri Pracht, Ettore's long-suffering company secretary at Molsheim, was regularly on the telephone imploring her to obtain a cheque from Ettore so that the bank could meet the weekly wages bill. It seems that collecting money was her principal activity at this time!

[52] Hodges, *The French Grand Prix*.

[53] Conway, 'Another little bit of history'.

Bankruptcy in the Offing

Against this background, it seems incredible that in 1938 Ettore bought, from the Radziwill family, a Château at Ermenonville to the north of Paris. He did so without telling the long-suffering Pracht who then somehow had to find the money to pay for it. Not that Bugatti ever lived there. Jules Goux, a retired works driver was installed on the premises though L'Ébé occupied it during and, for a time, after the war. Later her sister Lidia, by then the Countess de Boigne – she had married the Count in 1940 – lived there.

At the beginning of 1939 the French government was becoming sufficiently concerned by the state of Bugatti's finances – a state which might have endangered the record-breaking aircraft project – to set down its fears in a report, a copy of which was given to Bugatti. Dated February 1939, it stated that 'the general situation of this manufacture is still grave. He fears bankruptcy which will compromise all our efforts.'[54] Ettore's answer to his difficulties was that he should receive more railcar orders but the document noted that the *Société Nationale des Chemins de Fer* (SNCF) was reluctant to commit itself to purchasing more petrol, as opposed to diesel-fuelled machines. The report concluded: 'It is not possible that we allow this firm to disappear at a time when she can extract herself from the impasse in which she finds herself. Already 450 workmen have been paid off.'

The Last Season

In view of this deteriorating financial situation, the works competition programme was essentially a low key one, with no entries made in front-line Grands Prix. It was, nevertheless, reasonably successful and centred around three cars all of which were driven, with considerable skill in sports car races and hill climbs, by Jean-Pierre Wimille who, at the behest of Jean Bugatti, returned as works driver.

One of the two 57S45, readied for the 1937 French Grand Prix, may have formed the basis of a car which was, effectively, a Type 59 fitted with a 4.5-litre unsupercharged 50BII engine. The bodywork was new, a bulbous two-seater one with cycle wings and fitted with a detachable cowl which incorporated an elaborate grille and low-mounted integral headlamps. Then there was the definitive version of the 4.7-litre *monoposto*, built up around a new chassis frame and belatedly fitted with Lockheed hydraulic brakes. The bodywork was also new with a front grille in the manner of the W125 Mercedes-Benz of the 1937 season. The smallest capacity car of the trio was a supercharged 3.3-litre 57C-based Tank, built in 1939, which was the factory's sole entry in that year's Le Mans 24-hour race.

The Type 59 was run in 4.7-litre unsupercharged form in the Luxembourg Grand Prix in June, which Wimille won. It was entered as a 4.5-litre in the Comminges Grand Prix on 6 August, held less than the month before the outbreak of World War II, and it thus marked the last appearance by a works Bugatti in a significant pre-war event. *Motor Sport* pointed out that the Grand Prix used to be 'quite an important event in France . . . Then, possibly owing to the fact that the place is not far from the Franco–Spanish frontier, on the other side of which the civil war was being waged, the race was allowed to lapse.'[55]

There was, in fact, some doubt whether the race would be run in 1939 after this two-year gap but eventually twenty-three cars entered for the forty-lap, 438km (272-mile) race. Wimille was up against stiff opposition in the shape of two Darracqs and he made a poor start but managed to get past Lebègue, in one of them, and then caught Sommer in the other. A terrific duel ensued and he was re-passed by Sommer who later dropped back. But Lebègue managed to slip past the Bugatti on the final lap and won by a mere ⅖th of a second. The car that Wimille used in this final race of 1939 can today be seen at Mulhouse.

Wimille also had a busy season with the 4.7-litre

---

[54] Conway and Sauzay, *Bugatti Magnum*.

[55] *Motor Sport*, September 1939.

Jean Bugatti talks with Wimille in the 4.7-litre single seater in 1939. Its more enveloping front differs from that of the 3-litre of the previous year while hydraulic brakes are, at last, fitted. The event is the coupe de Paris race at Montlhéry on 7 May.

Wimille won the event and the Agaci Cup with Sommer in a 2.9 Alfa Romeo a close second.

Wimille putting up second fastest time of the day in the 4.7 at La Turbie in 1939.

single-seater which proved to be both fast and reliable. On 7 May at Montlhéry it ran in the *Coupe de Paris* event for racing cars and won by a mere 10 seconds at 137.7kph (85.3mph) from Sommer's 2.9-litre Alfa Romeo. Also campaigning the single-seater in hill climbs, Wimille put up second fastest time of the day at La Turbie. He brought the car to Britain to provide the Bugatti Owners' Club with its one and only works entry for its first international event at Prescott on 30 July. Jean Bugatti was also present to see him put up second fastest time of the day behind Raymond Mays' ERA, with a climb of 46.69 seconds which was, in its own right, an improvement on the 47.85 seconds set by Abécassis in his Alta in the previous year.

A Return to Le Mans

This visit to Britain occurred only six weeks after the Le Mans race in which, it will be recalled, Bugatti ran a 57C. The firm had not entered the 1938 event, Ettore arguing that if the victor exceeded the distance covered by the winning Tank in 1937, the works would run again in 1939. As it happened, the Delahaye that won in 1938 covered 3,180km (1,976 miles) in the 24 hours. This was 108km (67 miles) less than the 57G's 1937 record distance, but Jean managed to convince his father of the worth of a Bugatti presence in the 1939 event. Probably because of his parlous economic state, Ettore stipulated that only one car rather than two, be prepared.

Therefore on 5 January 1939, Robert Aumaître, chief mechanic in the racing department, was 'told to go to the assembly shop, to pick up a 57C chassis from the line, take it to the racing shop and prepare it for Le Mans'.[56] The last and final Tank was therefore supercharged, as opposed to its predecessors which were unblown. It was also longer as it retained the 3,300mm (10ft 10in) of the standard 57 while its lines varied somewhat from those built in 1936. Mechanical modifications were limited to the fitment of a higher 14/42 rear axle ratio, giving 228kph (142mph) at 5,000rpm and the engine which was, apparently, fitted with a different carburettor. It was said to develop 200bhp, which suggests that it was prepared to SC specifications. But compared with the two 57Gs that had run at Le Mans in 1937, fiscal constraints resulted in this relatively standard car being entered in 1939.

The 57C was to be driven by Wimille and Veyron and the event was scheduled to begin at 4 o'clock in the afternoon on Saturday 18 June though the Bugatti very nearly did not appear. On the first day of practice, everything went well. But during the second one, at 12.30 on Thursday night, Wimille arrived on foot at the pits with news that the car's engine had seized solid.

Jean favoured withdrawing and returning instead in 1940. Aumaître thought otherwise. He suggested that the works be telephoned and a mechanic, complete with a new set of pistons, be put on the Paris Railcar at Molsheim and then be met by a member of the racing team at the Gare de l'Est. In the meantime, the block would be repaired in the town of Le Mans. The plan worked. The engine was successfully rebuilt, and at 7 o'clock on the Saturday morning, Aumaître 'was on the road to Tours with the car running fine; no problems'.[57] Veyron then put in a further 100 or 200km (60 or 125 miles) and the 57C was ready for the 4 o'clock start.

Gérard's Delage went into the lead early on and Wimille was in fourth position though, by the eighteenth hour, he was running in second place but three laps behind the leader. Although the Bugatti was slowly gaining on the Delage, it seemed unlikely that there would be sufficient time for it to win, until soon after midday on the Sunday. It was then that the Delage pulled into the pits, the plugs were changed and it was off again but was back there on the following lap.

As *Motor Sport* reported: 'The Bugatti had already gained on one of its laps and now it was due round again. The big blue car came into sight,

---

[56] Robert Aumaître, 'The Le Mans 1939 Story' in *Bugantics*, vol. 48, no. 1 (spring 1985).

[57] Aumaître, *op. cit.*

Bugatti entered a single Tank-bodied 57C for Le Mans in 1939 and won the final race before the outbreak of the Second World War. Wimille (left) and Veyron are pictured after their triumph. This was the car in which Jean Bugatti was killed just eight weeks later.

swished up the straight and past the stationary Delage to take the lead.'[58] The 57C went on to win at 139.77kph (86.85mph), which was not only an improvement on the 1937 speed, but the car went further and covered a record 3354.65km (2084.54 miles) over the 24 hours. This gave Bugatti his second Le Mans victory and France's fifth of the inter-war years.

A Triumph for Jean

It is difficult to think of a greater endorsement of thirty-year-old Jean Bugatti's stewardship of the Molsheim factory than this Le Mans triumph. Since the departure, in 1935, of his friend Meo Costantini, who had hitherto been responsible for managing the works racing team, Jean had taken over that job as well as overseeing the design of the road cars, in addition to the demanding role of chief stylist. He also had to constantly travel to Paris to liaise with his father, who was always ready to veto a design initiative, particularly if he considered it to be too adventurous.

It was with bankruptcy looming that, in 1939, the 57's much needed Type 64 successor was nearing completion for it was destined to be unveiled at that year's Paris Salon. Inevitably, it was universally known at the works as '*la voiture de Monsieur Jean*'.

In 1934, an experimental saloon had been built, in some secrecy, by Jean and Costantini, and fitted with a badgeless oval radiator of the type which Ettore had vetoed on the 57. Similarly it employed the transverse leaf independent front suspension and alloy wheels, both akin to those used on the 53 four-wheel drive car. In an effort to quieten its Type 57 engine, the noisy timing gears were replaced by an inverted tooth chain. The car became the service vehicle for the racing team and it was only driven by Jean, Costantini and chief racing mechanic Robert Aumaître. He remembers that 'As the Patron must not hear of it, we named it *Crème de Menthe* so that we could even speak of it in front of him without him understanding anything.'[59]

The car was, by all accounts, faster than the 57 and made the run between Molsheim and the Paris' Prince of Wales Hotel in a mere four hours. Aumaître says that, from the beginning of 1937, 'M. Jean tested it mercilessly... Several times we did Molsheim to Strasburg in third gear at 5,800 rpm, there and back, thus twin 25km (15 miles) without trouble. It was unbreakable, agreeable to drive, silent at all revolutions, with first class braking. It had everything to please, even the name.'[60]

Then after three years and 250,000km (155,350 miles), the inevitable happened and the engine's timing chain broke, the valves touched and the engine was wrecked. It was dismantled in September 1937, just prior to that year's Paris Salon.

---

[58] *Motor Sport*, July 1939.

[59] Robert Braunschweig, ' "Crème de Menthe", an unknown Bugatti prototype' in *Bugantics*, vol. 43, no. 4 (winter 1980).

[60] *Ibid.*

The 57's Successor

Sadly, the Type 64, which was destined for the 1940 season, did not incorporate *Crème de Menthe* front suspension, it was no doubt banned by Ettore, and instead, retained the 57's rigid front axle and suspension. The increasingly outdated 3.3-litre fixed head straight-eight engine was effectively carried over though the timing chain had proved its worth and this took the place of the gears. For its operation, the teeth on the rear of the crankshaft were retained and meshed with an idler which also drove the chain.

The Bugatti-built box of the 57 was dispensed with and replaced with a Cotal epicyclic, electrically activated gearbox, which had been latterly offered on the 57. The chassis differed radically from the 57's conventional one. It was made of Duralumin and the box section frame was created by using the U-shaped members, one inside the other, and then riveted in place. Hydraulic brakes were perpetuated.

Two examples were prepared for the 1939 Paris Salon, a chassis and a two-door saloon. The radiator that Jean had introduced on the 57S was continued and the handsome two-door body was, in my opinion, very much in the same spirit as a celebrated British Grand Tourer, the Bentley 4.25-litre endowed with a handsome two-door body by Pourtout of Paris in 1938. It is thought that, in all, four Type 64 chassis were built. However, the model never entered production because of the outbreak of World War II and the Paris show was cancelled for the same reason. But by that time, Jean Bugatti, its creator, was dead.

Tragedy at Molsheim

It was on Friday 11 August, with the Bugatti factory closed for the annual holiday, that the racing department was preparing the winning Le Mans 57C. Jean-Pierre Wimille was to drive it at a Grand Prix to be held at the popular seaside resort of La Baule, on the Atlantic coast of France.

Robert Aumaître was working on the Tank and had already tested it to the east of Molsheim on a straight section of public road between Duttlenheim and Entzheim, *en route* to Strasburg, which the works used for such purposes. Afterwards Jean spoke to him to ascertain whether the car had exceeded 200kph (124mph). But Aumaître considered it was too dangerous to drive at that speed as 'the farmers were harvesting and there were many farm

Intended as the long-awaited replacement for the Type 57s, the Type 64, which perpetuated the 57S type radiator, was intended as the Bugatti for 1940. It was to have been powered by a Type 57 engine with quieter chain-driven camshafts and a Cotal gearbox was standardized. The chassis was a weight-saving duralumin. This car is now at Mulhouse.

carts on the road.'[61] Therefore, Jean decided to wait until there was little traffic and run the car at 9 o'clock at night, despite the fact that he was due to leave, with Noël Domboy, for Romania to talk about railcars, in the following morning.

He asked Aumaître to take the car out to a section of road between Duppigheim and Entzheim, warm it up and he would join him in his Type 57. Jean had dinner with his family at Molsheim and L'Ébé Bugatti recalled that he left them, saying, ' "I'll be back in 15 minutes" in answer to our warnings to take care.'[62]

Jean met up with Aumaître and the car. He had previously left his sixteen-year-old brother, Roland, at Duttlenheim, the Molsheim stable lad at the Duppigheim road and mechanic, Lucien Wurmser, at the Esso Station at Entzheim. Their brief was to police the road and prevent anyone straying on to it while the tests were under way.

Jean took the Tank towards Duppigheim but turned back because the car's dashboard lamp was not connected and he could not read the instruments. This fault had not been apparent in daylight, so Jean drove the car with Aumaître crouched alongside him to illuminate the rev counter by the light of a cigarette. They did at least four double runs. Then they stopped by the car for a cigarette and Aumaître took the wheel to take it back to the factory.

'Then he, almost at his car, suddenly made a sign to let him once more take the wheel.' Jean wanted to drive the car himself so that Aumaître could see how it rode at 200kph over a small nearby bump in the road. 'He went off turning in front of his brother Roland at Duttlenheim and, on the way back passed in front of me flat out, to crash into a tree 200 metres further on. The fuel tank having exploded, set fire to a corn mill and a tree; jumping in the car I arrived at the scene, searching for M. Jean, whom I found in the field on the left 20 metres from the roadway.'[63] Robert immediately drove with him to the hospital in Strasburg where he and a Sister Angélique got him to the operating theatre. 'It was here that an intern told me that M. Jean was dead.' He had, in fact, been killed instantaneously on hitting the tree.

After he had conveyed the terrible news to the Bugatti family at Molsheim, Lidia Bugatti asked Aumaître to telephone her father, who was staying in Belgium at the Château de Laekin, near Brussels. He did this at one o'clock in the morning and Ettore left immediately to be driven through the night in the *Coupé Napoléon* Royale. He arrived at the hospital between 6.30 and 7 o'clock on the Saturday morning.

It later transpired that Jean had crashed after avoiding a cyclist, nineteen-year-old Joseph Metz, who had somehow managed to get on to the road. According to a local Strasburg newspaper, published two days after the accident, on 13 August, he suffered 'two fingers broken and several other wounds not serious. His life was not in danger. The bicycle naturally was completely destroyed.' L'Ébé Bugatti says that the young man 'was haunted by the thought of having caused [Jean's] death and, three years later, he committed suicide'.[64]

It only emerged relatively recently[65] that Jean had previously been prosecuted for being involved in an accident, probably as a result of him driving too fast, with a cyclist who had died, further down the road at Lingolsheim on the outskirts of Strasburg. As a result, Jean's licence was suspended for one month. This may explain why, when confronted with a similar situation on the night of 11 August, he could have overreacted with such fatal consequences.

There is one further terrible element of irony to this tragedy. The La Baule Grand Prix, where Wimille was to have driven the 57C which Jean was testing when he was killed, was cancelled. It was to have been held on 3 September, and that was the day on which France and Britain declared war on Germany. The Second World War had begun.

---

[61] Robert Aumaître, 'The night of August 11, 1939' in *Bugantics*, vol. 46, no. 3 (autumn 1983).

[62] L'Ébé Bugatti, *The Bugatti Story*.

[63] Aumaître, *op. cit.*

[64] L'Ébé Bugatti, *op. cit.*

[65] H. G. Conway, 'Jean Bugatti' in *Bugantics*, vol. 46, no. 3 (autumn 1983).

11
Much More than a Car Factory

'One cannot write of Ettore Bugatti and confine oneself to this . . . aspect of what can properly be called his genius. In order to know the whole man, he had to be seen at Molsheim . . .'

The Duc de Gramont

Bugatti car manufacture lasted for a mere twenty-six years, from 1910 until 1914 and 1919 to 1939. During this period, production amounted to about 7,800 cars, which was around the same number of automobiles Henry Ford built in one and a half day's output at the height of the ubiquitous Model T's popularity. But Bugatti kept his business small because he wanted it that way and his works at Molsheim, in that quiet corner of Alsace, was much more than just a car factory. It was as if a mirror had been held up to reflect Ettore's interests and preoccupations where the apparently divergent worlds of engineering and a country estate were able to peacefully coexist.

For not only were Bugatti cars produced there, it was also home for their creator and his family; a small community where he could breed dogs and stable horses and pursue whatever interest took his fancy, sometimes, it should be said, to the exasperation of his closest colleagues. Little wonder that L'Ébé Bugatti has recalled that, 'a customer [who] came to take delivery of a car or to pick up a spare part . . . got the impression that he had discovered . . . the little fife of an Italian *signore* of the Renaissance who had strayed into the industrial age.'[1]

The Golden Age at Molsheim roughly followed the years of the arrival and ascendancy of the Type 35 in the mid-1920s and gradually declined from the early 1930s when Ettore began to spend less time there and car manufacture began to play second fiddle to railcar manufacture. Ettore's departure from Molsheim to Paris in 1936, and the labour problems that haunted industrial France at this time, effectively spelt an end to this automotive Arcadia.

It will be recalled that it was in December 1909, that Bugatti, accompanied by Baron de Vizcaya and Ernest Friderich, first visited the old dye works at Molsheim. Bugatti clearly liked what he saw and Friderich says that: 'It was at that old factory which was rented from the Frauleins Geisser for 5,000 marks [£245] per annum, that we installed our new plant.'[2]

The estate was located on the Molsheim to Obernai road about 1.6km (1 mile) south of the town. There was a substantial villa and wooded parkland, which fell within the parish of Dorlisheim, roughly bisected by a small stream which was, in fact, the infant River Bruche. Nearby was its old water mill. The park, which also contained a small circular pavilion close to the river, was entered through a pair of high wooden gates and surrounded by a fence of white palings but those nearest to the house disappeared over the years though their rendered pillars remained.

---

[1] L'Ébé Bugatti, *The Bugatti Story*.

[2] Friderich, 'How the firm of Bugatti was born'.

The Bugatti factory pictured prior to the building of the railcar shop in the open space in the right foreground in 1933. The town of Molsheim is to the left of the picture and the road runs to Obernai to the right. The roof of the Italianate gatehouse designed by Bugatti can be seen centre right, with the line of bicycle sheds for the firm's employees beyond. The wooded area on the extreme right of the picture is the edge of the villa's grounds. The chimney in the left background was for the boilers of the factory's electrical supply. The long building beyond the others, running parallel with the road, was the machine shop built after the First World War.

In these pre-war days and, indeed, in the early inter-war years, visitors arriving at the premises of Automobiles Bugatti entered the estate through a gateway, with the villa on their left and a range of buildings to the right. These were largely screened from the villa's occupants by the trees and this scatter of mostly brick structures, with tiled roofs, extended beyond the stream which was crossed by a small wooden bridge. Many of these had formed the old dye works, a wholly appropriate industry for the location because the manufacture of cotton was one of Alsace's principal industries. But during these early years they were used for storage, offices and car assembly.

W F Bradley has written that 'it never entered Bugatti's mind that he would erect a factory in the plains and build a house for himself and his family at St Odile or at some choice site in the Vosges mountains. His home was his factory; his factory was his home.'[3] When Bugatti first arrived at Molsheim, he was only able to rent one half of the villa. Presumably the Geisser sisters occupied the other portion but Ettore was to subsequently buy the rest of the house and, indeed, the entire estate. Then the drawing office staff, who had previously occupied one of the small separate buildings beyond the stream would be located, for a time, on the ground floor of the house itself.

When he began to build his own cars in Milan, Bugatti produced his own drawings, such as they were. Then when he moved to De Dietrich in 1902 he would probably have had the services of a drawing office to interpret his sketches and introduce the necessary details and dimensions for production. When he left that company in 1904 and became an independent engineer in Cologne, he employed a few draughtsmen, there were three of them by 1907, and some moved with him to Deutz and then on to Molsheim.

In addition to drawings he made on large sheets of good quality 1 metre (3.2 feet) wide cartridge paper, L'Ébé Bugatti recalls that her father's 'brain was always at work; thoughts and ideas which came to him through little details he had noticed during the day were jotted down on the backs of envelopes or scraps of paper, and these accumulated in his pockets – miniature archives, yet how important!' Draughtsmen would sometimes take weeks to interpret them into workable drawings.

As will have already been apparent, Bugatti had received no technical education. For his part, Jean-Albert Grégoire has said that he believed that the ideal engineer should combine a scientific culture with an artistic sense though:

> Bugatti was an artist pure and simple; his only scientific knowledge stemmed from ever growing experience plus a natural mechanical bent supported by the gift of observation. He did not believe in calculations, formulae and principles. He joked about pages covered with figures and integral calculations, which he called the sound holes of a violin. Fortunately he was wise enough to surround himself with engineers of talent whom he paid very well, while demanding of them complete anonymity.[4]

Like many self-taught engineers, Bugatti did not 'entirely approve of the technical training in state schools, believing it to be based too much on established principles and set rules, thereby curbing the creative imagination of young craftsmen'.[5] Nevertheless, the head of the design office, Felix Kortz, had gained his technical education in Switzerland, at the Zurich Polytechnic. This may be a reason for:

> . . . the relative complexity of [his] designs, which if stemming from Ettore's scheme layouts, resulted in detail drawings of cylinder block or complex engine castings covered with miniscule dimensions, delicately cored water pasages with walls of 5 or even 4mm thickness . . . and all those things that designers are told not to do to give the production shop a chance.[6]

Kortz had worked with Ettore at Deutz and, from there, followed him to Molsheim. Not that he had been given the title of chief draughtsman. There were no such formalities at the factory, apart from Ettore himself who was *Le Patron*. Kortz is thought to have remained in Alsace during World War I and rejoined Bugatti in 1919. His career with Ettore would, alas, be cut short in 1927 when he was killed in a car crash.

That the Bugatti works had, in these pre-World War I days, far more the appearance of a small estate than a car factory is echoed by account of a visit made by Louis Charavel who was later to race Bugattis in the 1920s under the pseudonym of 'Sabipa'. He bought an eight-valve Bugatti but the rear axle was damaged in an accident which necessitated a journey to Molsheim. When he arrived he

---

[3] Bradley, *Ettore Bugatti*.

[4] Grégoire, *Best Wheel Forward*.

[5] L'Ébé Bugatti, *The Bugatti Story*.

[6] Conway and Sauzay, *Bugatti Magnum*.

was bemused to find that all he could see at first 'were some stables, close to a country house. I was only half mistaken. The stables were being used as a workshop and, above them, in the loft, was an office reached by a short ladder.'[7]

This was the eyrie of Henri Pracht, Ettore's long-serving and prudent Alsatian accountant cum company secretary, and Charavel was just starting to explain the purpose of his visit when 'Bugatti came up. He was wearing a colonial helmet and a well fitting cream silk jacket with a blue border. But it was his shoes I noticed most – to my amazement, they had toes as gloves have fingers. He saw me staring at them, and said with a laugh, "They're much more comfortable like this. After all, when you want some gloves, you don't go and buy mittens. Why shouldn't it be the same with shoes?" '

L'Ébé Bugatti has explained that this distinctive footwear, which hailed from Germany, had 'divisions inside each toe cap for each toe and this was shown on the outside by the seams. They could only be worn, of course, with specially made socks.' Ettore then showed a no doubt still slightly bemused Charavel around. 'He took me to see the workshop where some thirty men were employed. Except for the cylinders which were made for him somewhere in the Cévennes, everything else was produced in his workshop.'

It was Bugatti's appearance that also made an impression on Roland Garros, on his first visit to Molsheim. His initial emotion was one of surprise. On passing through the gate, 'instead of the usual comings and goings of a car factory, even a small one, he saw Ettore dressed for riding, with a Tyrolean hat walking a small pony round the courtyard.'[8]

The works changed little until the outbreak of war and, as has already been recounted, Ettore and his family were absent from Molsheim for over four years, from August 1914 until January 1919. During the 1920s, as the business prospered, more buildings were constructed to cope with the extra production and others were built which were intended to increase the self-sufficiency of the works. These were mostly established beyond the River Bruche and allowed those structures that had originally served for car manufacture and assembly to revert to their original function as stables. There always seemed to be some building project underway at Molsheim in the 1920s.

Immediately after the war, the 16-valve cars were assembled in one of the original buildings beyond the stream and adjoining the park which was called *Le Fromage* because the method of production adopted resembled that of a circular cheese. The men, each holding a component part of the car, stood in a circle with a chassis frame at its centre. The individual then attached 'his' part to the frame, so building up a complete running chassis. It is not known whether this assembly method was perpetuated but it would appear to be the closest that Bugatti ever got to adopting mass production methods![9]

By the end of 1921 a new 100-metre (328ft) long building had been built well beyond the park which dwarfed all the existing ones. Single storied and illuminated by plenty of steel-framed windows, it served a multiplicity of purposes. Car production was transferred there, as was the drawing office. The same went for the racing department which was moved from a nearby L-shaped building. This extension of manufacturing facilities necessitated the construction of a secondary, and larger, entrance to the factory on to the Molsheim road.

Later a foundry was built beyond the stable block and, subsequently, towards the end of the decade, the bodyshop was expanded some distance behind what had become the '100-metre' machine shop.

This new building also served, for a period of several months, as home for Ettore and his family while the villa was being altered and redecorated. L'Ébé Bugatti says that, following the family's

---

[7] L'Ébé Bugatti, *op. cit.*

[8] *Ibid.*

[9] E Hallums, *The Quality of Work and the Quality of Art, A Study of Bugatti* (Mithras Press, 1979).
   There is a fascinating description of the early buildings at Molsheim, based on the recollections of two former Bugatti mechanics, Xavier Rohfritsch and George Lutz.

Ettore and his wife, Barbara, in the Type Garros, chassis 471.

return to Molsheim, 'the house was damaged and pillaged like the factory' and her mother's 'help and support were of the utmost value in restoring and modernizing the family home'. Her daughter says of Barbara Bugatti, of whom so little is known:

> She brought graciousness and gaiety to Molsheim, and her goodness, elegance, beauty and wit were like rays of sunshine to this lonesome corner of Alsace. My father used to say that she was always his greatest supporter and best counsellor . . . it was her great tact and encouragement which aided him in overcoming, with dignity and courage, the difficult periods with which his life was interspersed.

A member of the Milan aristocracy, Mme Bugatti was proud of her origins and Bradley recorded that 'There was hardly a workman's house with which she was unfamiliar . . . It was a cheering sight to observe this Italian lady, who alone of the entire family had preserved her Italian characteristics, moving among these hard headed Alsatians.'[10]

One of her initiatives at the villa was to create a bar decorated with panels which were inlaid between the shelves and created by an Alsatian artist named Charles Spindler. 'The bar itself was mahogany, and on it stood cocktail shakers, tankards, and a book of recipes of potent cocktails which were a joy to our guests.'

Adjoining the bar was a winter garden with glass walls 'giving views of trees and the purple heights of the distant Vosges mountains'. It also contained a palm tree with large cacti in each corner. In the centre of this indoor garden was a white marble statue about 900mm (3ft) high by Rembrandt Bugatti of a nude woman seated on a rock. This was a rare execution by a sculptor whose work was invariably of animal subjects cast in bronze. There were, however, three of these set against the glass walls: *Antwerp horsefair*, *Sick horses coming up from the mine*[11] and *The wounded hine*. The mosaic floor was covered with Persian carpets and a flower bed which was tended throughout the year. Furniture was mostly restricted to settees and cane chairs where people could sit and talk while sampling the pleasures of the nearby bar.

[10] Bradley, *Ettore Bugatti*.

[11] *The Antwerp horsefair* was exhibited at the Salon National in 1905 and survives in a private collection in England. *Sick horses coming up from the mine* may be another name for *'Les Vieux' Old Mine Horses* which was in the possession of the Bugatti family for many years. According to Mary Harvey, 'The depression which gradually overtook [Rembrandt's] life is already evident in this work . . . The sight of old Irish horses being taken off the ships at Antwerp to be sold for meat affected Bugatti deeply and his pity for these wretched creatures, barely able to stand, awaiting their fate, is conveyed with masterful understatement.' It is now in a private French collection.

The Bugatti villa at Molsheim viewed end on with the exterior of the winter garden readily apparent.

The villa's dining room was large with five tall windows along one side and there were two doors at one end. Ettore decided that the room was too dark and had the factory blacksmiths create a double chandelier for it to, of course, his own design. Indeed, Bugatti was 'always wanting to change the shape of household things which most of us accept without question'.[12] This included the beds which he designed and were specially made by the works carpenters. His ingenuity extended to the kitchens and he may even have created a machine out of spare parts from his cars for making pasta. The dough was compressed in a cylinder by a Royale piston, operated by a Type 46 steering wheel and Type 35 steering gear, gear while the cutting blades could be alternated to produce such Italian favourites as spaghetti, macaroni and tagliatelli.

Ettore's long-time friend, the Duc de Gramont, was a regular visitor to Molsheim and has recalled that Ettore 'kept a very good table and had several Italian cooks, for he liked Italian cooking best. All the ingredients that went into the dishes were specially chosen and came from all parts of France and Europe. He expected to be able to lunch or dine at any hour of the day.[13]

For W F Bradley, 'one of the delights of a visit to Molsheim was the gathering for lunch . . . uniting the male members of the family only. In addition to Ettore Bugatti and his son [Jean], Costantini would frequently join with such guests as happened to be passing through.' The rule was that the table should be of the best and would be complemented by 'the delicate wines of the region, supplemented by the liquers distilled on the premises'.[14] The latter was *mirabelle*, a plum brandy made by Ettore in the small pavilion on the park, and made of the best fruit which was picked for the purpose rather than having been bruised in falling from the tree.

A spirited conversation would ensue and although the subject of cars was invariably raised, the talk also embraced a diversity of topics. This might include fishing, and 'the latest tackle Jean had brought from London. To internal decorations and the disposal of some of Grandpa Carlo's masterpieces; to the arrival of a new hunter, to the possibilities of improving speed on water; to the

[12] Bradley, *Ettore Bugatti*.

[13] L'Ébé Bugatti, *The Bugatti Story*.

[14] Bradley, *Ettore Bugatti*.

quantity and quality of the kirsh then being produced by the distillery.'

The villa's kitchen was located in the basement, with its distinctive oval windows, and in the 1920s was converted, in part, into a cinema. Seats were arranged in groups with a few red upholstered boxes for special guests. There was also a glass-fronted area adjoining the projection room for members of the workforce. A programme of films, courtesy of the Strasburg Cinema, was shown every week, first silent ones though later the talkies arrived.

Initially when favoured customers visited Molsheim to collect their cars, or just to talk to *Le Patron*, he would entertain them at the villa. One such visitor, who was a guest of Ettore's in late 1923, was Raymond Mays, who had established a reputation for himself over the previous two seasons with his Brescia which he named *Cordon Rouge*. He was accompanied by Humphrey Cook in the latter's 30/98 Vauxhall and Bugatti greeted them cordially and ushered his English visitors into the house. But Mays was somewhat alarmed when 'suddenly, as if through a secret door, he disappeared from view. Then we heard a slight noise as if an electric button had been pressed, and silently one side of the room opened up to reveal a beautifully equipped bar with Ettore himself standing behind it.'[15]

As the decade progressed, more people made the pilgrimage to Molsheim, but, according to W F Bradley, accommodation there, which meant the Hôtel de la Gare and Hôtel Heim, run by Mme Heim and her son, was 'meagre and Strasburg being somewhat distant,' Ettore therefore decided to establish a small hostel on the road into the town. This would be for the use of his customers where they could stay free of charge. It was a charming two-storied turreted house and, appropriately, called the *Hostellerie du Pur Sang*.

There was a dining room, bar but only four bedrooms and Mme Bugatti was responsible for arranging and furnishing it. When the hotel opened, L'Ébé Bugatti became its supervisor and was responsible 'for all supplies of food and wine and maintaining the same standard of comfort as in her own home'. The day-to-day running of the *Pur Sang* was the responsibility of Fernand Hoffmann, who was also an excellent chef, and his wife.

It soon became apparent that there was insufficient room so further beds were provided in accommodation that had originally been designed for hens! Apparently Ettore thought that he would provide his guests with fresh eggs. Typically elaborate hen houses were constructed but then he decided against the idea . . .

Bradley has said that Bugatti 'spoke pure French, without any trace of Italian accent, nor was there anything to suggest that he had spent a number of years in Germany, or that he had been influenced by the strong Alsatian *patois*'. Ettore's sartorial preferences have already been mentioned but Bradley says that, if a car needed necessary work, 'his cream gloves, his immaculate dust coat, his well cut riding breeches and his yellow jersey did not prevent him dropping on his knees or crawling beneath the machine to assure himself that everything was as he desired it.'

Jean-Albert Grégoire had given us this description of Bugatti in the 1920s: 'With his loud voice, high colour and bubbling vitality, a light brown bowler at the back of his head, he looked rather like one of the "horsey" set that had stumbled into the car world.'[16]

Bugatti's passion for horses was such that the Duc de Gramont says that 'I think that he liked them better than he did the cars.' Because of this commitment, 'The stables were perhaps the most extraordinary sight at Molsheim; they were kept as spick and span as the finest drawing room . . . The stalls were impeccable and were divided by plaited straw matting.' Ettore would entertain his guests by taking them to the nearby riding ring while stable lads brought the best horses forward and walked them round. This display would always be completed with 'four Shetland ponies galloping round the sawdust ring like clowns.'[17]

[15] Raymond Mays, *Split Seconds* (Foulis, 1951).

[16] Grégoire, *Best Wheel Forward*.

[17] L'Ébé Bugatti, *The Bugatti Story*.

Ettore in his element at Molsheim on one of his many thoroughbred horses.

L'Ébé Bugatti has said that there were about fifteen horses stabled at Molsheim. Ettore's own mount was a large Irish grey named Brouillard (smoky) and his dedication to all matters equestrian was also reflected by the fact that he also built up an impressive collection of about forty horse-drawn vehicles of various ages. It was a theme echoed in Bugatti's *fiacre* coachwork which was such a distinctive feature of some Bugatti car bodies. He even went to the extent of building his own carriages. There were four in all and one coach, which was incomplete at the time that Bugatti left Molsheim for Paris in 1936, was constructed for him to drive from Alsace, across the Alps, to his Milan birthplace. It fortunately survives, along with examples of harnesses and associated leatherwork produced at Molsheim, which is of the finest possible quality.

Then there was Totosche, a Sicilian donkey and a present from Vincenzo Florio in 1929, after Bugatti had won the Targa Florio for five years in succession. Bradley says that 'It is typical of Florio that he should have sent the slowest animal in Sicily to the maker of the fastest vehicles . . . seen there.' The donkey came complete with a hand-painted Sicilian peasant cart and seems to have been one of the few animals to have had the run of the factory and its grounds.

Bugatti shared his passion for horses with Lidia, his second daughter, who rode regularly. She also had some artistic ability and was responsible for producing drawings which were used to illustrate some of the all too few advertisements for Bugatti cars. Inevitably, the only theme was an equestrian one. She was rather more outgoing than her sister L'Ébé, and often entertained Ettore's guests at Molsheim. L'Ébé herself never married. She was close to her mother and although there was some talk about an engagement to a family friend, a Jewish banker, Ettore vetoed the liaison.

Jean was growing to manhood in the 1920s though Bradley says that 'as soon as he was old enough to work, he was sent into the factory and was expected to report at 7 a.m. together with the other workmen and to share their life in every respect.' What he never received was any formalized engineering education as Ettore no doubt believed that this would stifle his undoubted and precocious artistic talent. Jean was thus given a similar practical training as his father while getting to know the workforce that one day he would have to manage.

The baby of the family was Roland, born in 1922. His lessons came from an English governess named Esmee Yeulett, who had gone to France in 1920 to learn the language, and had been engaged by the Cozette family at Tours. There were motoring

Bugatti's love of horses was enthusiastically endorsed by Lidia, his younger daughter.

Lidia once again on horseback with Jean Bugatti and a Type 40A for which he had designed the coachwork.

Young Roland Bugatti, complete with bicycle in the 1920s.

assocations here because M. Cozette's brother made carburettors and superchargers. Esmee remained at the villa for seven years and left in 1936 when her charge was fourteen. Roland thus became the only member of the Bugatti family to be truly bilingual and would be able to switch, with ease, from French to English.

Ettore seems to have had a distant and rather formalized relationship with his children which probably echoed his own upbringing. This is shown in letters exchanged between him and Jean, in which he adopted the formal *vous* to address his eldest son and the latter responded by writing to his father as *Monsieur*. Roland seems to have had a similarly detached relationship. Griffith Borgeson, who knew him for nine years, has written that he would refer to Ettore 'as *Le Patron*, or The Boss, depending whether he was speaking French or English. Occasionally he would use *mon père* or "my father". Perhaps only two or three times . . .

did Roland refer to *Papa*. "Daddy" never.'[18] Reputedly, the only person outside his family to address Ettore with the more familiar *tu* was his unpaid racing manager and friend, Meo Costantini, who had taken up permanent residence at the *Pur Sang*.

A great social occasion at Molsheim was *Le Shoot*. Hares, rabbits and pheasants would be hunted on the large plain to the north of the factory bounded by the Altorf woods to the south which were about 19km (12 miles) away. Despite the hunt being such a regular feature of the Bugatti family's life, L'Ébé says that her father was 'not really keen on going shooting, neither was Jean; but they liked the friendly atmosphere and the party occasions'. Ettore would follow the others, some distance behind, on horseback. Then would come a mule fitted with a special harness carrying his four different guns.

When the shoot was in progress, its participants were serviced by an extraordinary mobile dining room which could accommodate twenty to twenty-five people. This vehicle had originally been conceived by Ettore for use by his family and friends on the occasion of the Type 35's debut at the 1924 French Grand Prix. It was built at Molsheim on a four-wheeled chassis, double wires at the rear and singles at the front, and was towed to the proceedings behind a lorry. Always known at the factory as the *roulette* (caravan), it was metal-framed and built of oak.

Inside, the walls were decorated with marquetry panels made by local artist, Charles Spindler, whose work, it will be recalled, also featured in the villa in the shape of the celebrated bar. The principal one showed a flight of storks while other smaller scenes displayed aspects of Alsatian life. The caravan was self-contained with its own electricity supply and running water was provided by a tank located on the roof.

After lunch, during which liberal quantities of wine were consumed, the shoot continued and, in the early evening, the party might repair to the bar of the *Pur Sang* where Fernand would produce fortifying and substantial whiskies and sodas. Then

[18] Borgeson, *Bugatti by Borgeson*.

The Hostellerie du Pur Sang at Molsheim where guests at the factory stayed. The wooden building for the mechanics' accommodation can just be seen through the trees.

there might be the sound of breaking crockery from the adjoining room as a result of Jean Bugatti having hurled an old shell case at a nearby dresser[19] and this would signal the start of an orchestrated 'smash up' in which items of furniture, crockery and glasses would be thrown about to the detriment of the windows and chandelier. After about two hours of this mayhem, the destruction would cease. Fernand then tidied up the place until the next time . . .

Some dogs would accompany the hunt and at one time there were no less than sixty at Molsheim with two men employed to look after them. Typically, Ettore conceived a special machine so that four could apparently be exercised simultaneously. The canine response to the constraints of such a device is not recorded! Bugatti also briefly attempted to breed wire-haired terriers and even brought a prize-winning dog named Epping Forest from Britain for the purpose.

One visitor to Molsheim, who had an unforgettable encounter with Ettore's dogs, was Henry de Hane Segrave. He was at the factory in 1920 to collect what he believed to be the 16-valve car in which Friderich had won that year's *voiturette* race at Le Mans. During his weekend stay, Ettore showed him around and Segrave soon heard the barking of dogs so 'asked to see the kennels, but his host was dubious, *Ils sont très féroces mes chiens* [my dogs are very ferocious] he said, but de Hane, fond of animals and reputed to "have a way with dogs" begged to be shown them.'

The two approached the kennels but, on nearing them, Bugatti shouted a warning 'at which several garden hands swiftly approached, armed with pitchforks, and surrounded de Hane! At the same moment several large hounds of the Dobermann Pinscher species . . . rushed up with much baying and growling, their obvious desire to tear de Hane to pieces being frustrated only by the ring of pitchforks around him. Bugatti smiled sardonically, then said, "Come Monsieur Segrave, we will now see the car", and de Hane left the stables with relief.'[20]

As the business prospered, Ettore would enhance many of his factory buildings with an Italianate air. One such was a colonnaded gatehouse to which visitors would first report to be issued with their special passes to enter the works. Perhaps the best known of them was a museum

---

[19] This was the experience of Col. Eric Giles and his brother when they attended a shoot at Molsheim in 1935. They were in the forefront of the Bugatti movement in Britain and the Bugatti Owners' Club. See Conway, *Bugatti Magnum*.

[20] Cyril Posthumus, *Sir Henry Segrave*, (Batsford, 1961).

A Type 43 as announced in 1927 pictured in front of the small museum at Molsheim dedicated to Rembrandt Bugatti's sculpture. A typical factory door, complete with its distinctive hinges, can be seen on the left.

which displayed his brother Rembrandt's sculptures in its windows and was used as a background for photographs of Bugatti cars. It nominally had such a role and was also used for a variety of purposes, including, on one occasion, a leather shop. All the important buildings at the factory were fitted with purpose-made doors of varnished oak, hung with distinctive bronze hinges and fitted with large finger plates. When the French journalist, Maurice Phillipe, visited Molsheim in the 1920s, he recorded that 'there was not a trace of finger marks on the copper door plate. "Locks made by Bugatti", the proprietor pointed out.'[21] And Ettore had the key.

It was the factory's extraordinary cleanliness that also impressed visitors. Phillipe recorded that 'an employee did nothing else but keep the paths and the workshop floors clean.' It was this aspect of the works that so impressed Bess Duller, the wife of George Duller who combined the career of a successful hurdling jockey with that of a racing driver. When he took his Bugatti back to Molsheim for a tune-up his wife went with him. 'What a factory! Do you know you could have eaten off those floors, I've never seen anything so perfect in my life', she remembered.[22]

Once, in an effort to keep costs down, the prudent Henri Pracht made the mistake of reducing

[21] L'Ébé Bugatti, *The Bugatti Story*.

[22] Jonathan Wood, 'Mrs Duller' in *Classic Cars*, December 1976.

the amount of cleaning rags used but soon reverted to the number previously allocated. Pracht informed Phillipe, 'The workmen know what M. Bugatti is like, and they tell him they haven't enough rags to keep everything as clean as it ought to be. And that's a matter where M. Bugatti always gives way!'

Ettore can be seen to have played hard but he was also an extremely conscientious worker. He began at 8 o'clock in the morning though sometimes did not finish until 8 or 10 in the evening. His daughter L'Ébé has recalled that 'He never used a slide rule, and only rarely a drawing board . . . He sometimes spent many hours in the drawing office at night, for he thought that the silence helped him work out problems which had cropped up during the day; but this sometimes led him far from his starting point.'

He was popular with is workforce and knew many of them by their christian names. However, Noël Domboy has said that 'what annoyed *Le Patron* most was to be misunderstood when giving orders or directives.'[23] He would also never hesitate to discipline an employee whom he considered to be not working properly and on one particular occasion, he witnessed one of his men using a hammer incorrectly. L'Ébé has said that he 'gave the man the rough edge of his tongue . . . His wrath was long remembered, and it had good effect.'

Domboy has also reflected that, at the factory, Bugatti's 'changing moods were legendary and, when he was about to vent his displeasure, frequently justified, the staff would disappear on a variety of pretexts. Only those who had no choice but to remain in place were exposed to attack, frequently followed by more friendly words because his anger was always short-lived.'

L'Ébé Bugatti has provided us with a vivid memory of her father in his factory: 'It was amazing to see the way he would pick up and feel a car part on his way through the workshops, holding it in his gloved hands.' This, she says, was because he did not want to leave a mark on the metal. He used to finger 'pieces of machinery just as he felt a horse's leg after a ride or when he made his daily visit to the stables'.

Not only did Ettore design cars but he was also responsible for many of the machine tools in use at Molsheim. In pre-World War I days, for instance, he designed a multi-spindle to simultaneously drill twelve holes in the eight-valve car's aluminium crankcase. Perhaps the most famous of them was the lovely enamelled vices used at the factory which bore the oval Bugatti badge and were tools of great precision in addition to their undoubted aesthetic quality. There were, in fact, two types. One was an essentially conventional unit but there was a more sophisticated version which included an integral surface plate which served as a base for a dial indicator graduated to a hundredth of a millimetre.

After the war, in about 1922, Ettore began producing his own nuts and bolts. The latter had square heads while the hexagonal nuts were soon ingeniously fitted with their own integral washers. This made the nuts less likely to vibrate loose, a valuable asset on any car and particularly one involved in competitive activities.

Throughout his working life, Bugatti continued to patent inventions. W F Bradley has said that there were about a thousand in all. This is a fairly accurate figure because, in 1942 Casanova, his patent agent, prepared a list of 925 with thirty-seven more in the offing though they were not all concerned with motor cars. Jean-Albert Grégoire has recounted that they included 'a mechanical razor, a butterfly nut for inner tubes, an unbreakable windscreen, ultra light bicycle and motor cycle frames, tip up seats, safety locks, automatic filler caps, a casting rod for fishermen . . .'

When it came to the design process, one of Ettore's idiosyncrasies, to which Hugh Conway has drawn attention, was that 'He believed he could get his designs right first time.'[24] This contrasted with his great contemporary, Henry Royce, who was wedded to the development process and might built a number of experimental cars and then

---

[23] Conway and Sauzay, *Bugatti Magnum*.

[24] Conway, *Bugatti, 'Le pur-sang des automobiles'*.

refine any shortcomings that became apparent, which was the antithesis of Ettore's approach!

Ettore Bugatti's relationship with his drawing office has already been mentioned and, as the decade progressed, its staff also grew in number. Felix Kortz's role has been chronicled earlier and, on his death on 1927, his place was taken by thirty-four-year-old Édouard Bertrand who was to become an 'engine man' and was latterly in charge of tool design.

Although born in Karlsruhe, Bertrand's family was of French Protestant origins; he had joined Bugatti in 1924 and, after working on such simple assignments as making drawings of car tools, he progressed to more responsible jobs. In fact, 'he was one of those rare draughtsmen who would work quickly with a soft pencil to outline a proposal he wanted to make, the results being technically clear and of some aesthetic quality.'[25]

Noël Domboy considered that the Italian, Alpinola Turina, from Mantua was a 'great engineer'. The latter arrived at Molsheim in the 1920s, departed and then rejoined the firm to work on a steam engine which Bugatti was developing for the French railways. Adolphe Nuss, who joined Bugatti in the 1920s, was responsible for detailing the 16-cylinder Type 45 and 47 and he later worked on the railcars. It has already been recorded that the Italian, Antonio Pichetto, was recruited in 1931 to work on the Type 53 front-wheel drive car. Essentially a chassis engineer, he would make a key contribution in that regard to the Type 57, the Type 59 and the sports racing 57G. Noël Domboy himself went to Bugatti in May 1932 and, although he was initially involved on the Type 57 chassis, he soon began to work closely with Ettore on railcar design.

Henri Muller, who was Ettore's architect, was also pressed into service in the drawing office and it was he who made all the stress calculations on the first railcar which was the four-engined *Président*. However, Domboy says that, 'it was not his fault if the weights he had been given turned out to be too optimistic and that the chassis had to be strengthened after the first trials.'[26] The production version therefore weighed 1,778kg (35 tons) rather than 1,270kg (25 tons) of the prototype.

As the 1930s progressed, Jean Bugatti would take the place of his father in working with this team although his first involvement with the creative process was in the all-important area of coachwork design. In this field he worked closely, and secretly, in a special shed at Molsheim with Joseph Walter, a talented draughtsman, who has said that the ideas were Jean's and it was his role to interpret them. Not only is it difficult to fault the proportions of Jean's coachwork but the detail is equally impressive, so Walter deserves every credit for his role in the creation of the finished product.

In August 1932 the Bugatti family was enlarged by the arrival of Ettore's parents, Carlo and Thérèse, following the death of their fifty-two-year-old daughter, Deanice, in the Strasburg hospital. It will be recalled that they had previously lived at Pierrefonds near Paris and Carlo, in particular, soon settled into a routine at Molsheim. At 11 o'clock every morning he would set off, his dog at his heels, for the town. He had become a familiar figure in Molsheim, for he soon established himself as a 'regular' in the Hôtel de la Gare and enjoyed his daily jug of beer.

The couple did not, however, live at the villa but instead took up residence in the charming wooden southern chalet of the Château de St Jean which Ettore had bought, along with its land. The château was located just to the south of the factory on the main road towards Dorlisheim. Not that the château itself was occupied. It was used for the occasional reception and its imposing façade provided a suitably impressive background against which to photograph Bugatti cars. Its adjoining gate house was similarly employed.

This rather pointless purchase illustrates the profligate way in which Ettore spent his money, when he was making it, though his later impromptu acquisition of the Ermenonville château indicates he was capable of similar impulses, even when he was not! L'Ébé Bugatti has said that, 'He

---

[25] Conway and Sauzay, *Bugatti Magnum*.

[26] Letter in *Bugantics*, vol. 44, no. 1 (spring 1981).

made plenty of money . . . almost without noticing it. The coffers . . . filled at a rate which would seem bewildering today though they emptied just as fast.'

In 1928 Ettore commissioned from Bréguet a copy of one of its most complicated carriage clocks, no. 2973, which dated from about 1815.[27] The work took three years to complete and was to cost Bugatti 60,000fr. (£485) on its completion in 1931. He also built up a collection of arms which included a medieval crossbow at one chronological extreme and an octagonally barrelled Winchester Model 94 rifle at the other. Bugatti's interests also embraced maritime matters and the Duc de Gramont has remembered that he kept at Molsheim 'a 15-metre [50ft] boat under a hangar next to the stables, and he was always carrying out modifications for it, from keel to mast-head. It was there for years'.[28]

The coming of the depression in the 1930s slowed these excesses somewhat and when René Dreyfus, who as a works driver, arrived at Molsheim in 1933, his impression was that Ettore:

> . . . was always broke. Living on the grand scale to which he had quite happily become accustomed himself meant that ready cash was not always that ready. And if it was, he might use it instead on some new invention or a new idea he had, or something he had designed which had nothing whatsoever to do with the business in hand'.[29]

Although Dreyfus always got paid, some months Henri Pracht would ask him to accept a chassis in lieu of salary or prize money. He would then drive it to Paris and the Bugatti showrooms in the Avenue Montaigne where Robert Benoist would sell it for him.

On a happier occasion, Ettore presented him with one of the rare horseshoe-shaped watches that he gave to his favoured customers, racing drivers and friends. They were made for him in Switzerland in the late 1920s by the Mido Watch Company of Bienne. Approximately fifty were produced and today are much sought after collectors' items, so much so that in 1983 Mido produced a limited edition of a hundred in 18 carat gold, each signed by Michel Bugatti, Ettore's son by his second marriage.

One of Dreyfus's first impressions of the Molsheim factory was that 'Everyone rode bicycles . . . you might see a few cars about . . . but mostly it was bicycles, dozens upon dozens of them.' Bugatti was no exception, either on a machine of his own design or in his purpose-built Type 56 electrically powered vehicle.

As a driver, Dreyfus would see much of the aristocratic Meo Costantini and he was able to observe the close friendship that had sprung up between him and Jean Bugatti. 'They were good, good friends. Meo was approaching forty, Jean was in his twenties and, in addition to liking Costantini, he was a little anxious to be like him too. He wanted to emulate his *savoir faire*, his class – and he succeeded.'[30]

One of Costantini and Jean's favourite occupations was to drive the 26km (16 miles) to Strasburg, though not on a clear night but 'whenever the fog rolled in'. Jean was fast establishing a reputation as something of a daredevil while Meo had, of course, raced Bugattis prior to becoming team manager and had twice won the Targa Florio held on its formidable, twisting circuit. Their objective in the city was Julia's bar at the Hôtel de la Maison. But for Dreyfus such journeys were rarely a pleasure as 'Very often the road was used for logging trucks. We couldn't see them until we were nearly upon them . . . somehow we always managed to swerve just in the nick of time. I always arrived at Julia's a bit shaken. Jean and Meo thought the adventure a great delight.'

Such pranks were recounted to Bugatti owner, T P Cholmondeley Tapper, when he visited Molsheim in 1934. On being shown around he was surprised to see a museum which contained many old Bugattis, including one of the 1923 Tanks, the

[27] Letter from Cecil Clutton in *Bugantics*, vol. 49, no. 2 (summer 1986).

[28] L'Ébé Bugatti, *The Bugatti Story*.

[29] Dreyfus with Kimes, *My Two Lives*.

[30] *Ibid*.

Sculpter Paul Troubetskoy at Molsheim in the mid-1930s. He is seen here talking to Lidia who is working on a frieze which was going to be placed at the front of the factory but was never positioned because of the strikes of 1936. Just out of the picture is Troubetskoy's Lancia Lambda.

16-cylinder wheel and the newly sidelined four drive ones. Also present, and in running order, were the three Millers which Ettore had traded with Leon Duray and 'these were sometimes taken out by Jean and Costantini' though their two-speed gearboxes, designed for Indianapolis, occasionally gave way. In one instance they 'were having an unofficial race together on the nearby Route Nationale from Strasburg. Costantini was forced to run off the road at high speed in to a ploughed field due to another weak point, the rather ineffectual braking system.'[31]

Similarly, when Cholmondeley Tapper arrived at the factory, he was amused to see Jean 'passing in and out testing various cars. He would approach the factory gates at tremendous speed while two men rushed to open them and signal him through if the road was clear and in no time he would disappear in a cloud of dust.' He was no doubt on his way to lap the roughly triangular 25 or so kilometres (15 miles) factory test route. It was known locally as the Bugatti Circuit and included the nearby small town of Obernai, plenty of mountain roads and with a return through the town of Molsheim. But whereas everyone else covered this course clockwise, Jean sometimes went in the opposite direction and other factory drivers might encounter him coming towards them at breakneck speed, just to test their reflexes . . .

Perhaps the highlight of Cholmondeley Tapper's visit was a run he had with Jean Bugatti in a Type 50. He found him 'an excellent driver' but as Ettore had banned him from racing Bugattis, 'the only opportunity he had of exercising his skill was during such test runs.' On their return through the town of Molsheim they encountered a large flock of geese. 'There was an interval of swift repartee between Jean and the herdsman' but fortunately the geese escaped with nothing worse than severe shock. 'At the end of the run we came down the road in front of the factory at a hundred miles an hour [160kph], braked sharply and were waved in by the gatemen. I was glad to be back safe and sound.'[32]

His arrival at the factory had been rather more tranquil and Cholmondeley Tapper was delighted to find that Ettore had thoughtfully provided 'special chairs so that visitors at the factory gatehouse could sit in the sun if kept waiting'. He also noticed that in his office Bugatti displayed, framed on the

---

[31] Cholmondeley Tapper, *Amateur Racing Driver*.

[32] Ibid.

Ancient and modern: Ettore at Molsheim on a penny farthing and the Tank in which Wimille and Benoist won Le Mans in 1937.

Like his father, Ettore Bugatti also designed furniture and here he is relaxing in the fruits of his own labours. These chairs were provided for customers who had to wait awhile at the factory.

wall, the 1902 letter from de Dietrich, addressed to his father, which heralded the start of his extraordinary career.

T P did not meet Ettore on his first visit but did so on a second one when his Type 37A engine was being rebuilt. Adrien Paul, Bugatti's commercial director, showed him where to park and then a few minutes later, 'I was surprised to see *Le Patron* himself approaching, bowler hatted and riding a bicycle.' On greeting him, Ettore 'plunged immediately into a discussion of the career of the 37A since I had been driving it, and particularly wanted to hear from me descriptions of all the major events in which I had taken part. I was amazed to find that he already knew a great deal about my driving exploits.'

Cholmondeley Tapper stayed overnight in Molsheim at the Hôtel de la Gare near the railway station and, during the night, he was reassured to hear

The disused entrance to the old foundry photographed in the post-war years.

the scream of his engine under test at the factory. On the following morning, he saw old Carlo Bugatti at the hotel for, in addition to his daily beer, he 'would sit vigilantly on guard over his son's racing cars whenever they waited in the yard to be dispatched by train to a race or purchaser'.[33] The truth was that, as the decade progressed, these dispatches became fewer and, as has already been

[33] Cholmondeley Tapper, *Amateur Racing Driver.*

Molsheim pictured in 1985. The familiar Bugatti ovals have been removed from the gateposts of what is now a factory in the Messier-Hispano-Bugatti group.

recounted, by 1939 the financial affairs of Automobiles Bugatti were in disarray.

Then came the terrible affairs of the night of 11 August with the death of Jean Bugatti. Ettore must have been devastated by the loss and L'Ébé has said that her mother had only recently undergone serious surgery. It was 'feared that the blow to her might be too much; but she showed admirable courage, and was in fact an example to us all'. Jean had been planning to marry a Mexican night club *chanteuse* named Riva Reyes, and her distress can only be imagined.

There was a marriage in the Bugatti family in the following year when, in March, Lidia married the Comte de Boigne, though Carlo died in the following month and he was buried alongside his wife, who had been laid to rest five years previously, in the family vault at Dorlisheim church. Although a handful of cars would be produced after the war, it was the end of Bugatti car production at Molsheim.

The distinctive doors and hinges, as used at Molsheim, have been imaginatively incorporated into the Bugatti Trust's study centre, which was opened at Prescott in 1990. It contained a magnificent archive, to which this author owes a great debt of gratitude, and its curator, David Sewell, welcomes students of the marque from the world over.

12
Reversals, then Renaissance

'Bugatti is back in business . . . the revived name will adorn around fifty specially built cars a year for selected customers.'
Autocar & Motor, 20 January 1988

The last eight years of Ettore Bugatti's life were overshadowed by personal unhappiness, war and, subsequently, his own poor health. The outbreak of hostilities meant a second wartime sojourn in Paris and, although he continued to design cars, perhaps fortuitously, none of them progressed beyond the prototype stage because they reveal how his creative powers had become ossified. He died, only seven years after his father, in August 1947.

Despite the Molsheim factory reopening after the war under Roland Bugatti's direction, automobile production effectively ceased and only a handful of updated Type 57s were built, along with an ill-fated Grand Prix car. The firm was mostly dependent on sub-contract work but, in 1963, it was taken over by Hispano-Suiza and today remains in the ownership of its SNECMA (Société Nationale d'Études et de Construction de Moteurs d'Aviation) successor. Then, in late 1987, came news that

Ettore Bugatti pictured in his later years.

a new privately funded company, Bugatti Automobili SpA, had been founded to revive the marque. A purpose-built factory has been established in Ettore's native Italy and the new Bugatti, appeared in the autumn of 1991.

Ettore was fifty-eight in 1939, the year in which the war broke out on 3 September, a mere twenty-three days following Jean's death. In that month, the Ministry of Air ordered that such vulnerable facilities as Bugatti's factory be transported away from a possible line of an enemy advance. The contents of the plant, and its workforce, were therefore uprooted and moved as far from the German border as possible and for Bugatti that meant the city of Bordeaux, over 800km (497 miles) to the west. The Ministry had already awarded Ettore important sub-contract work for the manufacture of crankshafts, camshafts and propeller shafts for Hispano-Suiza V12, 12X and 12Y aero engines.

The considerable transport costs were to be borne by the Ministry, but Bugatti was responsible for the preparation and packing and he later claimed this had cost him 'millions of francs' while 'breakages, the destruction of installations were never taken into account... these amounted to 15 per cent of the value of the machinery, which meant 20 to 25 million francs [£113,314 to £141,643] in 1939.'[1]

The Ministry charted special trains to transfer the material to the west, with much of the exercise being undertaken under the cover of darkness. Not only did the crates contain essential machine tools but also many of the historic cars at Molsheim, even down to Ettore's Type 10 prototype[2] and all the works drawings.

The Move to Bordeaux

The Bordeaux premises, located at 6, Boulevard Alfred Danney, was a far cry from the order and cleanliness of Alsace. It was a group of buildings vacated by a transport company named the CIMT and Noël Domboy has recounted that they were 'decayed and the floors were soil'.[3]

In view of the influx of the 1,500 strong workforce, temporary accommodation had to be bsilt to house them. Once there, Bugatti took it upon himself to speed the crankshaft manufacturing process by designing and building a special lathe for the purpose.

In the four years prior to the outbreak of the war, Ettore had been living in Paris but, with the move to Bordeaux he was, for a time, reunited with his family. They all moved to the Château de Roquefort near Sauveterre-de-Guyenne, to the south-east of the city.

The factory was in production for a relatively short time because, on 10 May, the German army invaded the Low Countries of Holland, and swept through Belgium and into France which fell on 22 June. That month, the workforce at Bordeaux grew with the arrival of Bugatti personnel from Paris. This meant Noël Domboy and his associates, who had been based at Bugatti's rue Boissière establishment, along with men from a drawing office and machine shop facility in the rue du Débarcadère which had been established for the design and manufacture of the de Monge-designed Bugatti aeroplane.

Bordeaux fell within the area governed by the Vichy government and, in October 1940, the Bugatti factory was bombed by the RAF, which did little damage, though the accounts and design office had to move and were transferred, under the direction of Adrien Paul, to premises occupied by a lutemaker in the rue Condillac.

By this time, the move back to Alsace had begun. At the end of the July, *Obersturmbannführer* Hans Trippel had appeared at the Bordeaux factory with the request that the personnel and tools be returned to Molsheim where he was already in occupation. Back in 1934 Trippel had began building an amphibious car at Homburg in the Saar. This four-wheel drive vehicle, designated the SG6, was

---

[1] L'Ébé Bugatti, *The Bugatti Story*.

[2] The Type 10 prototype later disappeared in Bordeaux but surfaced in Europe after the war and was sold to the USA and restored by Harrah's Auto Collection.

[3] 'Bugatti's Addresses' in *Bugantics*, vol. 45, no. 4 (winter 1982).

originally powered by a four-cylinder Adler engine though was later replaced by a six-cylinder 2.5-litre Opel unit.

In 1936 Trippel had demonstrated his vehicle to Hitler, who immediately seized on its military potential, and Trippel was provided with state funds to develop his invention. In 1937 came a front-wheel drive SG7 which reverted to Adler power. Then came the war and, in 1940, Trippel arrived at Bugatti's empty Molsheim factory in what had become the German province of Elsass and renamed it Trippelwerke GmbH.

Ettore says that 'My refusal to obey the pressing demands of the German authorities to return to Alsace resulted in their taking over the machinery and all the material at Bordeaux and expelling me from the factory.'[4] He had been in a position to make such a stand because, as an Italian citizen, he was, ostensibly, an ally of Germany since, on 11 June, Mussolini had sided with Hitler to declare war on France and Britain. With his role at Bordeaux effectively at an end, Bugatti therefore decided to return to Paris at the end of 1941, though his wife and the family stayed at the Château de Roquefort.

There still remained the problem of the ownership of Molsheim for, says Bugatti, 'the authorities announced that my property in Alsace would be put up for sale to meet my overdraft at the bank in Strasburg and debts to various suppliers.' This would have left him with some 35 million francs (£198,300), a figure that did not begin to reflect its true value which, Bugatti wrote, 'was estimated at 334 million francs [£1.9 million] on 1 December 1941'. Eventually, a compromise was agreed and Ettore 'was obliged, in the face of threats, to accept a sum of 150 million francs [£850,000], against the surrender of my factory. This compulsory sale destroyed nearly half a century's work. Still, the balance is owing to me by virtue of a requisition order.'

With Molsheim effectively in German hands, it was then sold, at the instigation of the German Air Ministry, to Trippel for RM7.5 million (£652,173).

[4] L'Ébé Bugatti, *op. cit.*

Trippel at Molsheim

In the meantime, Trippel had to cede part of the facility to the *Kriegsmarine*, which separately set up Maschinenfabrik Molsheim to manufacture torpedoes. The factory was then divided in two to keep the factions apart, a state of affairs that continued until the spring of 1942 when the wall was demolished and the business came under the control of the German Air Ministry for which the torpedoes had been produced. New machine tools were introduced and some of Bugatti's were moved to Germany.

By this time the monthly production target was 150 torpedoes and twenty amphibious vehicles and it is thought that approximately 1,000 examples of the Trippel SG6 were built at Molsheim between 1940 and 1944. There was also an SG7 version of 1943, powered by an air-cooled V8 Tatra engine. In addition to this output, snow ploughs were manufactured for use on the Russian front, Hitler having invaded that country in 22 June 1941. A prototype powered sledge was built, along with covered sledges for transporting the German wounded.

Officials of the Trippel company occupied the Bugatti villa during the conflict and had the basement converted into a sophisticated air raid shelter, complete with its own air purification plant. After the war it was found that the caravan, which had been such a feature of the Molsheim hunting parties, was missing. It was subsequently found, in a badly damaged state, about 48km (30 miles) away near Narzwiller, at Struthof which was the only German concentration camp to have been established on French soil.

Back in Paris Ettore was established at his house at 20 rue Boissière. It will be recalled that Ettore and his wife had moved there during World War I and he had occupied it virtually continuously since leaving Molsheim in 1936. A traditional and handsome building from the *Belle époque* era, it had large apartments, one on each floor, and the family had lived on the first. Ettore's office was on the ground floor and led on to the street.

In the meantime, work had been proceeding on a new generation of Bugatti cars. The detail design work was undertaken on a second floor office in this building by Noël Domboy and Antonio Pichetto. The mainstream project was a new 1,500cc car which, in Ettore's own words, would have been capable of 'seating four comfortably, will have the performance of the Type 57. It will be much lighter and a lot smaller.'[5]

Cars for the Post-War Era

Then there was the automotive extreme to the gargantuan Royale, the Type 68 with a minuscule 370cc four-cylinder supercharged engine. It would, said Ettore, have a 'performance similar to the Type 57 . . . Its petrol consumption will be very low. This assembly was tried out in a motor cycle so as not to reveal my true intentions . . . It has a road speed of over 140kph (nearly 90mph).'

Ettore was also to reveal that he envisaged a luxury car programme, 'a Grand Sport de luxe with an engine of 4 litres capacity' and, incredibly, 'a Royale, but much lighter than the previous Bugatti Royale. These cars will only be manufactured in small numbers.'

None of these models ever entered production though, in some instances, prototypes were built, of which the most significant was the Type 73 which Ettore envisaged being available in a number of variants. For this model, he reverted to the four-cylinder theme which had been absent from the Molsheim model range since the demise of the Type 40A in 1933. It seems likely that work on the project began prior to World War II and was, perhaps, a tacit recognition by Bugatti that his days of pre-eminence in Grand Prix racing were over and he was intending to follow Maserati and Alfa Romeo and return to the field of *voiturette* racing.

It has to be said, however, that the competition version, the Type 73C, was a world away from the respective 6 CM and Alfetta and its chassis was, in essence, rooted in the 1920s.

The series began with Type 73. This was conceived as a sports car, powered by a supercharged, twin overhead camshaft 70 x 95mm four-cylinder engine of 1,460cc[6] with, for the first time on a Bugatti, a detachable cylinder head. Another innovation was the fitment of wet cylinder liners, which greatly simplified the aluminium block casting, a feature which had been popularized by the *Traction Avant* Citroën of 1934. The camshafts in the aluminium cylinder head were driven by gears from the front of the engine and the four valves per cylinder were inclined at 70 degrees. The supercharger was front-mounted and fed the inlet manifold via an ungainly pipe that swept up to the right of the engine. The barrel crankcase echoed Miller practice and the crankshaft ran in split aluminium 'cheeses' which must have considerably slowed the assembly process.

Ettore made no attempt to perpetuate the 57's single plate clutch and unit construction gearbox and instead reverted to his own multiplate unit and the all-synchromesh gearbox, with its central change, though this was separately mounted. The rear axle had echoes of the Brescia unit. The 73's chassis was essentially conventional Molsheim though had cruciform bracing behind the engine and Ettore did make some effort to cheapen his traditional front axle assembly. Again, quarter-elliptic rear springs were used but hydraulic brakes, courtesy of the Type 57, were employed.

A Backward Look

The 73A was a passenger car. Its chassis was essentially the same though the engine was different, in that it was a supercharged single cam 76 x 82mm unit of 1,480cc with three valves per cylinder and the customary fixed cylinder head. The cast iron block differed from the 73's and extended downwards to support the main bearings in the Royale manner. Its 73B derivative shared its dimensions though was fitted with a detachable cylinder head and other refinements.

---

[5] 'Letter to Col. Eric Giles' published in *The Autocar*, 2 February 1945.

[6] John Barton, 'Probe, Probare Type 73' in *Bugantics*, vol. 52, no. 3 (autumn 1989). A detailed article on a shadowy area of Bugatti history.

By way of a contrast to producing the largest car in the world, Bugatti also had the idea of building one of the smallest, the 370cc supercharged twin overhead camshaft four valves per cylinder Type 68. A fluid flywheel was employed, complete with brake to permit the silent engagement of bottom gear. Exhibited at the 1946 Paris Show, the Type 68 never entered production.

The diminutive Type 68 prototype can today be seen at Mulhouse.

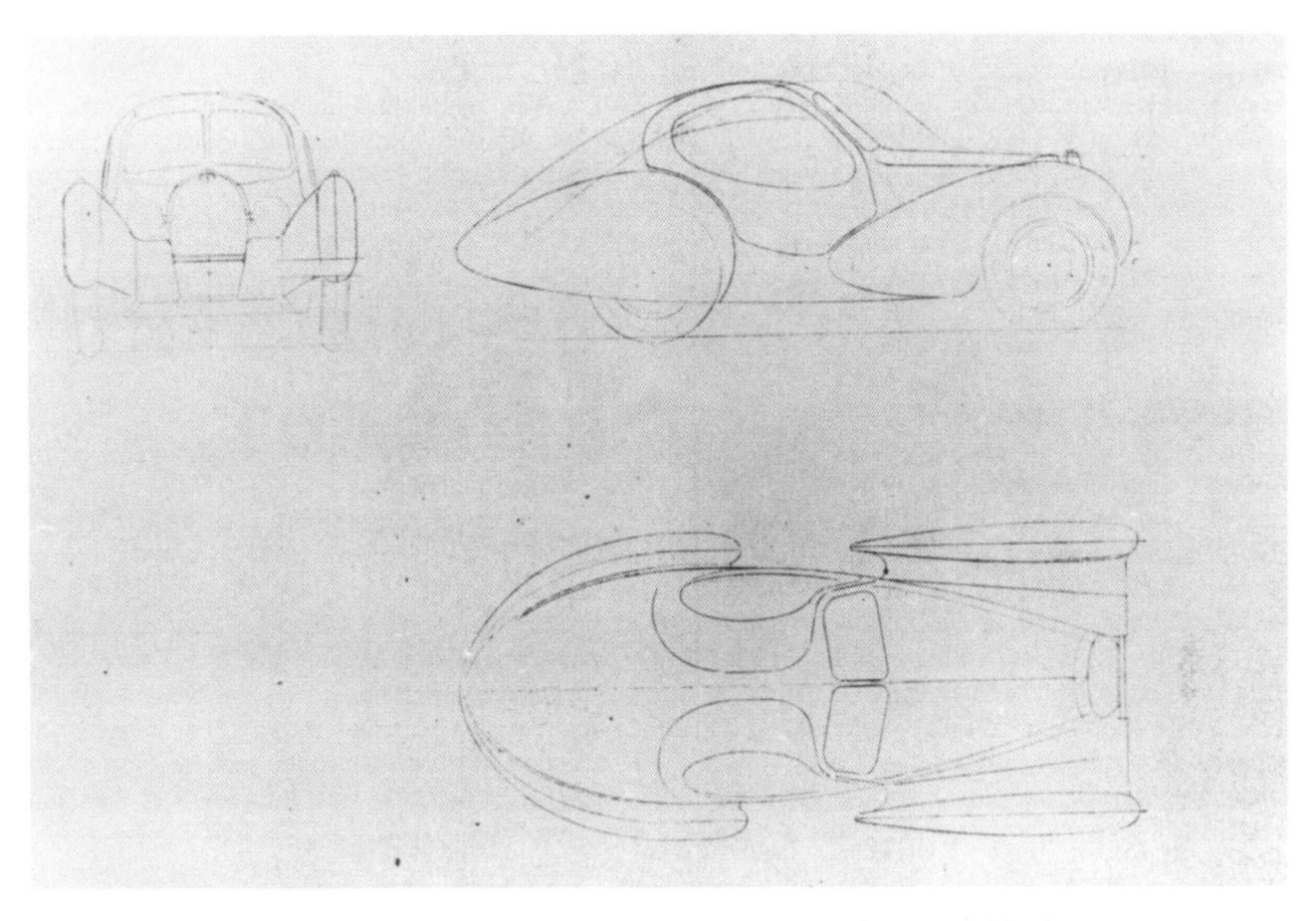

There were clearly thoughts about producing a sporting version of the Type 68, a sort of poor man's Atlantic.

The competition version of the model was the Type 73C which had a similar engine to the Type 73 though the gear lever was cranked to the right. There was a transmission brake fitted behind the gearbox.

The 73C also differed from the 73 in having a modified Type 59 steering box, while the chassis was much lighter and more related to the 59 and the spring hangers were almost identical to those of this earlier racing car. The 73C was envisaged as a single-seater though two surviving drawings show it with different fronts: one has the traditional Bugatti radiator, and the other with a cowl similar to that used on the 3-litre single-seater.

So much for the Type 73 but, as already mentioned, Ettore also conceived a miniature car intended to meet the austerity of the post-war years. Its design had been completed by the time that Noël Domboy returned to Paris in mid-1941. He has recalled that: 'in the interim, Édouard Bertrand and I had designed the four-cylinder 330cc [sic] supercharged Type 68, becoming, on paper only, a 350cc engine intended to fit both a small two-seater, the prototype of which was built but which never saw the road, and a motor cycle of which a prototype was also built.'[7]

This car, with its tiny 48.5 x 50mm engine dimensions, giving 370cc, would have probably been one of the smallest four-cylinder cars in the world. It is sobering to reflect that the Fiat 500's engine was over half as big again at 569cc! The Type 68, however, was a supercharged twin overhead camshaft unit with four valves per cylinder which echoed the Brescia of the 1920s, though intended

[7] Domboy.

357

to run at 7,500rpm. Unlike the Type 73, there was no ugly induction pipe as the carburettor was bolted directly to the front-mounted blower. The gearbox was in unit with the engine and had a silent bottom gear. The prototype, to which Domboy referred, was fitted with an open two-seater body and today can be seen displayed at Mulhouse. A stylish coupe was also contemplated but never built.

More Paris Purchases

It seems likely that during the war, Ettore imagined that he would not be returning to Molsheim. But he would still need a manufacturing facility to produce the 73 family of cars and the 68, so he bought the La Licorne works at 169–176 rue Armand Sylvester, Courbevoie, Seine for that purpose. This was a make that had begun life as the Corre in 1901 and had evolved into the La Licorne, which means Unicorn in English. Castings, pressings and stampings would be the responsibility of the nearby La Fournaise works which Bugatti had also purchased.

The task of building the Type 68 and 73 prototypes was assigned to his rue du Débarcadère works, which Bugatti rented from the Zenith Carburettor Company. In addition to this automobile work, it was also involved in the creation of one of Ettore's maritime projects; a motor torpedo boat with a stepped hull built of a weldable aluminium alloy called Duralinox. It was to have been powered by eight supercharged 50B engines, grouped in lines of four, though Domboy says that the blocks were fragile, despite a number of reworks by Édouard Bertrand.

However, the concept was flawed in that unlike conventional torpedo boats which, for example, launched their weapon to the starboard, and would then swerve to port to avoid it, on the Bugatti craft the tube was located in the prow and, with a projected speed of 40 knots, it ran the risk of overtaking the torpedo which it had just fired. The hull lingered on for a time after the war and was then scrapped.

The Bugatti Yacht

Ettore was becoming increasingly preoccupied with maritime matters because, during the Occupation, he bought the small Deconnick boat yard on the Seine at Maison-Laffite, which was 21km (13 miles) to the north-west of Paris. Bugatti was clearly looking to the post-war years because, prior to the outbreak of hostilities, he had literally thrown caution to the winds and commissioned a large yacht from the Macario shipyard at Trouville on the Normandy coast. Its keel was laid down in 1939, the year in which, it will be recalled, Bugatti was fearing bankruptcy.

The yacht was approximately 295m (90ft) long and gave Ettore an opportunity to apply many of the patents he had filed over the years for such marine applications as windlass, capstan, rudder controls, propeller shafts and the like. There was a foremast and mainsail and the latter was fitted with a Bugatti-designed roller reefing gear. Ettore was also responsible for the rudder and steering mechanism, the winches to raise and lower the anchor and the pumps.

Inevitably, Ettore did not spare himself when it came to the furniture and interior fittings which he designed for the six cabins. This even included the wash hand basins and these were made at Molsheim. The same went for the saloon's oval table which had to be remade a number of times because its contours had to follow the lines of the hull. Despite the vessel's dependence on wind power, there were two auxiliary Ford V8 engines.

The yacht was launched in Easter 1940 but the war meant construction ceased in May of that year. Then M. Macario towed it across the Seine estuary to Le Harve but, with the fall of France, it was removed across the English Channel to Britain. It spent the war moored in the Solent and, after hostilities, was returned, somewhat the worse for wear, and came to rest at the Maison-Laffite yard.

Wooden trawlers were built there during and after the war and Ettore also had plans to produce a wooden double-skinned *youyou* (dinghy) from the same location. It was to be powered by a single-cylinder 60 x 60mm, 170cc Bugatti-designed engine, allotted the Type 75 designation. Production was scheduled for 1948 and it was intended to manufacture the boat in three sizes of 2.8, 3.3 and 4 metres but the project died with Ettore. At least one example was built and its immaculate sectioned hull can be seen at Mulhouse.

Business as Usual?

The Bugatti Paris showrooms in the Avenue Montaigne remained open throughout much of the war as the company's head office. It was here that the firm's directors met, though they appear to have had little success in restraining Ettore! Its most celebrated member was Rear Admiral Guéguen, former director general of the Navy, and his two colleagues were M. Bonnefond-Craponne and M. de Maistre. At some stage during hostilities these premises were demolished and Bugatti's head office was transferred to an apartment at 30 Avenue Hoche which had previously been occupied by a Jew who had fled in the wake of the German advance.

W F Bradley has said that these premises were 'the centre of a dangerous school of resistance and sabotage'[8] and the group included the former Bugatti driver Williams, whom Bradley claims, was betrayed by a traitor in the summer of 1943. He was captured and never seen again. The same fate befell driver Robert Benoist who had also been Bugatti's Paris-based sales manager in pre-war days. His valiant activities for the Résistance resulted in his being captured. He was sent to the Buchenwald concentration camp where he was one of thirty-six prisoners hanged on 12 September 1944, only eight months before the end of the war in Europe.

As far as Bugatti's own family was concerned, at the end of 1941 they moved from the Château de Roquefort and were to live at the Château at Ermenonville for the remainder of the war. Barbara Bugatti was an invalid by this time, for she had never recovered from the surgery she had undergone for cancer just before the war. Nursed by her son, Roland, she died on 21 July 1944 at the age of sixty-two.

A New Mme Bugatti

Over two years later, on 12 October 1946, Ettore married for a second time. His bride was twenty-five-year-old Geneviève Marguerite Delcuze and, like his first marriage, the children had come first. A daughter, Thérèse Georgette Fernande, had been born at Neuilly-sur-Seine on 12 December 1942 and a son Michel on 12 July 1945.

The war did not prevent Ettore from pursuing his passion for riding. He had brought his Irish grey Brouillard to Paris but, while in Bordeaux, he had bought another. However, once in Paris, he had forgotten the name and whereabouts of the dealer. The luckless Adrien Paul was then commanded to find both dealer and horse and he proceeded to do so . . . on a bicycle. Eventually he discovered the individual, and the horse, and the latter was duly transported to the French capital. Its arrival permitted Bugatti to use a horse-drawn carriage, or sometimes he rode his Irish grey. Any visitor to Ettore's office would have seen two pig skin saddles he had bought from Hermès sitting on trestles and covered with towelling.

Noël Domboy remembers that he shared his own office with eight large packing cases containing the harness which had once hung in the Molsheim saddlery. One Saturday morning he, and others, were assembled by Ettore to 'open all these to find two sets of harness in black leather, the blinkers of which carried the EB motif in silver. He nearly wept when he found that some of the horsehair on the collars had been eaten by bugs.'[9]

In 1944 Bugatti heard that Jean-Pierre Wimille's mother was in financial difficulties, so he bought her a villa on the French Riviera. This was at Beaulieu-sur-Mer near Nice but he rarely used this retreat, which was called the Villa Tunis, though it remained in the ownership of the Bugatti family

---

[8] Bradley, *Ettore Bugatti*.

[9] 'Bugatti's Addresses', in *Bugantics*.

until Hispano-Suiza bought the firm in 1963. Then in June 1944 came D-Day and Paris was liberated on 23 August. In October, the de Gaulle administration was recognized by the United Nations and elections followed a year later. This was, inevitably, a difficult period and Bradley has written that there was 'great turmoil of suspicion, strife and vengeance . . . In the white heat of passion many an employer was denounced . . . Even Bugatti was not immune to these attacks.'[10]

After the Liberation

Ettore had imagined that, after the liberation, his property would be returned to him 'but instead it was claimed by the State and re-equipped for production by the State.'[11] In this hostile climate, Bugatti, who for much of the inter-war years had alone upheld French prestige on the racing circuits of the world, was forced to prepare a document, dated 3 April 1945, 'to justify my past as a car manufacturer'. In these circumstances, it must have gladdened his heart to witness Wimille win the first major motor racing event of the post-war years. Driving the 4.7-litre single-seater, he won the over 3-litre race of the *Grand Prix de la Libération*, held in the Bois de Boulogne on 9 September 1945.

These difficult circumstances must have spurred Ettore to grasp the nettle of his Italian citizenship, so he applied to become a naturalized Frenchman and received the papers on 25 February 1946. He had waited so long, says L'Ébé, because 'he thought it in poor taste to change what nature had made him . . . he looked upon Italy and France as a son of well-bred parents thinks of his father and mother. It was a noble conception but brought much worry and difficulty into his life and twice contributed to his ruin.'

This standpoint has been confirmed by the famous French motoring journalist, Charles Faroux, who recalled that:

[10] Bradley, *Ettore Bugatti*.

[11] L'Ébé Bugatti, *The Bugatti Story*.

Ettore Bugatti . . . told me many times during the forty years that I . . . followed the course of his life: 'I love France deeply but I was born an Italian and I will remain an Italian; my children born in France or Alsace are French, thus all future Bugattis will be French – this is essential . . .'[12]

There was no alternative but for Bugatti to take legal action to recover his factory from the *Administration des Domaines*, a state office. A hearing was held at Saverne in Alsace and Ettore lost. He appealed and the case was heard at the Court of Appeal in Colmar in April 1947. It was a traumatic occasion and L'Ébé Bugatti has recounted that 'counsel for parties hostile to Bugatti made violent speeches which were a great emotional shock to him and caused a nervous breakdown.'[13] Thankfully, the words of Ettore's advocates, M. Rengade, a legal and financial consultant from Paris, M. Weiller from the Strasburg bar and M. Sinay, his opposite number from Colmar, won the day. The judgement was dated 11 June 1947 and Molsheim was returned to him as the rightful owner though he was never to see it again because, by that time, he was seriously ill.

The Death of Ettore

The hearing had taken its toll. L'Ébé says that he caught a chill on his return by car to Paris by which time he was 'exhausted and in low spirits'. This led to pneumonia and he was then left half paralysed by an obstructed artery. He lingered on for three months in this condition, and 'despite the care of specialists, he gradually sank into a coma and died on 21 August 1947 at the American Hospital at Neuilly.'[14] He was sixty-six. A memorial service was held in Paris and another at the parish church at Dorlisheim and he was buried in the family vault there.

[12] Grégoire, *Best Wheel Forward*.

[13] L'Ébé Bugatti, *op. cit.*

[14] *Ibid*.

Ettore's death created a new set of problems and, as W F Bradley has said: 'The Bugatti family decided that the life work of "the patron" should not die. There were five direct descendants, not one of whom could take an active part in the revival: Mademoiselle L'Ébé Bugatti, the eldest daughter; her sister Lidia . . . the son Roland, obliged to reside in the south for health reasons; the two young children of a second marriage.'[15] The sixth all-important individual was the new Mme Bugatti, whom had been left a 50 per cent stake in the business. The uneasy alliance inevitably soon divided itself into two camps with the children of Ettore's first marriage on one side and his second, young wife on the other.

In the event, overall responsibility for the Bugatti factory would be vested in Roland but the day-to-day running of the business went to Pierre Marco, a Molsheim veteran of thirty years' standing, who was appointed general manager. Ironically he was 'an old enemy of Ettore's . . . was partial to the new Madame Bugatti's interests. Every issue seems to have developed into a tug of war between the two factions, with L'Ébé often opposing them both.'[16] With such a background, it is perhaps inevitable that the post-war history of Automobiles Bugatti should be such an unhappy one.

First Molsheim had to be rebuilt. Prior to his departure, Hans Trippel had been determined that the works would not fall intact into allied hands, so he had sabotaged the machine shop. 'Cincinnati millers, Landis grinders, Gleason gear cutters lay overturned in the dust filled machine shops. A bomb here and there and the offices went up in flame and smoke.'[17]

The Factory Revived

When Marco returned to Molsheim on 1 August 1947, he found 600 French marines in occupation and the factory in ruins and no less than 3,000 wrecked machine tools. 'In the great pile we picked out turret lathes, broaching machines, multi purpose drillers the whole gamut of metal working machines all useless,' Marco told Bradley, 'In one corner of what would be a new machine shop there was a huge wreckage of costly magnesium cylinder blocks and an experimental Duralumin chassis.'[18]

Despite Ettore's death, there was a Bugatti stand at the 1947 Paris Salon which opened at the Grand Palais on 23 October. There, a Type 73 chassis and engine were exhibited and, incredibly, some orders and deposits were placed for the 73C racer. One came from Serge Pozolli while Bart Loyens wanted two.

But such was the chaos at Molsheim that it soon became clear that the model could not immediately enter production, if ever, though one 73 road car, a two-door saloon, was completed and has ended up at Mulhouse. At some stage, the remaining Type 73 chassis were dismantled and moved from the rue due Débarcadère to Molsheim, and the customers appropriately reimbursed.

It was to be over three years before the revelation came that 'new' Bugattis were on the way. On 2 March 1951 *The Autocar* reported that 'two new models in production before the end of the year is the latest hope of the Bugatti company.' One was the Type 101, which was effectively a mildly updated version of the 3.3-litre Type 57. This was despite the fact that, in the interim, the French government had passed a law which penalized cars of over 16CV, which is about 3 litres. This meant that a 12–15CV model would cost the equivalent of £23 a year to tax, and above that capacity a swingeing £73.

This spelt death to those *Grandes Routières* that had survived the war, of which the 17CV Type 101 was one. It may well have been for this reason that a 1,500cc sports car, the Type 102, was contemplated. This employed what sounds like a derivative of the Type 73 engine with 78 x 80mm dimensions,

---

[15] Bradley, 'Bugatti revival' in *The Autocar*, 8 February 1952.

[16] Borgeson *Bugatti by Borgeson*.

[17] Bradley, *op. cit.*

[18] *Ibid.*

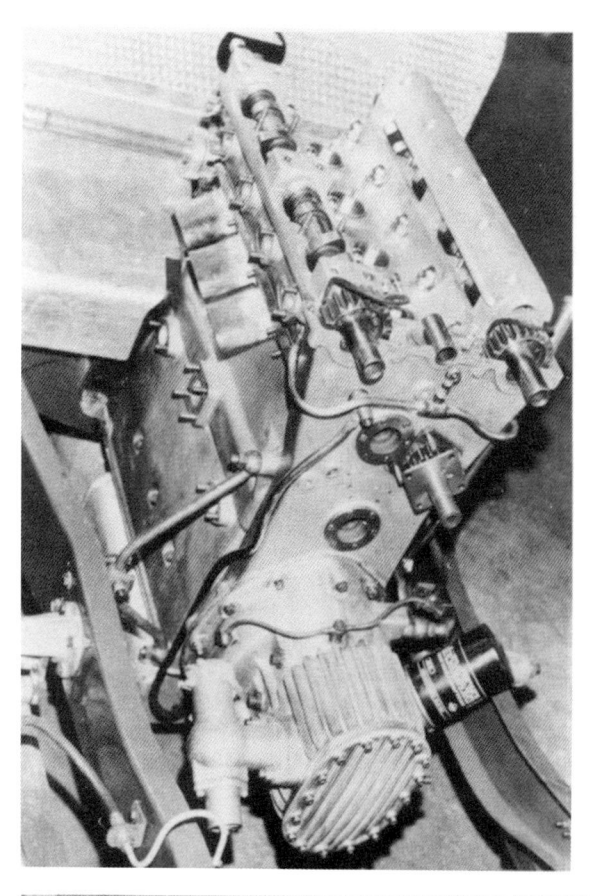

The Type 73C was the competition version. This 1.5-litre unit has an aluminium detachable cylinder head, with four valves per cylinder, a wet liner alloy block. As can be seen, the camshafts are gear driven.

One of the two 101s displayed by Bugatti at the 1951 Paris Salon, still with fixed head 3.3-litre straight-eight engine and right-hand drive. The coachwork was by Gangloff. Not surprisingly there were no takers.

chain driven twin overhead camshafts, fixed cylinder head, wet or dry sump lubrication and a separate gearbox. It had transverse leaf independent front suspension.

Bradley saw an example during a visit he made to Molsheim in 1952 though, despite his declaration that; 'the 102 is a thoroughbred' it sounds as though it was far from it and was never seen. The Type 101, by contrast, was and appeared at the 1951 Paris Salon.

The Type 101 Arrives

Prior to the event, *L'Auto Journal* reported that: 'M. Marco triumphantly announced that the Bugatti is at last ready to resume its place on the world automobile market.'[19] The 101's mechnicals were little changed from its predecessor's and the car was available in unblown and supercharged form. It was therefore a re-creation of the Type 57 and 57C respectively though, on the latter version, the twin choke Weber carburettor was mounted on top of the blower with associated changes to the manifolding. A Cotal electric gearbox was employed – there was said to be a five-speed Bugatti-designed unit in the offing but it never appeared. Lockheed eight-shoe hydraulic brakes were fitted.

[19] Eaglesfield and Hampton, *The Bugatti Book*.

If the chassis was outdated, then the same could not be said of the coachwork. It was by Bugatti's long-time associate, Gangloff, and was modern, full-width and visually impressive. Two cars, a two-door saloon and a convertible, were exhibited but there could be little demand for such an anachronism. The 101 was a costly, cart-sprung model, which had been outdated in 1939, with an obsolete, fixed head eight-cylinder engine that was both expensive to run and tax. For its price, which was over £3,500, a British purchaser could have bought a Bristol 401 with some change to spare.

Neither 101s were sold at Paris and both were retained by the factory. There were, at least, seven built and one chassis was never bodied. The Paris Show Gangloff supercharged 101C saloon was built on a 1937 Type 57 chassis (57454) though the prototype (101500), had been fitted with a four-door saloon by Guilloré of Paris which also bodied 101502 as a two-door saloon and this survives in Switzerland.

Today, the four-door car is at Mulhouse, as is a convertible (101503) which is similar to the 1951 Paris Show car. That particular 101 (101501) remained at the works, and was little used until 1959 when it was taken to America by Gene Cesari and then sold to Dr Edwin Rucker and subsequently Godfrey Howard and P Ceresole. It is now in Germany.

A two-door coupe by Van Antem of Paris was

This Type 101 is chassis number 101502, and the two-door saloon bodywork is by Guillore of Courbevoie.

This last Type 101 chassis, 101506, which bore the word Fini. *It left Molsheim 1961 and this body was designed by stylist Virgil Exner and completed in 1965.*

taken to America by Robert Stanley and purchased by the Harrah's Auto Collection. It was bought by Jacques Harguin Deguy of California in 1986 who resprayed it red and black. The 'missing' 101 is chassis 101505.

The final 101 chassis is number 101506. It remained at Molsheim until 1961 when it was bought in a bodyless state by E Allen Henderson of Marlboro, New Jersey. He then disposed of it to L Scott Bailey, founder editor of the respected American publication, *Automobile Quarterly* who then passed it on to the celebrated former Chrysler stylist, Virgil Exner. Ghia in Turin had already produced many Exner one-offs during his Chrysler years and it was responsible for bodying this chassis to their client's design.

The 101 was shortened by 460mm (18in) and Exner came up with a full-width open two-seater body which uneasily retained the Bugatti horseshoe radiator flanked by two deeply recessed rectangular headlamps. Many of the panels were heavily louvred and the car was displayed at the 1965 Turin Show. Appropriately, the chassis is said to have been stamped *Fini*.

The Ill-Fated Type 251

These years of the Bugatti family's post-war stewardship were outwardly clouded by the débâcle of the Type 251, which was the last Grand Prix car to have been built at Molsheim. To put this unhappy project into perspective, we must first retrace our steps and see how the firm survived in the difficult early post-war years.

Initially, it had undertaken the manufacture of blank shells for the navy and underwater cable cutters. This work was of relatively short duration but then came a contract from the French State Railways for spare parts for its Bugatti railcars, together with the establishment of a service facility at Molsheim. The SNCF was, for a time, Bugatti's largest customer. Then there were looms for the Gantois company and castings for Holwegg. A

Roland Bugatti, who was to play a role in the creation of the ill-fated Bugatti 251 of 1955, pictured at Molsheim in the pre-war 4.7-litre single seater.

machining contract from Citroën followed in 1949 and this lasted until 1951.

But by 1952 the railway was beginning to make its own servicing arrangements and this work began to decline. Fortunately, by this time the firm had discovered the lucrative military market. Since 1946 the armed services had become embroiled in a costly war in French Indo-China and, in 1951, Bugatti obtained a valuable contract for the manufacture of tank engines. This grew to the extent that, by 1953, it accounted for 78 per cent of the firm's commercial activities.[20]

Meanwhile, in 1951, Ettore's widow had married again. Her new husband was the Rizla 'cigarette paper king', René Bolloré, who was keen that the Bugatti name should, once again, appear on the race tracks. This meant no less than the revival of

[20] Jacques Ickx in *L'Équipe*, 19 November 1954, in Conway, *Bugatti Magnum*.

the marque by not only creating a Grand Prix team but also a roadgoing sports car derived from the racer which was designated the Type 251. A new 2.5-litre Grand Prix formula was due to start in 1954, so it seems likely that this decision was taken in 1953 when defence orders were running high. Plans were laid to produce a batch of six cars.

With Jean's death, the factory had no creative mainspring, but now thirty-one-year-old Roland moved forward to make his contribution. There was no question, of course, of him designing the car and his only real excursion in the field of car technology had been an unsuccessful attempt to convert the Type 73 engine to rotary valve operation. The task was assigned to the celebrated Italian engineer, Gioachino Colombo, who was, appropriately, Milanese and had the highly successful 158/159 Alfetta to his credit, as well as the post-war 1.5-litre V12 supercharged Ferrari.

Roland's input to the design would appear to

have been that its engine was mounted behind the driver, in memory of his father's regard for the Auto Unions of pre-war days. This was, in itself, a revolutionary feature in an era of front-engined racers but the 251's engine was also positioned transversely behind the driver. The eight's layout echoed that of the pre-war 8C Alfa Romeo, in that the twin overhead camshaft and its ancillaries were driven by gearing between the two blocks.

It was, in effect, two four-cylinder engines bolted together – oh, how Ettore would have approved – and Colombo spoke of the crankshafts being assembled in different planes to alter the engine's torque characteristics to suit specific circuits. One of these fours would later form the basis of the projected sports car, which was designated the Type 252. The blocks were of magnesium alloy, with wet liners, while four twin choke Weber carburettors were fitted. Ignition was by twin Marelli magnetos, which ignited the 16mm sparking plugs. With dimensions of 75 × 68.8mm, the eight had a capacity of 2,430cc.

Power was transferred from the centre of the crankshaft, via spur gearing, to the integral five-speed gearbox with Porsche-type synchromesh. The drive thereafter passed through the differential to the half shafts of the De Dion rear axle.

Roland Makes a Contribution

Colombo designed a welded tube space frame type chassis but unorthodoxy reared its head when it came to the suspension. The rear axle slid on a centre trunnion and was located by twin trailing radius rods. On each hub was a vertical pushrod which was attached to a bell crank mounted on the upper chassis members. Each, in turn, was connected to the long coil spring damper units which were anchored on the opposing side of the chassis and the two therefore criss-crossed the frame. But at Roland's insistence, independent front suspension, which had been the norm on every front-line Grand Prix car since the mid-1930s, was dispensed with and the rear layout was effectively duplicated at the front. The car could not be a true Bugatti, he maintained, unless it was fitted with a hollow and rigid front axle . . .

Roland was, apparently, able to establish a good rapport with Colombo, and Alfa Romeo historian and designer, Luigi Fusi, has recalled that although he hardly spoke any Italian, 'he amazed and delighted us by speaking excellent Milanese. But what was really astonishing was that it was the Milanese of our parents and grandparents. It was a fossil language which this young stranger spoke fluently.'[21]

Much more progressive was Colombo's specification of disc brakes and the 251 was the first continental Grand Prix car to be fitted with them, their worth having been proved in Britain by Connaught, BRM and Vanwall. The radiator was front-mounted and coolant was run down the main chassis long irons to the rear engine. There were echoes here of Bugatti's abortive 1903 Paris to Madrid racer. Pannier fuel tanks were fitted either side of the driver. The Bugatti had only a 2,184mm (7ft 2in) wheelbase which was to endow the finished car with a squat, stubby appearance.

Colombo finished his work, which was all undertaken at his Milan office, in June 1954 but on 20 July of the following month the French signed a truce with the Communists at Geneva and the war in the Far East was over, along with the defence contracts on which the Bugatti factory was so dependent. By this time, the 251 project is reputed to have absorbed the equivalent of £60,000.

Work pressed ahead, nevertheless, and the construction of the first car began at Molsheim. On Colombo's advice, the factory employed Ferrari's former chief mechanic, Stefano Meazza, to oversee its manufacture. This took the remainder of 1954 and much of 1955 and the engine ran for the first time 29 July of that year. On a modest 7.5:1 compression ratio it was said to have produced 230bhp at 8,000rpm and the completed car had its initial public airing at the Entzheim airfield near Molsheim on 4 October 1955 with Marco at the wheel.

Over a month later, on 21 November, Roland's

---

[21] Borgeson, *Bugatti by Borgeson*.

The eccentric mid-engined 251, unveiled at Entzheim airfield near Molsheim on 4 October 1955. Maurice Trintignant is at the wheel of this first car.

Roland Bugatti at the wheel of the second 251. Note the different nose and there were also suspension modifications.

friend, Bugatti enthusiast and celebrated French racing driver, Maurice Trintignant, drove the 251 for a press showing. This first car was intended to be little more than a works hack on which to put some miles to bed in the components. It accordingly ran on low octane fuel rather than the usual F1 alcohol cocktail.

Testing continued of this single car, numbered 001. It proved to be a twitchy performer and a second chassis was therefore prepared with a longer wheelbase. It was not numbered. The works was, however, hoist with its own petard because the publicity that had followed the November unveiling resulted in the ACF pressing Bugatti to run a car in the 1956 French Grand Prix, which was to be held at Rheims on 1 July.

Débâcle at Rheims

The two cars were taken to the circuit for practice but both proved disappointing. Trintignant's best practice time was 2min 41.9sec which was over 17 seconds slower than Fangio's Lancia-Ferrari in pole position. In addition, handling of both 251s was positively alarming. They wandered from side to side of the track along the straights, their noses had a tendency to rise at high speed and the cars behaved unpredictably over bumps. Trintignant decided to use 001, though it was fitted with drum brakes in place of the pioneering discs.

The story of the race is soon told. Not only did the car lack power but reports mention its, by then, characteristic weaving at speed. Trintignant never managed higher than twelfth position, though he did engage in a spirited duel with the equally pedestrian Gordini. On the eighteenth lap he came into the pits because dust was entering the unprotected carburettor inlets which had jammed the controls. That represented the end of the story and the cars were never again seen in action.

Roland briefly persisted with modifications to the front suspension and large drilled spring brackets and vertical coil springs were introduced on the second car before it was abandoned. Work also ceased on a further four chassis and their attendant engines which would have made the full complement of six.

It was much the same story with the 1.5-litre 252 sports car. Michelotti was commissioned to produce a styling study and a prototype was built but when it was tested, its driver, Pierre Macoin, was invariably alarmed to find himself stationary by the side of the road with steam, oil or petrol pouring

The prototype of the 252 sports car at Mulhouse is the only reminder of what might have been a new generation of Bugatti sports cars.

The Type 252's 1.5-litre four-cylinder twin overhead camshaft engine, effectively half the 251's unit.

out of the bonnet.[22] To avoid any unwelcome publicity, it was decided to fit the prototype 252 engine under the bonnet of the single Type 73 saloon to ensure anonymity. But when the engine worked, the rear axle broke ... By 1958 there were even thoughts about fitting it in a Peugeot 403 but this never materialized.

Marco retired soon after this and the engine was redesigned along with the chassis. Work on the design of the underframe, which had all-round MacPherson strut suspension, began in August 1959 and was not completed until 1962, by which time the Colombo-designed engine was reintroduced! But the project was then abandoned in view of the fact that the Bugatti name was associated with large cars on which there were higher profit margins. This prompted a final car, the 451 of 1962.

This concept, essayed by Noël Domboy, remained a paper project but was clearly intended as a Ferrari challenger with a front-mounted 80 x 74mm 4.5-litre V12 engine and rear-mounted gearbox, clutch and differential. It would have had a space frame chassis with all-independent MacPherson strut suspension and a 2.75m (9ft) wheelbase.

Hispano-Suiza Takes Over

The 451 was, however, overhauled by events. The factory's financial position had deteriorated badly by the early 1960s to the extent that Fritz Schlumpf, who was building up an extraordinary collection of Bugattis at nearby Mulhouse, approached the firm and offered to buy, for the equivalent of £50,000, the eighteen historic cars still retained by the works. Contracts were exchanged on 11 April 1963 but three months later, in July, came the Hispano-Suiza take-over and the family penned a farewell message, dated 22 July, to the Molsheim workforce. Domboy has recounted that he 'was called into an office to meet the new directors, Mr Baumgarten and Mr Robert Blum ... and Gaston Surel, the Manager at Molsheim ... Mr Baumgarten said, "*L'automobile c'est fini*." And so it was!'[23]

The purchase meant that there were no links

---

[22] Barton, 'Probe, Probare Type 73, Part II', in *Bugantics*, vol. 52, no. 4 (winter 1989).

[23] 'The Bugatti Type 451' in *Bugantics*, vol. 44, no. 1 (spring 1981).

The Schlumpf brothers, Hans, left and Fritz who built up the world's largest collection of Bugattis at Mulhouse. Now taken over by the French government, the museum is today is one of the most popular tourist attractions in France.

between the Bugatti family and Molsheim for the first time for fifty-four years and, by 1980, all Ettore's children by his first marriage would be dead. Sixty-five-year-old Lidia Bugatti died in 1972 and Roland in 1977, aged only fifty-four. L'Ébé, who was the oldest, died in 1980 at the age of seventy-six. None had any children, though the second generation of Bugattis has. Thérèse married in 1965 and her daughters are named Manuela and Geneviève.

Thérèse, Ettore's daughter by his second marriage, pictured at 'The Amazing Bugattis Exhibition', staged at the Royal College of Art, in 1979.

Michel's wedding came two years later and his offspring are Emmanuel and Caroline Bugatti.

One spin-off from the change in ownership was that the order went out to remove any 'unwanted junk' from the factory and the firm's M. Seyfried then invidiously had to ring up the individuals who had ordered 73Cs back at the 1947 Paris Show and asked whether they wanted to take delivery of their cars in kit form, sixteen years after they had

Michel Bugatti, born in 1945, at a Bugatti gathering in Europe.

ordered them! This perhaps explains why all five examples survive today.

In 1968 Hispano-Suiza was itself taken over by SNECMA,[24] the nationalized French aircraft combine, and this consolidated an earlier switch at Molsheim to the manufacture of aircraft components. The works found itself producing undercarriages and parts for Concorde and Caravelle aircraft. The famous oval Bugatti signs from the factory entrance subsequently disappeared and, from 1 January 1978, the firm became a part of SNECMA's new Messier-Hispano-Bugatti division, a state of affairs that continues to this day.

A New Bugatti

It seemed that M. Baumgarten had the last word as far as any Bugatti car was concerned but, at the end of 1987, the American *Automotive News* broke the story that Bugatti Automobili SpA had been registered with a view to producing a new Bugatti car. It later emerged that this new company had been registered in Modena, Italy on 14 October

Another model announced at the 1946 Paris Show was the Type 73 though, after Bugatti's death in 1947, the cars were dismantled and the parts returned to Molsheim. It is thought that there were sufficient components for seven cars, two Type 73/73As and five 73Cs, of which this is '73001'. It was only after the Hispano-Suiza takeover of Bugatti in 1963 that the dismantled cars were disposed of and this example was built up by specialist Jean de Dobeleer in Brussels. It is now in Japan.

[24] The Gnome-et-Rhône company was given the title Société Nationale d'Études et de Construction de Moteurs d'Aviation on its compulsory nationalization on 29 August 1945. In 1946, SNECMA also took over the already nationalized aviation section of the Renault company, along with SECM (Lorraine) and the GEHL oil engine company.

1987, with a capital of £85,000 though this increased as the project got underway. A 65 per cent holding came from the Luxemburg-based Bugatti International SA, with the 35 per cent balance owned by former Lamborghini manager, Paolo Stanzani. A factory site had been purchased near Modena and it was intended to produce fifty cars a year.

In July 1988, news broke that the car was to be powered by a 3.5-litre V12 engine and, in October, Stanzani revealed more details of the new Bugatti. It would have a projected top speed of 321kph (210mph) and be in competition with Jaguar, BMW and Aston Martin and other members of the 200-mile-an-hour club. Stanzani told *Autocar & Motor* that the new car:

> . . . must maintain the flavour of an elegant, classy car, typical of the old Bugattis. I don't want a car that looks like a Group C car. It must perform well everywhere, on road or on the track, but it must also be right for visiting the opera. I don't want enormous spoilers and dramatic air intakes – just a functional graceful body.[25]

The V12 engine would have five valves per cylinder and no less than four turbo chargers, one for every three cylinders to cope with the American emissions legislation. Stanzani was anticipating 550bhp at 9,000rpm. The engine would, of course, be rear-mounted and the Bugatti would be a four-wheel drive car. Stanzani later revealed that the car would be endowed with:

> . . . two personalities, like Dr Jekyll and Mr Hyde. The road personality will have maximum torque at low revs and a soft suspension. The track persona will develop its torque at higher revs and will have a stiffer suspension. To change from Dr Jekyll to Mr Hyde you won't need a mechanic's help. You'll just drive to the track, push a couple of buttons and, *voilà*![26]

In the meantime a new, purpose-built factory has sprung up at Campogalliano about 10km (6miles) from Modena. It has cost £1.25 million, is dominated by a circular glass tower which houses the technical department and is computer controlled to follow the sun and thus gain the maximum amount of natural light.

The project has not been without its fair share of corporate upheavals. In the summer of 1990, a behind-the-scenes power struggle between the company's low-profile president, Romano Artioli, and Stanzani boiled over and, on 13 July, the latter was removed from his position as managing and technical director.

The official opening of the plant took place on 15 September 1990, which would have been Ettore's 109th birthday. A link between Molsheim and the new establishment was forged when a flame was lit from the foundry at the Alsace works and conveyed to Compagalliano in a 1934 Type 57. It was one of a collection of fifty or so Bugattis which had assembled at Molsheim for the run to Italy on 8 September.

Twelve hundred guests gathered to attend the opening ceremony that was attended by Michel Bugatti. Although the engine was on display, the prototype was covered by a sheet with the car's announcement intended for exactly a year later, which would have been Ettore's 110th birthday.

There then followed a trip to Ettore Bugatti's newly discovered birthplace at Castello Sforzesco in Milan, where our story began one hundred and ten years ago . . .

But the firm was reported to be dissatisfied with the lines of the new Bugatti, the work of the celebrated stylist, Marcello Gandini, which was said to be too similar to that of the Diablo and Cizetta supercars.

So a small team, headed by Giampaolo Benedini, the architect responsible for the new Campogalliano factory, redesigned the EB 110's lines and the resulting car was unveiled to very mixed reviews at its Paris launch on 14 September 1991. Production is scheduled to begin this year (1992) though Romano Artioli is reported to be encouraging stylists to come up with concept cars which may be a reflection of the reception the prototype received. But whatever the outcome, there can be little doubt that Bugatti is back!

---

[25] *Autocar & Motor*, 19 October 1988.

[26] *Autocar & Motor*, 1 March 1989.

Appendix 1

The Bugatti Owners' Club

The oldest Bugatti club in the world is not, as might be expected, French but English. The Bugatti Owners' Club was formed in the heyday of the marque in 1929 and is still in robust health today. The idea for such an organization originated with one D B Madeley, who seems to have been one of the few people to have claimed to have owned a Crossley Bugatti! He wrote to the motoring press to suggest the idea and, as a result, a trio of enthusiasts met on 18 December 1929 over a pint of beer to discuss the idea. There were two pipe smokers – one was Madeley himself and the other Colonel Godfrey Giles – while the third member of the triumvirate was one T Ambrose Varley who

Hugh Conway and his Type 43 in a Bugatti Owners' Club meeting at Prescott.

preferred a long cigarette holder. It was out of this smoke, as Colonel Giles later put it, that the Club was born. The inaugural meeting was immortalized in a charming line-drawing of three half emptied glasses of beer, complete with smoking accoutrements. It has appeared on the first page of the Club's excellent magazine, *Bugantics*, that arrived in June 1931, ever since. As a result of this gathering, the first meeting of the club was held at Colonel Giles' house in Regents Park and he became chairman, Madeley was treasurer and Varley secretary. The latter's term of office was destined to be relatively short and Giles later recounted that his last memory of Varley was of him 'rushing down to his house to retrieve the Club's books etc., before they also vanished'. Kenneth Bear replaced him as secretary but, in 1932, Godfrey Giles's brother Eric, took over and held the post until 1946. In 1952 he became the Club's president, a position he held until his death in 1988. The Club was delighted when Ettore Bugatti agreed, in 1931, to become its Patron and he was to remain so until he died in 1947. His place was then taken by Earl Howe, who had previously been president.

The Club's associations with the factory were underlined when Jean Bugatti joined in 1930 and he attended its dinners in 1934 and from 1937 to 1939.

The club was diligent in providing events for its members. A speed trial, which Jean Bugatti attended, was held at Lewes in 1933, hillclimbs were held at Chalfont St Giles and Aston Clinton and, in 1937, the Club staged a race meeting at Donington in conjunction with the Vintage Sports Car Club, which had been formed just three years previously.

The same year witnessed the Club obtaining its own hill climb course at Prescott about 8km (5 miles) from Cheltenham in the glorious Cotswold countryside near Cleeve Hill. Prescott consisted of a house and surrounding estate which had been owned until 1871 by the Earl of Ellenborough. In the early 1930s it belonged to an elderly couple called Royds who were friends of the parents of Tom Rolt, later to delight a wider public as L T C Rolt with his magnificent books on industrial history, including celebrated biographies of George and Robert Stephenson and Isambard Kingdom Brunel. Rolt had welcomed any excuse to visit the Royds in his GN 'because motoring up their drive was such an exciting exercise. It included one bend so acute that it could not be negotiated on one lock . . . this and other corners were completely blind by dense shrubberies, it was perhaps fortunate that the Royds owned no car.'[1]

But the house and estate were sold in 1936 and bought as a speculative venture by the Gloucestershire Diary Company based in nearby Cheltenham. This so alarmed Rolt that he brought the higher echelons of the VSCC to Prescott because, at the time, the club was dissatisfied with Aston Clinton and he envisaged that the drive to Prescott house would make an ideal hill climb course. Regrettably the club did not possess sufficient funds to buy the estate and Rolt has recalled that 'It was Sam Clutton who saved the day. He knew that the wealthier and more established Bugatti Owners' Club were as dissatisfied as we were with Aston Clinton and were looking for some other course to take its place.' Clutton proposed that the idea should be sportingly passed on to the BOC on the understanding that, if Rolt's initiative came to fruition, the VSCC would be able to hold a hill climb meeting there once a year. In truth the Bugatti Owners' Club also lacked the funds but, to their eternal credit, the Giles brothers stepped in and bought the estate which was then rented to the BOC for a nominal fee.

Some £2,000 were raised for the hill to be resurfaced by the Beaufort Quarry Company and, on 15 May 1938, the first meeting was held there. Apart from the war years, when the army occupied the estate, this most picturesque of venues has been in use ever since as one of the country's leading hill climbs. In 1960 the course was extended by the addition of a new loop at Orchard Corner which lengthened the climb from around 1,000 yards to 1,127. However, the VSCC continue to use the old course for its annual meeting.

The Bugatti Owners' Club's headquarters is, very appropriately, at Prescott. Since 1932 the BOC

[1] L. T. C. Rolt, *Landscape with Machines*, Longman, 1971.

has been open to Bugatti enthusiasts as well as owners and full details can be obtained from:

Mrs Sue Ward,
Secretary,
The Bugatti Owners' Club,
Prescott Hill,
Gotherington,
Cheltenham,
Gloucestershire GL52 4RD.

Bugatti Clubs on Both Sides of the Atlantic

I am grateful to Andre Rheault for providing details of the American Bugatti Club which was founded in 1960 for the purpose of establishing ties between Bugatti owners in America for their mutual benefit.

Eligibility for membership requires present or past ownership of a Bugatti car. The present membership numbers 220 in the USA and 70 overseas in some eighteen different countries. Given the relatively small number of members and large size of the country, events are infrequent but do take place throughout the year. These are both social – dinners, luncheons and the like – and automotive, in the form of hill climbs and rallies.

Pur Sang is the ABC's high quality quarterly publication and is edited by Andre Rheault, who is also the Club's Registrar. He maintains a 'Register of All Bugattis' in North America which contains detailed information on some 400 plus cars.

Any present or previous Bugatti owner wishing to join the American Bugatti Club should write to the secretary:

Mr Laurence Deutsch,
400, Buckboard Lane,
Ojai, CA 93023,
USA.

On continental Europe, the oldest of the clubs is the Bugatti Club Nederland, established in 1955. It was during an old car run to Utrecht, which culminated with dinner at the Zandvoort racing circuit, that five Bugatti owners decided to form a Dutch Bugatti club. They were B Laming, W M Pieters, van Ramshorst, G F M F Prick and R Andersen. This small but active club regularly organizes meets and rallies including the Grand Rally International at Ermenonville–Le Mans which was to lead to the Bad Honnef Rally organized by the Bugatti-Club Deutschland.

That organization was founded a year after the Dutch one, in 1956. It was established by Bugatti owner Kurt Kieffer, very appropriately, at that year's German Grand Prix. The Club's aim is to maintain the Bugatti tradition and to promote comradeship amongst drivers as well as for the care of current cars. Former Bugatti Grand Prix drivers, Louis Chiron and Maurice Trintignant, were welcomed as honorary members.

On 31 October 1967 came the Club Bugatti France. This was founded by five enthusiasts, Paul Badre, Pierre Bardinon, Wladimir Granoff, Jess Pourret and Phillipe Vernholes. Well-attended rallies are organized.

Appendix 2

Bugatti Type Numbers 1899–1962

1	1899	Prinetti and Stucchi four-engined quadricycle
2	1901	3-litre four-cylinder passenger car for Milan Exhibition
3	1903–1904	De Dietrich*
4	1903–1904	De Dietrich*
5	1903–1904	De Dietrich*
6	1905–1907	Hermes for Mathis*
7	1905–1907	Hermes for Mathis*
8	1907	Overhead camshaft engine in Hermes-type chain-driven passenger car chassis
8A	1909–1910	9.9-litre Deutz passenger car with chain drive
8B	1909–1910	6.4-litre Deutz passenger car with chain or shaft drive
8C	1909–1910	4.9-litre Deutz passenger car
9A	1909–1910	6.4-litre Deutz passenger car
9B	1909–1910	4.9-litre Deutz passenger car
10	1909	1.2-litre 'Le Petit Pur Sang' passenger car
11	1910	Deutz

Start of Bugatti Car Production at Molsheim

12	1910	Prince Henry sports car with chain drive
13	1910–1914	1.3-litre eight-valve engine with 2m wheelbase passenger and racing car
	1914–1926	16-valve engine with 2m wheelbase
14	1910	Engine and radiator
15	1911–1913	eight-valve engine in 2.4m wheelbase passenger car
16	1911	(Prototype of four-cylinder 855cc passenger car which became Bébé Peugeot)+
17	1912–1913	eight-valve engine in 2.55m wheelbase passenger car
18	1912–1913	5-litre chain-driven racing car
19		Not known
20	1912	1.5-litre Peugeot prototype passenger car
21	1910–1911	(3.5-litre Deutz passenger car)+
22	1914 & 1920	eight-valve engine with 2.4m wheelbase passenger car
	1920–1926	16-valve engine with 2.4m wheelbase passenger car
23	1914 & 1920	eight-valve engine in 2.55m wheelbase passenger car
	1921–1926	16-valve engine in 2.55m wheelbase passenger car
24		Not known
25	1914	As Type 22 but with eight-valve 68 x 108mm engine
26	1914	As Type 23 but with eight-valve 68 x 108mm engine
27	1914	Thought to be 68 x 100m 16-valve car
28	1921	3-litre eight-cylinder prototype passenger car

* It is not possible, at this stage, positively to identify these individual models.
+ Speculative attribution. *See* text.

376

Type	Years	Description
29	1922	1.5-litre version of Type 30. Not built.
30	1922–1926	2-litre eight-cylinder passenger car
31		Not known
32	1923	2-litre eight-cylinder 'Tank' racing car
33	1923	2-litre eight-cylinder passenger car with gearbox in rear axle. Not built.
34	1925	35.4 litres 16-cylinder aero engine
35	1924–1927	2-litre eight-cylinder Grand Prix racing car
35A	1925–1930	2-litre car similar to above with Type 38 engine
35T	1926–1927	2.3-litre version of Type 35
35C	1927–1930	Supercharged version of Type 35
35B	1927–1930	Supercharged version of Type 35T
37	1925–1931	1.5-litre four-cylinder racing and sports car
37A	1927–1931	Supercharged version of above
38	1926–1929	2-litre eight-cylinder passenger car
38A	1927–1929	Supercharged version of above
39	1926–1927	1.5-litre version of Type 35
39A	1926–1927	Supercharged version of above
40	1926–1931	1.5-litre four-cylinder passenger car
40A	1931–1933	1.6-litre version of above
41	1927–1933	La Royale. 12.8-litre eight-cylinder passenger car
42		Marine engine
43	1927–29	2.3-litre eight-cylinder supercharged sports car
43A	1929–1935	Roadster version of above
44	1927–1930	3-litre eight-cylinder passenger car
45	1928	3-litre 16-cylinder racing car
46	1929–1932	5.3-litre eight-cylinder passenger car
46S	1930–1932	Supercharged version of above
47	1929	Grand Sport version of Type 45
48	1931	994cc four-cylinder engine for Peugeot 201X
49	1930–1932	3.3-litre eight-cylinder passenger car
50	1930–1933	4.9-litre eight-cylinder twin overhead camshaft passenger car
50B	1936	Racing engine, B1 4.7-litre supercharged
	1936	B11 4.5-litre unsupercharged
	1938	B111 3-litre supercharged
51	1931–1935	Eight-cylinder 2.3-litre supercharged Grand Prix racing car
51A	1932–1936	1.5-litre version of above
52	1927–1931	Miniature electrically powered version of Type 35 for children
53	1932	Eight-cylinder 4.9-litre four-wheel drive racer
54	1931–1933	4.9-litre Grand Prix racing car
55	1931–1935	2.3-litre eight-cylinder supercharged sports car
56	1931–19	Electric-powered runabout
57	1933–1939	3.3-litre eight-cylinder passenger car
57C	1937–1939	Supercharged version of above
57S	1936–1938	Sports version of Type 57
57SC	1937–1938	Supercharged version of above
57G	1936	3.3-litre 'Tank' sports racing car
57S45	1937	4.5-litre sports racing car
58	1934	Diesel version of Type 41 engine for railcar use
59	1933–1934	3.3-litre Grand Prix racing car
60		4.1-litre aero engine. Not made.
61		Railcar gearbox and bevel drive
62		Dry sump gearbox and pump for Type 59
63		Chassis with 50S, for Sport, engine
64	1939	Eight-cylinder 3.3-litre passenger car. Not built.
65	1940	Single-cylinder engine
66	1938	Type 50B-type aero engine
67	1939	V-type aero engine. Not built.

68	1946	Four-cylinder 360cc passenger car
69		Not known
70		Hydraulic coupling of four engines
71		18.5-litre eight-cylinder marine engine
72		13cc single-cylinder bicycle engine
73	1946	Sports car with supercharged 1.5-litre twin overhead camshaft engine
73A		Passenger car with 1.5-litre supercharged four-cylinder single overhead camshaft engine
73C		1.5-litre four-cylinder twin overhead camshaft supercharged racing car
74		Four-cylinder steam engine
75		Single-cylinder marine engine for dinghy
76		Eight-cylinder lorry engine
78		Eight-cylinder touring engine
79		Not known
80	1949	Reverse gear for 12-cylinder La Licorne engine
101	1951	3.3-litre eight-cylinder passenger car
102		1.5-litre four-cylinder sports car. Not built.
251	1955	2.4-litre eight-cylinder Grand Prix racing car
252		1.5-litre four-cylinder sports car
451	1962	4.5-litre 16-cylinder project. Not completed.

Selected Bibliography

Works on Bugatti

Barker, R. 1971. *Bugatti*. Ballantine Books.
Borgeson, G. 1981. *Bugatti by Borgeson*. Osprey.
Bugatti, L. 1967. *The Bugatti Story*. Souvenir Press.
Bugatti Owners' Club. 1957 to date. *Bugantics*.
Bradley, W. F. 1959. *Ettore Bugatti, Portrait of a Man of Genius*. Motor Racing Publications.
Conway, H. G. 1959. *The Automotive Inventions of Ettore Bugatti*. The Newcomen Society.
Conway, H. G. 1984. *Bugatti*. Octopus Books.
Conway, H. G. 1987. *Bugatti 'Le pur-sang des automobiles'*. Foulis.
Conway, H. G. and Sauzay, M. 1989. *Bugatti Magnum*. Foulis.
Conway, H. G. 1983. *Grand Prix Bugatti*. Foulis.
Conway, H. G., von Fersen, H. H., Jedding, H., Malhotra, R., von Salden, A. and Spielmann, H. 1983. *Die Bugattis Carlo, Rembrandt, Ettore und Jean*. Christians.
Dumont, P. 1985. *Bugatti Thoroughbreds from Molsheim*. EPA.
Eaglesfield, B., Hampton, C. W. P. 1954. *The Bugatti Book*. Motor Racing Publications.
Hallums, E. 1979. *The Quality of Work and The Quality of Art: A Study of Bugatti*. Mithras Press.
Harvey, M. 1979. *The Bronzes of Rembrandt Bugatti*. Palaquin Publishing.
Haslam, M., Garner, P., Harvey, M., Conway, H. G. 1979. *The Amazing Bugattis*. The Design Council.
Jarraud, R. 1977. *Bugatti Doubles Arbres*. Editions de l'automobiliste.
Kestler, P. 1977. *Bugatti: Evolution of a Style*. Edita.

Related Books

Anselmi, A. T. 1977. *Isotta Fraschini*. Interauto.
Bird, A. 1960. *The Motor Car 1765–1914*. Batsford.
Bird, A. 1967. *Early Motor Cars*. George Allen and Unwin.
Boddy, W. 1960. *Montlhéry The Story of the Paris Autodrome, 1924–1960*. Cassell.
Borgeson, G. 1966. *The Golden Age of the American Racing Car*. Norton.
Bradley, W. F. *Targa Florio*. Foulis.
Cholmondeley Tapper, T. P. 1954. *Amateur Racing Driver*. Foulis.
Clutton, C., Posthumus, C., Jenkinson, D. 1956. *The Racing Car Development and Design*. Batsford.
Court, W. 1966. *The Power and the Glory*. Macdonald.
Dees, M. L. 1981. *The Miller Dynasty*. Barnes Publishing.
Dreyfus, R. with Kimes, B. R. 1983. *My Two Lives*. Aztex.

Georgano, G. N., editor. 1982. *The Complete Encyclopedia of Motor Cars*. Ebury Press.
Georgano, G. N., editor. 1977. *The Encyclopedia of Motor Sport*. Ebury Press.
Grégoire, J. A. 1954. *Best Wheel Forward*. Thames and Hudson.
Hodges, D. 1967. *The French Grand Prix*. Temple Press.
Hodges, D. 1963. *The Le Mans 24 Hour Race*. Temple Press.
Hodges, D. 1964. *The Monaco Grand Prix*. Temple Press.
Hough, R. 1957. *Tourist Trophy*. Hutchinson.
Hull, P. and Slater, R. 1964. *Alfa Romeo*. Cassell.
Karslake, K. 1950. *Racing Voiturettes*. Motor Racing Publications.
Laux, J. M. 1976. *In First Gear*. Liverpool University Press.
Mays, R. 1951. *Split Seconds*. Foulis.
Montagu, Lord, 1969 and 1971. *The Lost Causes of Europe* (Volumes I and II.) Cassell.
Montagu, Lord, 1966. *The Lost Causes of Motoring*. Cassell.
Nicholson, T. R. 1966. *The Vintage Car*. Batsford.
Nye, D. 1989. *Famous Racing Cars*. Patrick Stephens.
Nye, D. 1974. *Motor Racing Mavericks*. Batsford.
Pomeroy, L. 1959 and 1965. *The Grand Prix Car* (Volumes I and II.) Motor Racing Publications and Temple Press.
Posthumus, C. 1977. *Classic Racing Cars*. Hamlyn.
Posthumus, C. 1965. *The German Grand Prix*. Temple Press.
Posthumus, C. 1980. *The Roaring Twenties*. Blandford.
Sedgwick, M. C. 1974. *Fiat*. Batsford.
Segrave, H. O. D. 1928. *The Lure of Speed*. Hutchinson.
Sheldon, P. 1987 and 1990. *A Record of Grand Prix and Voiturette Racing* (Volumes 1 and 2.) St Leonards Press.
Sloniger, J. and von Fersen, H-H. 1965. *German High-Performance Cars*. Batsford.
von Fersen, H-H. 1976. *Autos in Deutschland 1885–1920*. Motorbuch.

Magazines

The Autocar
Automobile Quarterly
Classic Cars
Classic and Sportscar
The Motor
Motor Sport
Veteran and Vintage Magazine

Index

Figures in *italics* refer to illustrations.

Agnelli, Giovanni, 20, *50*, 175
Alfa Romeo, 166, 172, 175, 181, 198, 206, 211, 213, 225, 233, 239, *240*, 280, 290, 308, 310, 314, 326, 355
Alfonso XIII, King, 250, 251, 264
Alvis, 134, 203, 230, 289
Artioli, Romano, 372
Aumaître, Robert, 329, 330, 331
Austin, B.H., *122*, 123
Austin Seven, 87
Auto Union, 228, 278, 307, 310, 312, 314, 317, 326, 366
Autocar, The, 13, 23, 34, 40, 114, 121, 132, 156, 175, 181, 184, 200, 248, 249, 272, 276, 309, 361
Autocar & Motor, 352, 372

Baccoli, Michele, 113, 114, *114*, 116, 118
Ballot, 138, 139, 140, 146, 147, *149*, 194
Béchereau, Louis, 151, 153
Benoist, Robert, 312, 314, 315, 317, 318, 319, 322, 323, 324, 325, 359
Bentley, 223, 279, 331
Bentley, WO, 128, 129, 223
Bernardi, Count Enrico, 20, 28, 29
Bertrand, Edouard, 346, 357
Bianchi, 21, 27
Binder, Henry, 270, 273
Bird, Sir Robert Bt., 196, 220
Birkin, Henry 'Tim', 243, 244
Blériot, Louis, 59
Bollée, Amédée, 32, 47
Bolling, Raynal C, 99, 100
Bolloré, René, 365
Borgeson, Griffith, 28, 76, 168, 193, 246, 342
Bouriat, Guy, 223, 239
Bradley, W F, 13, 59, 67, 74, 75, 76, 79, 81, 84, 88, 90, 94, 95, 100, 102, 107, 112, 114, 115, 148, 184, 193, 194, 198, 223, 225, 241, 247, 250, 262, 285, 287, 334, 337, 338, 339, 340, 345, 359, 361, 363
Bréguet, Louis, 108, 109
Brilli Peri, Count Gastone, 210, 244
Brivio 'Tonino', 314, 315
Brooklands, 129, 187, *189*, 228, *315*
Brooks, Robert, 266, 269
Bulleid, Oliver, 284, 285
Burlington Carriage Company, 35, 43

Bugatti, Barbara, 30, 40, 41, *57*, 66, 337, *337*, 339, 359
Bugatti, Carlo, 10–15, *11, 16*, 17, 25, 26, 32, 69, 93, 94, 143, 295, 346, 350, 351
Bugatti, Ettore, *18, 26, 27, 28, 33, 36, 50, 82, 147, 173, 198, 337, 340, 349, 352*
  early life in Milan, 10–30
  racing, 1898–1901, 22–25
  career with De Dietrich, 32–40
  enters car in 1903 Paris to Madrid race, 38, *38, 39*
  with Mathis, 41–4
  with Deutz, 45–59
  moves to Molsheim, 62
  returns to Italy, 92
  designs aero engines in Paris, 95–102
  wins at Brescia, 118
  produces first Grand Prix car, 144
  designs Type 35, 164–72
  creates Royale, 243–8
  attitude to organized labour, 286, 287
  returns to Paris, 287
  at Molsheim, 333–51
  transfers factory to Bordeaux, 353
  moves back to Paris, 354
  creates post-war car range, 355, 357
  remarries, 359
  death, 360
Bugatti, Genevieve, 359, 361
Bugatti, L'Ébé, 10, 29, 40, 41, 59, 92, 94, 96, 255, 265, 267, 268, 270, 281, 285, 286, 287, 332, 333, 335, 336, 339, 340, 342, 346, 349, 351, 360, 361, 370
Bugatti, Jean, 61, *82*, 92, *117, 198*, 203, 204, 205, 208, 221, 223, 224, 230, 231, 233, 236, *236*, 254, 255, 256, 257, 259, 262, 282, 289, 295, 299, 323, *328*, 330, 331, 332, 340, *341*, 347, 348, 351, 374
Bugatti, Lidia, 40, *82*, 163, 203, 204, 255, *256*, 331, 332, 340, *341, 348*, 351, 370
Bugatti, Michel, 347, 359, 370, *371*
Bugatti, Rembrandt, 15, 16, *16*, 17, 50, 65, 93, 96, 109, 110, 249, 337, 344
Bugatti, Roland, 38, 69, 110, 241, 288, 332, 340, 342, *342*, 352, 361, 365, *365*, 366, *367*, 370
Bugatti, Thérèse, 359, 370, *370*
Bugatti, Thérèse (wife of Carlo), 15, 93, 346
Bugatti aero engines: Eight, 94, *95*, 96, 97
  U16, 98–109, *101, 103, 104, 105, 106, 108*
  Type 34, 165, 245, 246

Bugatti cars: 3 litres, 325, *325, 326*
   4.7 litres, *319, 320, 328, 365*
   Brescia, 119, *120, 121, 122, 128, 129, 130*
   EB 110, 9, 372
   Eight valve, *67, 68, 69, 70, 71, 72, 73*
   Prince Henry, 70, *78*, 89
   Sixteen valve, *71, 72, 133, 137*
   Type 2, 26, *26, 27, 27,* 28, *28,* 29, *29*
   Type 8, 45–51, *48, 49, 50*
   Type 10, 45, 56, 57, *57, 58,* 59, *59*
   Type 13, 62, 63, 64, 74, *74,* 112, 113, *117, 119*
   Type 15, 64, *65, 66,* 70
   Type 17, 64, 69, 70
   Type 18 (Garros), 72, 76–87, *79, 81, 82, 83, 84, 85, 86,* 93, 94, 97, 112
   Type 22, 70, 90, 112 *122,* 125, *133*
   Type 23, 70, *72,* 112, *131, 133, 134, 135, 136, 137*
   Type 25, 90
   Type 26, 90
   Type 27, 90, 97
   Type 28, 141, 142, *142,* 143, *143,* 245
   Type 30, 131, 144, *145,* 146, *147, 148, 151, 152, 153,* 160–3, *161, 162*
   Type 32, *155, 159,* 165
   Type 35, 18, 164–72, *166, 169, 173, 176, 177, 180, 181, 188, 190, 191, 192*
   Type 35A, *178,* 179, *179, 184*
   Type 36, 181, 183
   Type 37, 144, 199–202, *199, 200, 201, 202, 203*
   Type 38, 206–7, *206*
   Type 39, 181, 182, 183, *185, 186, 189*
   Type 40, 144, 202–6, *203, 204, 205*
   Type 41 (Royale), *58,* 114, 243–77, *248, 249, 250, 251, 252, 253, 254, 255, 256, 257, 259, 260, 261, 263, 268, 271, 274, 275, 276*
   Type 43, 194, 207–12, *207, 208, 209, 210, 211, 344, 373*
   Type 44, *205,* 214–15, *215, 216*
   Type 45, 212–14, *212, 213*
   Type 46, 216–20, *218, 219*
   Type 47, 214, 231, 236
   Type 49, 215–16, *217, 218*
   Type 50, 220–25, *219, 221, 222, 223, 224*
   Type 50B, 318, 319
   Type 51, 192, *195,* 196, *196, 197*
   Type 52, *137,* 239, 241, *241,* 242
   Type 53, 229–35, *229, 231, 232*
   Type 54, 225–9, *225, 227, 228*
   Type 55, 235–9, *236, 237, 238, 240*
   Type 56, 242–3, *242,* 257
   Type 57, 278, 288–94, *286, 290, 291, 293, 294, 305, 306*
   Type 57C, 293, *293,* 295, *306, 327, 330*
   Type 57G, 320–4, *321, 322, 323, 324, 330, 349*
   Type 57S, 297, *297,* 298, *299,* 301, *301, 302, 303, 304*
   Type 57S Atlantic, 14, 295, 296, *296, 298, 299*
   Type 59, 278, 307–319, *309, 311, 313, 314, 315, 316, 317,* 322, 325
   Type 64, 290, 330, 331, *331*
   Type 68, *356,* 357, *357,* 358
   Type 73, 355, *362, 371*
   Type 101, 361, *362,* 363, *363,* 364, *364*
   Type 102, 361
   Type 251, 365, 366, *367,* 368
   Type 252, 368, *368,* 369, *369*
   Type 451, 369
Bugatti Owners' Club, 262, 300, 329, 373–5
Bugatti Railcars, 280–5, *281, 283, 284, 286*
Bugatti Trust, 87, 90, 300, *351*
Burlington Carriage Co., 35, 37, 40, 43
Burton, Jack Lemon, 66, 269, 272, 276, 277, 316

Campbell, Malcolm, 187, 201, 209, 211, *241,* 303
Cappa, Giulio, 47, 147, 230, 231
Carol, King of Rumania, 270
Charavel, Louis (Sabipa), 187, 202, 251, 335, 336
Chassagne, Jean, 172, 175, 176
Chayne, Charles A, 274, 276
Chiron, Louis, *188,* 191, 196, 221, 223, 224, 233, 239, 318
Cholmondeley Tapper, T P, 281, 347, 348, 349
Citroën, 87, 156, 290
Citroën, Andre, 239, 261
Classic and Sportscar, 266, 301
Classic Cars, 273
Clément-Bayard, 56, 57, 94
Colombo, Gioachino, 365, 366
Conelli, Count Carlo, 189, 202, 212, 223
Conway, Hugh, 34, 70, 107, 125, 153, 246, 345, *373*
Corsica, *301, 302, 303, 304*
Costantini, Meo, 116, 172, 174, *174,* 180, 181, 184, 185, 187, 188, 210, 230, 237, 251, 312, *313,* 330, 342, 348
Craig, Ian, 316
Crossley Bugatti, 125, 126, 128
Cunningham, Briggs, 265, 266, 267, 268, 269, 270
Cushman, Leon, 129, *130*
Czaykowski, Count Stanislaus, 226, 227, 228, 229

Davis, S C H 'Sammy', 164, 175
de Alzaga Unzue, Martin, 151, 153, 154, 170, 171
de Cystria, Prince Bernard, 152, 154, 159
De Dietrich, 31, 32, 40
De Dietrich, Baron, 32
De Dietrich cars, 34, 35, *35, 36, 37*
De Dion Bouton, 19, 20, 28
de Gramont, Duc, 96, 280, 333, 338, 339
Delage, 84, 138, 160, 173, 177, 182, 187, 188, 189, 279, 292
Delaney, Luke Terence, 35, 37, 38, 44
Delaunay–Belleville, 44, 96, 97, 98
d'Erlanger, Leo, 206, 211
Deutz, 44, 45
Dutz cars: Prince Henry, 54
   Type 8, 48, *48, 49,* 50, 51, *51, 52, 53*
   Type 9, 51, 52, *53*
   Type 21, 52, *54,* 57, 63
de Vizcaya, Baron Agustin, 60, 61, 64
de Vizcaya, Ferdinand, 61, 182
de Vizcaya, Pierre, 61, 113, 114, 118, 145, 146, 149, 150, 151, 158, 160, 172, 175, 176, 177, 180, 181, 182, 185, 208
Diatto, 95, 97, 125, 126
Divo, Albert, 189, 190, 212, 233

Domboy, Noël, 259, 281, 282, 332, 345, 346, 353, 355, 357, 359, 369
Don, Kaye, 226, 228
Dreyfus, René, 191, 202, 230, 232, 233, 234, 280, 281, 311, *311*, 312, 313, 314, 315, 347
Duesenberg, 103, 106, 107, 111, 138, 139, 140, 141, 144, 161, 221, 251, 279
Dunlop, 169, 175, 176, 223, 225, 235
Dutilleaux, Louis, *210*, 211
Duray, Leon, 192, 193, 194, *211*, 234

Eccles, Lindsay, 312, 316
Esders, Armand, 254, 257, 269
Espanet, Dr Gabriel, 65, 81, 84, 91, 93, 94, 95, 96, 98, 102, 103, 107, 244
Evans, Denis, *209*
Eyston, George, 188, 266

Faroux, Charles, 256, 360
Fiat, 20, 23, 24, 27, 41, 47, 55, 74, 79, 80, 147, 148, 150, *159*, 160, 164, *165*, 167, 173, 175, 213, 279
Fletcher, Alec Rivers, 233, 234
Foresti, Giulio, *149*, 182
Foster, Captain Cuthbert, 261, 262, 263, 272, 276
Friderich, Ernest, 44, 45, 57, 59, 60, *60*, 63, 67, 69, 72, 74, *74*, 75, 80, 81, 84, *85*, 87, 89, 92, 93, 102, 103, 111, 113, 114, *116*, 118, *119*, 145, 148, *155*, 158, 159, *159*, 172, 175, 287, 333
Fuchs, Dr Josef, 259, 261, 267, 274

Gaggenau, 46, 47
Gangloff, 65, 216, 217, 291, 301
Garnier, Leonico, 172, 175
Garros, Roland, 81, 83, 84, 94, 110, 245
Giles, Eric, 374
Giles, Colonel Godfrey, 83, 303, 373
Goux, Jules, 182, 184, 185, *186*, 187, 188, 251
Grands Prix: Alsace, 1926, 183
  Belgian, 1925, 181; 1933, 312; 1934, 315; 1935, 318
  Boulogne, 1924/25, *132*
  British, 1926, 187; 1927, 189, *189*
  Cork, 1938, 325, 326, *326*
  French, 1908, 56; 1911, 73–5, *74*; 1912, 80, 81; 1914, 87; 1921, 144, 146; 1922, 147–50; *148, 149*; 1923, 158–60, *157, 159*; 1924, 172–6; 1925, 181, 182; 1926, 184–7; 1927, 188–9; 1928, 190; 1929, 191; 1931, 196; 1932, 196; 1933, 228, 312; 1934, 314, *314*, 315; 1935, 317; 1936, 320, 322; 1937, 324, 325; 1938, 326; 1956, 368
  German, 1928, 234, 244, 252; 1929, 191
  Italian, 1922, 150; 1923, 154; 1924, 176; 1925, 182; 1926, 187, 188
  Monaco, 1929, *190*, 191; 1930, *191*; 1931, 196; 1933, *197*, 198; 1934, 313, *313*, 314; 1936, 319, *319*
  Monza, 1929, 193, 194; 1931, 226; 1933, 228
  San Sabastian, 1924, 176; 1926, *186*, 187
  Spanish, 1926, 187; 1927, 189; 1929, 191; 1933, 312, *313*; 1934, 315; 1935, 318
  Swiss, 1936, *320*
Grégoire, Jean-Albert, 44, 199, 235
Gressley, Sir Nigel, 284, 285
Gulinelli, Counts, 24, 25, 26, 31

Hall, Eddie, *122*
Hampton, Peter, 65, 66, *66, 238*, 300, 304
Harrah, William, 266, 273
Harrah Auto Collection, 266, 273
Henry, Ernest, 76, 139, 147
Henry, Frederick, 270–3
Hermes, 42, *42*, 43, 44, *45, 48*
Hispano-Suiza: aero engines, 98, 139, 279, 353, 369, 371
  cars, 138, 139, 244, 248, 264, 279
Howe, Earl, 209, 211, 226, 233, 234, 262, 295

Indianopolis, 500 Miles Race: 1914, 76, 84, 85
  1915, 87
  1919, 138, 140
  1920, 168
  1923, 151, 168, 170
  1932, 235
Isotta Fraschini, 25, 41, 46, *46*, 55, 56, *56*, 57, 80, 117, 140, 264
Itala, 230

Jano, Vittorio, 172, 196, 239
Jarrot Engines, 162, 179
Jarrot, Charles, 23, 37
Jarrott and Letts, 37, 123, 124
Junek, Elizabeth, 150, *150*, 160, *161, 185*, 190, 244
Junek, Vincent, 150, 160, 166, 176, 244

Karslake, Kent, 55, 74, 111, 114
Kellner, 256, *257*, 268, *268*
Kestler, Paul, 254, 305
King, Charles Brady, 103, 104, 105, 106, 107, 109
Kurtz, Felix, 335, 346

Labourdette, 81, *83*
Labric, Roger, 324
Lancia, Vincenzo, 24, 27, *50*
Le Mans: 1923, 206
  1930, 206
  1931, 223
  1932, 239, *240*
  1937, *322*, 323, *323*, 324
  1939, 329, 330, *330*
Le Zèbre, 87
Lefrère, Major EGA, 123, 124
Lehoux, Marcel, 226, 229
Lewis, Hon. Brian, 315, 316
Loiseau, Frederic, 205, 208
Lurani, Count Johnnie, 210, 223

Maggi, Count Aymo, 210
Marco, Pierre, 118, 146, 158, 160, 361
Marshall, Bertram, 129, *129*, 131
Marquis, Johnnie, *86*, 87
Martini, 55
Maserati, 214, 225, 226
Materassi, Emilio, 188, 189, *189*
Mathis, Émile, 33, 34, 38, 40, 41, 42, 44, *50*, 60, 62, 95
May, Austen, 233, 234
Maybach, William, 42, 47

Mays, Raymond, 119, *120*, 121
Mercedes, 28, 29, 30, 33, 41, 42, 43, 47, 56, 76, 78, 147, 148, 177
Mercedes-Benz, 228, 278, 289, 307, 310, 314, 317, 320, 326
Michelin, 176, 223, 279, 280
Mille Miglia, 1928, 210, 211
Miller, 154, 160, 168, 169, *171*, 192, 196, 234
Miller, Harry, 106, 154, 168, 169, 170, 171, 172, 192, 230
Minoia, Ferdinand, 184, *186*, 187
Mischall, Emil, 113, 114
Moglia, Edmund, 124, 183
Molsheim, 13, 60, 61, 112, *334*, 333–51, *338, 340, 341, 344, 348, 349, 350, 354*, 361, 372
Monès–Maury, Jacques, 124, *128*, 129, 145, 146
Montlhéry, 181, 182, *188*, 193, 226, *314, 321*
Motor, The, 67, 73, 79, 90, 114, 175
Motor Sport, 215, 237, 238, 302, 307, 327, 329
Motor Shows: Berlin, 1906, 46; 1907, 51; 1911, 65
   London, 1921, 141, 142; 1929, 216; 1932, 257; 1935, 296
   Paris, 1910, 65, *67*; 1919, 125; 1921, 141, 142; 1925, 246; 1928, 252; 1929, 216; 1931, 256, 257; 1932, 223; 1938, 305, 331; 1939, 305, 331; 1946, 356; 1947, 361; 1951, *362*, 363
Muller, Henri, 346
Musée national de l'Automobile, 44, 89, 160, 242, 267, 277, 285, *356*
Mussolini, Benito, 250, 279

NAG, 47
NSU, 62
Nuvolari, Tazio, 126, 196, 198, 210, 225, 227, 313, 314, 317

Offenhauser, Fred, 106, 171
Oldfield, Barney, 87
Opel, 62

Packard, 103, 107, 192, 246
Panhard Levassor, 19, 24, 30, 95
Park Ward, 261, 262, *263*
Paul, Adrien, 349, 353, 359
Peck, D Cameron, 265, 266, 268, 273
Petit, Émile, 187
Peugeot, 76, 81, 139, 164
Peugeot, Bébé, 13, 70, 76, 87, *88*, 89–90, *89*
Peugeot, Robert, 76, 89
Phillipe, Maurice, 344, 345
Pichetto, Antonio, 230, 289, 355
Pirelli, 24
Pomeroy, Laurence, 76, 79, 183
Pope, Richard, 299, 301, 304
Porsche, Ferdinand, 34, 77, 78
Pracht, Henri, 326, 327, 336, 344, 347
Price, Barrie, 300
Prince Henry Trials, 54, 70, 77, *77*, 78, 80
Prinetti and Stucchi, 17, *17*, 19, 20, 21, *22*, 24, 25, 27, 75

Rabag Bugatti, 126, 127, *127*
Racing, *See under* Grand Prix *and* Voiturette
Ricordi, Giuseppe, 26, 27, *117*
Roberts, T A, 312
Rolland-Pilain, 47, 73, 75, 158, 160

Rolt, L T C, 374
Rost, Maurice, 223, 224
Rothschild, Lord, 14, 298, 299, 300
Royce, Henry, 345

Salmson, 187
Schlumpf, Hans and Fritz, 44, 242, 267, 277, 369, *370*
Scholte, Llewellyn, 112, 172, 280
Scott-Moncrieff, David, 251, 270
Segrave, Henry de Hane, 129, 147, 155, 160, 170, 343
Shelsley Walsh, *231*, 233, 234, *236*
Siko, Odette, 206
Sorel, Lt.-Col. Wyndham, 218, 220, 262, 280, 302
St John, Geoffrey, 235
Stanzani, Paolo, 372
Stefanini, Giuseppe, 46, 55
Sunbeam, 87, 147, 148, 159, 160, 173, 175, 177

Talbot-Darracq, 124, 127, 128, 129, 137
Targa Florio: 1923, *136*
   1925, 179, 180
   1926, 184, *201*
   1927, 188
   1928, 190, 191
   1929, 191
   1930, 191
Tourist Trophy: 1922, 127–9
   1928, 211, 212
   1935, 295
   1936, 301
Trintignant, Maurice, *367*, 368
Trippel, Hans, 353, 354
Turin Motor Museum, 24, 230, 242

Vanden Plas, 301, 304
Vanvooren, 223
Varzi, Achille, 192, *197*, 198, 223, 224, 226, 227, *228*, 229, 237, 312, *313*, 314
Vendromme, Stella, 280, 326
Veyron, Pierre, 227, 324, 329, *330*
Villiers, Amherst, *120*
Voisin, Gabriel, 94, 156–7, 279
Voisin, *157*
Voiturette: 1500 Trophy, 1922, 128–9, *128, 129, 130*
   Brescia, 1921, 116–18, *119*
   Coupes des Voiturette, 1908, 55, 56; 1914, 90
   Le Mans, 1920, 112–18, *114, 116*
von Zeppelin, Count Ferdinand, 92, 94

Walter, Joseph, 346
Wanderer, 62, 89
Weinberger, Ludwig, 259, 261
Weymann, C T, *216, 219*, 223, 252, *253*, 254
Williams, *190*, 191, 212, 226, 227, 299, 300, 320, 321, 322
Wimille, Jean-Pierre, 226, 267, 314, *317*, 318, 320, 322, 324, 325, 326, 327, *328*, 329, *330*, 332, 359

Young, James, 217, 218

Zborowski, Count Louis, 154, 160, 173, 175